Multicriteria Optimization

Matthias Ehrgott

Multicriteria Optimization

Second edition

With 88 Figures and 12 Tables

Springer

Dr. habil. Matthias Ehrgott, Associate Professor
The University of Auckland
Department of Engineering Science
School of Engineering
Private Bag 92019
70 Symonds Street Level 3
Auckland 1001
New Zealand
m.ehrgott@auckland.ac.nz

Originally published as volume 491 in the series:
Lecture Notes in Economics and Mathematical Systems

Cataloging-in-Publication Data
Library of Congress Control Number: 2005924730

ISBN 3-540-21398-8 Springer Berlin Heidelberg New York

Bibliographic information published by Die Deutsche Bibliothek.
Die Deutsche Bibliothek lists this publication in the Deutsche Nationalbibliografie; detailed bibliographic dat is available in the Internet at http://dnb.ddb.de

Springer is a part of Springer Science+Business Media

springeronline.com

Printed in Germany

Hardcover-Design: Erich Kirchner, Heidelberg

SPIN 10996973 42/3153-5 4 3 2 1 0 – Printed on acid-free paper

Preface

Life is about decisions. Decisions, no matter if made by a group or an individual, usually involve several conflicting objectives. The observation that real world problems have to be solved optimally according to criteria, which prohibit an "ideal" solution – optimal for each decision-maker under each of the criteria considered – has led to the development of multicriteria optimization.

From its first roots, which where laid by Pareto at the end of the 19th century the discipline has prospered and grown, especially during the last three decades. Today, many decision support systems incorporate methods to deal with conflicting objectives. The foundation for such systems is a mathematical theory of optimization under multiple objectives.

Fully aware of the fact that there have been excellent textbooks on the topic before, I do not claim that this is a better text, but it has a considerably different focus. Some of the available books develop the mathematical background in great depth, such as Sawaragi *et al.* (1985); Göpfert and Nehse (1990); Jahn (1986). Others focus on a specific structure of the problems covered as Zeleny (1974); Steuer (1985); Miettinen (1999) or on methodology Yu (1985); Chankong and Haimes (1983); Hwang and Masud (1979). Finally there is the area of multicriteria decision aiding Roy (1996); Vincke (1992); Keeney and Raiffa (1993), the main goal of which is to help decision makers find the final solution (among many "optimal" ones) eventually to be implemented.

With this book, which is based on lectures I taught from winter semester 1998/99 to winter semester 1999/2000 at the University of Kaiserslautern, I intend to give an introduction to and overview of this fascinating field of mathematics. I tried to present theoretical questions such as existence of solutions as well as methodological issues and hope the reader finds the balance not too heavily on one side. The text is accompanied by exercises, which hopefully help to deepen students' understanding of the topic.

The decision to design these courses as an introduction to multicriteria optimization lead to certain decisions concerning the contents and material contained. The text covers optimization of real valued functions only. And even with this restriction interesting topics such as duality or stability have been excluded. However, other material, which has not been covered in earlier textbooks has found its way into the text. Most of this material is based on research of the last 15 years, that is after the publication of most of the books mentioned above. This applies to the whole of Chapters 6 and 7, and some of the material in earlier chapters.

As the book is based on my own lectures, it is well suitable for a mathematically oriented course on multicriteria optimization. The material can be covered in the order in which it is presented, which follows the structure of my own courses. But it is equally possible to start with Chapter 1, the basic results of Chapters 2 and 3, and emphasize the multicriteria linear programming part. Another possibility might be to pick out Chapters 1, 6, and 7 for a course on multicriteria combinatorial optimization. The exercises at the end of each Chapter provide possibilities to practice as well as some outlooks to more general settings, when appropriate.

Even as an introductory text I assume that the reader is somehow familiar with results from some other fields of optimization. The required background on these can be found in Bazaraa *et al.* (1990); Dantzig (1998) for linear programming, Mangasarian (1969); Bazaraa *et al.* (1993) for nonlinear programming, Hiriart-Uruty and Lemaréchal (1993); Rockafellar (1970) for convex analysis, Nemhauser and Wolsey (1999); Papadimitriou and Steiglitz (1982) for combinatorial optimization. Some results from these fields will be used throughout the text, most from the sources just mentioned. These are generally stated without proof. Accepting these theorems as they are, the text is self-contained.

I am indebted to the many researchers in the field, on whose work the lectures and and this text are based. Also, I would like to thank the students who followed my class, they contributed with their questions and comments, and my colleagues at the University of Kaiserslautern and elsewhere for their cooperation and support. Special thanks go to Horst W. Hamacher, Kathrin Klamroth, Stefan Nickel, Anita Schöbel, and Margaret M. Wiecek. Last but not least my gratitude goes to Stefan Zimmermann, whose diligence and aptitude in preparing the manuscript was enormous. Without him the book would not have come into existence by now.

Preface to the Second Edition

Much has happened in multicriteria optimization since the publication of the first edition of this book. Too much in fact for all the contributions in the field to be reflected in this new edition, which – after all – is intended to be a textbook for a course on multicriteria optimization. Areas which have seen particularly strong growth are multiobjective combinatorial optimization and heuristics for multicriteria optimization problems. I have tried to give an indication of these new developments by adding "Notes" sections to all chapters but one. These sections contain many references to the literature for the interested reader. As a consequence the bibliography has more than doubled compared to the first edition. Still, heuristics feature only in the very last section and metaheuristics are not even mentioned.

There are a number of other changes to the organization of the book. Linear and combinatorial multicriteria optimization is now spread over five chapters, which seems appropriate for material that covers roughly half the pages. It also reflects the way in which I have usually taught multicriteria optimization, namely a course on general topics, containing material of the first five chapters, and a course on linear and combinatorial problems, i.e. the second half of the book. I have therefore tried to make the second part self contained by giving a brief revision of major definitions.

Some reorganization and rewriting has taken place within the chapters. There is now a section on optimality conditions, previously distributed over several chapters. Topics closely related to the weighted sum method have been collected in Chapter 3. Chapter 4 has been extended to include several scalarization techniques not mentioned in the first edition. Much of the material on linear programming has been rewritten, and scalarization of multiobjective integer programs has been added in Chapter 8.

Of course, I have done my best to eliminate errors contained in the first edition. I am grateful to all students and colleagues who made me aware of them, especially Dagmar Tenfelde-Podehl and Kathrin Klamroth, who used the book for their own courses. There will still be mistakes in this text, and I welcome any suggestions for improvement. Otherwise, I hope that you approve of the changes and find the book useful.

Auckland, March 2005 *Matthias Ehrgott*

Notation

These are some guidlines concerning the notation I have used in the book. In general, calligraphic capitals denote sets, latin capitals denote matrices (or some combinatorial objects) and small latin or greek letters denote elements of sets, variables, functions, parameters, or indices. Superscripts indicate entities (such as particular vectors), subscripts indicate components of a vector or matrix. Due to a limited supply of alphabetical symbols, I have reused some for several purposes. Their usage should be clear form the context, nevertheless I apologize for any confusion that may arise.

The following table summarizes the most commonly used symbols.

Notation	Explanation
$\mathcal{X}$	feasible set of an optimization problem
$\mathcal{Y} := f(\mathcal{X})$	feasible set in objective space
$\mathcal{C}$	cone
$x = (x_1, \ldots, x_n)$	variable vector, variables
$y = (y_1, \ldots, y_p)$	vector of objective function values
$f = (f_1, \ldots, f_p)$	vector of objective functions
$g = (g_1, \ldots, g_m)$	vector of constraint functions
$A \in \mathbb{R}^{(m \times n)}$	constraint matrix of an LP
$C \in \mathbb{R}^{(p \times n)}$	objective matrix of an MOLP
$b \in \mathbb{R}^m$	right hand side vector of an (MO)LP
y^I	ideal point
y^N	nadir point
y^U	utopian point
$\lambda \in \mathbb{R}^p$	vector of weights
$y^1 < y^2$	$y^1_k < y^2_k$ for $k = 1, \ldots, p$
$y^1 \leqq y^2$	$y^1_k \leq y^2_k$ for $k = 1, \ldots, p$
$y^1 \leq y^2$	$y^1 \leqq y^2$ but $y^1 \neq y^2$
$\mathbb{R}^p_>$	$\{y \in \mathbb{R}^p : y > 0\}$
$\mathbb{R}^p_\geq$	$\{y \in \mathbb{R}^p : y \geq 0\}$
$\mathbb{R}^p_\geqq$	$\{y \in \mathbb{R}^p : y \geqq 0\}$

Contents

Preface v

Notation ix

Contents xi

1 **Introduction** 1
1.1 Optimization with Multiple Criteria 1
1.2 Decision Space and Objective (Criterion) Space 4
1.3 Notions of Optimality 7
1.4 Orders and Cones 8
1.5 Classification of Multicriteria Optimization Problems 16
1.6 Notes 19
Exercises 20

2 **Efficiency and Nondominance** 23
2.1 Efficient Solutions and Nondominated Points 23
2.2 Bounds on the Nondominated Set 33
2.3 Weakly and Strictly Efficient Solutions 38
2.4 Proper Efficiency and Proper Nondominance 50
2.5 Notes 59
Exercises 62

3 **The Weighted Sum Method and Related Topics** 65
3.1 Weighted Sum Scalarization and (Weak) Efficiency 68
3.2 Weighted Sum Scalarization and Proper Efficiency 71
3.3 Optimality Conditions 80
3.4 Connectedness of Efficient and Nondominated Sets 86
3.5 Notes 91

Exercises 93

4 Scalarization Techniques 97
4.1 The ε-Constraint Method 98
4.2 The Hybrid Method 101
4.3 The Elastic Constraint Method 102
4.4 Benson's Method 106
4.5 Compromise Solutions – Approximation of the Ideal Point 111
4.6 The Achievement Function Method 121
4.7 Notes 122
Exercises 124

5 Other Definitions of Optimality – Nonscalarizing Methods 127
5.1 Lexicographic Optimality 128
5.2 Max-Ordering Optimality 132
5.3 Lexicographic Max-Ordering Optimization 135
5.4 Notes 146
Exercises 148

6 Introdcution to Multicriteria Linear Programming 151
6.1 Notation and Definitions 152
6.2 The Simplex Method and Biobjective Linear Programs 160
Exercises 170

7 A Multiobjective Simplex Method 171
7.1 Algebra of Multiobjective Linear Programming 171
7.2 The Geometry of Multiobjective Linear Programming 184
7.3 Notes 193
Exercises 195

8 Multiobjective Combinatorial Optimization 197
8.1 Basics 198
8.2 Problems with Explicitely Given Feasible Set 209
8.3 Scalarization of Multiobjective Integer Programs 214
8.4 Notes 218
Exercises 220

9 Multiobjective Versions of Polynomially Solvable Problems 221
9.1 Algorithms for the Shortest Path Problem 221
9.2 The Minimum Spanning Tree Problem 236
9.3 Matroids 247
9.4 The Assignment Problem and the Two Phase Method 253

9.5 Notes 267
Exercises 269

10 Multiobjective Versions of Some $\mathcal{NP}$-Hard Problems 271
10.1 The Knapsack Problem and Branch and Bound 271
10.2 The Travelling Salesperson Problem and Heuristics 279
10.3 Notes 288
Exercises 290

Bibliography 291

List of Figures 307

List of Tables 311

Author Index 313

Subject Index 315

1

Introduction

In this book, we understand the solution of a decision problem as to choose "good" or "best" among a set of "alternatives," where we assume the existence of certain criteria, according to which the quality of the alternatives is measured. In this introductory chapter, we shall first give some examples and distinguish different types of decision problems. Informally, we shall understand optimization problems as mathematical models of decision problems. We introduce the concepts of decision (or variable) and criterion (or objective) space and mention different notions of optimality. Relations and cones are used to formally define optimization problems, and a classification scheme is introduced.

1.1 Optimization with Multiple Criteria

Let us consider the following three examples of decision problems.

Example 1.1. We want to buy a new car and have identified four models we like: a VW Golf, an Opel Astra, a Ford Focus and a Toyota Corolla. The decision will be made according to price, petrol consumption, and power. We prefer a cheap and powerful car with low petrol consumption. In this case, we face a decision problem with four alternatives and three criteria. The characteristics of the four cars are shown in Table 1.1 (data are invented).

How do we decide, which of the four cars is the "best" alternative, when the most powerful car is also the one with the highest petrol consumption, so that we cannot buy a car that is cheap as well as powerful and fuel efficient. However, we observe that with any one of the three criteria alone the choice is easy. □

Table 1.1. Criteria and alternatives in Example 1.1.

		Alternatives			
		VW	Opel	Ford	Toyota
Criteria	Price (1,000 Euros)	16.2	14.9	14.0	15.2
	Consumption $\left(\frac{l}{100km}\right)$	7.2	7.0	7.5	8.2
	Power (kW)	66.0	62.0	55.0	71.0

Example 1.2. For the construction of a water dam an electricity provider is interested in maximizing storage capacity while at the same time minimizing water loss due to evaporation and construction cost. A decision must be made on man months used for construction as well as mean radius of the lake, and also it must respect certain constraints such as minimal strength of the dam. Here, the set of alternatives (possible dam designs) allows infinitely many different choices. The criteria are functions of the decision variables to be maximized or minimized. The criteria are clearly in conflict: A dam with big storage capacity will certainly not involve small construction cost, for instance. □

Example 1.3. As a third example, we consider a mathematical problem with two criteria and one decision variable. The criteria or objective functions, which we want to minimize simultaneously over the nonnegative real line, are

$$f_1(x) = \sqrt{x+1} \text{ and } f_2(x) = x^2 - 4x + 5 = (x-2)^2 + 1, \tag{1.1}$$

plotted in Figure 1.1. We want to solve the optimization problem

$$\text{“}\min_{x \geq 0}\text{”}(f_1(x), f_2(x)). \tag{1.2}$$

The question is, what are the “minima” and the “minimizers” in this problem? Note that again, for each function individually the corresponding optimization problem is easy: $x_1 = 0$ and $x_2 = 2$ are the (unique) minimizers of f_1 and f_2 on $x \in \mathbb{R} : x \geq 0$, respectively. □

The first two examples allow a first distinction of decision problems. Those decision problems with a countable number of alternatives are called *discrete*, others *continuous*. In this book, we will be concerned with both continuous and discrete problems.

Comparing Examples 1.1 and 1.3, another distinguishing feature of decision problems becomes apparent: In Example 1.1 the alternatives are explicitly

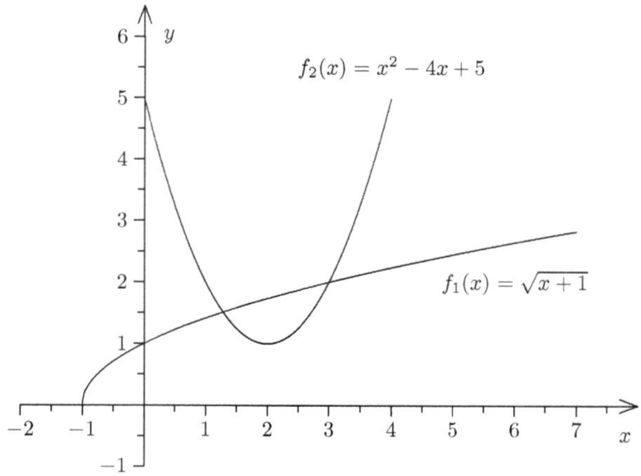

Fig. 1.1. Objective functions of Example 1.3.

given, whereas in 1.3 the alternatives are implicitly described by constraints ($x \geq 0$). Thus, we may distinguish the following types of decision problems, based on the description of the set of alternatives.

- Problems with finitely many alternatives that are explicitly known. The goal is to select a most preferred one. Multicriteria decision aid deals with such problems. We will only have one short section on such problems in this book (Section 8.2).
- Discrete problems where the set of alternatives is described by constraints in the form of mathematical functions. These problems will be covered in Chapters 8 to 10.2.
- Continuous problems. The set of alternatives is generally given through constraints. These are the objects of interest in Chapters 2.

Historically, the first reference to address such situations of conflicting objectives is usually attributed to Pareto (1896) who wrote (the quote is from the 1906 English edition of his book, emphasis added by the author):

> We will say that the members of a collectivity enjoy *maximum ophelimity* in a certain position when it is *impossible to find a way of moving from that position very slightly in such a manner that the ophelimity enjoyed by each of the individuals of that collectivity increases or decreases.* That is to say, any small displacement in departing from that position necessarily has the effect of increasing the ophelimity

which certain individuals enjoy, and decreasing that which others enjoy, of being agreeable to some and disagreeable to others.

Applying this concept in our examples, we see that in Example 1.1 all alternatives enjoy "maximum ophelimity," in Example 1.3 all x in $[0, 2]$, where one of the functions is increasing, the other decreasing. In honor of Pareto, these alternatives are today often called *Pareto optimal solutions* of multiple criteria optimization problems. We will not use that notation, however, and refer to *efficient solutions* instead (see 2.1 for a formal definition). Large parts of this book are devoted to the discussion of the mathematics of efficiency.

1.2 Decision Space and Objective (Criterion) Space

In this section, we informally introduce the fundamental notions of decision (or variable) and criterion (or objective) space, in which the alternatives and their images under the objective function mappings are contained.

Let us consider Example 1.1 again, where – for the moment – we consider price and petrol consumption only for the moment. We can illustrate the criterion values in a two-dimensional coordinate system.

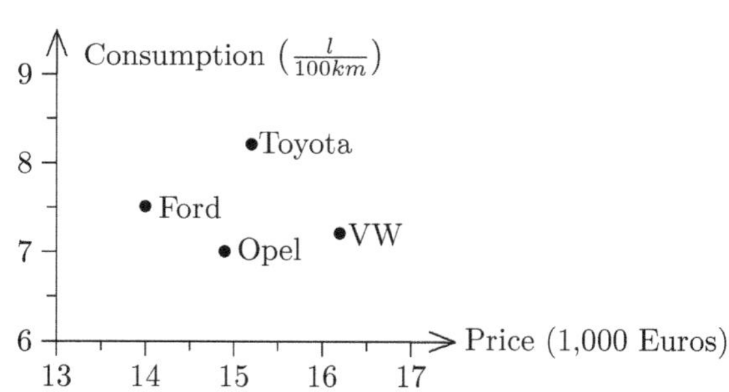

Fig. 1.2. Criterion space in Example 1.1.

From Figure 1.2 it is easy to see that Opel and Ford are the efficient choices. For both there is no alternative that is both cheaper and consumes less petrol. In addition, both Toyota and VW are more expensive and consume more petrol than Opel.

We call $\mathcal{X}$ = {VW, Opel, Ford, Toyota} the *feasible set*, or the set of alternatives of the decision problem. The space, of which the feasible set $\mathcal{X}$ is a subset, is called the *decision space*.

If we denote price by f_1 and petrol consumption by f_2, then the mappings $f_i : \mathcal{X} \to \mathbb{R}$ are criteria or objective functions and the optimization problem can be stated mathematically as in Example 1.3:

$$\text{“}\min_{x \in \mathcal{X}}\text{”}(f_1(x), f_2(x)). \tag{1.3}$$

The image of $\mathcal{X}$ under $f = (f_1, f_2)$ is denoted by $\mathcal{Y} := f(\mathcal{X}) := \{y \in \mathbb{R}^2 : y = f(x) \text{ for some } x \in \mathcal{X}\}$ and referred to as the image of the feasible set, or the feasible set in criterion space. The space from which the criterion values are taken is called the *criterion space.*

In Example 1.3 the feasible set is

$$\mathcal{X} = \{x \in \mathbb{R} : x \geq 0\} \tag{1.4}$$

and the objective functions are

$$f_1(x) = \sqrt{1+x} \text{ and } f_2(x) = x^2 - 4x + 5. \tag{1.5}$$

The decision space is $\mathbb{R}$ because $\mathcal{X} \subset \mathbb{R}$. The criterion space is $\mathbb{R}^2$, as $f(\mathcal{X}) \subset \mathbb{R}^2$. To obtain the image of the feasible set in criterion space we substitute y_1 for $f_1(x)$ and y_2 for $f_2(x)$ to get $x = (y_1)^2 - 1$ (solving $y_1 = \sqrt{1+x}$ for x). Therefore we obtain $y_2 = ((y_1)^2 - 1)^2 + 4 - 4(y_1)^2 + 5 = (y_1)^4 - 6(y_1)^2 + 10$. The graph of this function (shown in Figure 1.3) is the analogue of Figure 1.2 for Example 1.1. Note that $x \geq 0$ translates to $y_1 \geq 1$, so that $\mathcal{Y} := f(\mathcal{X})$ is the part of the graph to the right of the vertical line $y_1 = 1$.

Computing the minimum of y_2 as a function of y_1, we see that the efficient solutions $x \in [0, 2]$ found before correspond to values of $y_1 = f_1(x)$ in $[1, \sqrt{3}]$ and $y_2 = f_2(x) \in [1, 5]$. These points on the graph of $y_2(y_1)$ with $1 \leq y_1 \leq \sqrt{3}$ (and $1 \leq y_2 \leq 5$) will be called *nondominated points.*

In Figure 1.4 we can see how depicting the feasible set $\mathcal{Y}$ in criterion space can help identify nondominated points and – taking inverse images – efficient solutions. The right angle attached to the efficient point $(\hat{y}_1, \hat{y}_2)$ illustrates that there is no other point $y \in f(\mathcal{X})$, $y \neq \hat{y}$ such that $y_1 \leq \hat{y}_1$ and $y_2 \leq \hat{y}_2$. This is true for the image under f of any $x \in [0, 2]$. This observation confirms the definition of nondominated points as the image of the set of efficient points under the objective function mapping.

In the examples, we have seen that we will often have many efficient solutions of a multicriteria optimization problem. Can we consider these as “optimal decisions,” in an application context such as, e.g. the dam construction problem of Example 1.2. Or, in the car selection problem, do we have to buy all four cars after all? Obviously, a final choice has to be made among efficient

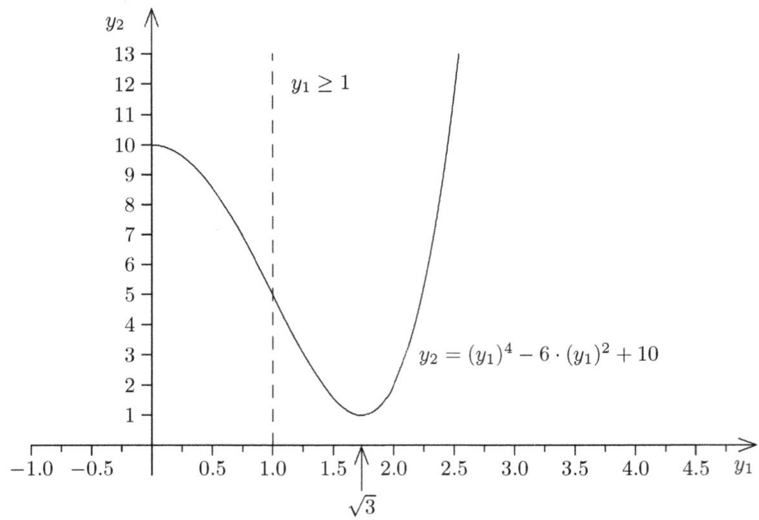

Fig. 1.3. Criterion space in Example 1.3.

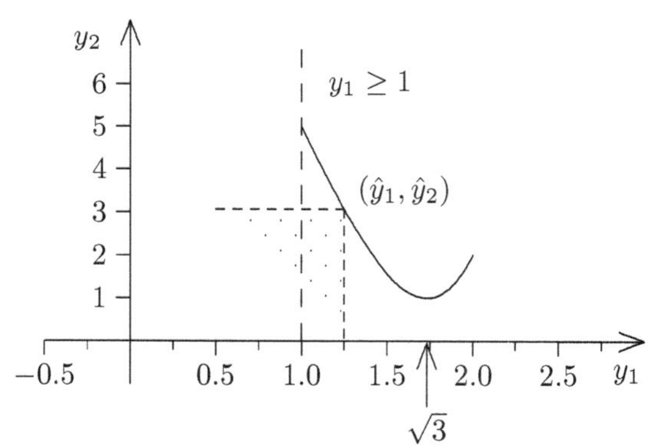

Fig. 1.4. Nondominated points in Example 1.3.

solutions. This aspect of decision making, the support of decision makers in the selection of a final solution from a set of mathematically "equally optimal" solutions, is often referred to as multicriteria decision aid (MCDA), see e.g. the textbooks of Roy (1996), Vincke (1992), or Keeney and Raiffa (1993).

Although finding efficient solutions is the most common form of multicriteria optimization, the field is not limited to that concept. There are other possibilities to cope with multiple conflicting objectives, as we shall see in the following section.

1.3 Notions of Optimality

Up to now we have written the minimization in multicriteria optimization problems in quotation marks –

$$\begin{aligned} &\text{“}\min\text{''}(f_1(x),\ldots,f_p(x)) \\ &\text{subject to } x \in \mathcal{X} \end{aligned} \tag{1.6}$$

– for good reason, since we can easily associate different interpretations with the “min.” In this and the following sections we discuss what minimization means.

The fundamental importance of efficiency (Pareto optimality) is based on the observation that any x which is not efficient cannot represent a most preferred alternative for a decision maker, because there exists at least one other feasible solution $x' \in \mathcal{X}$ such that $f_k(x') \leq f_k(x)$ for all $k = 1, \ldots, p$, where strict inequality holds at least once, i.e., x' should clearly be preferred to x. So for all definitions of optimality we deal with in this text, the relationship with efficiency will always be a topic which needs to be and will be discussed. Some other notions of optimality are informally presented now.

We can imagine situations in which there is a ranking among the objectives. In Example 1.1, price might be more important than petrol consumption, this in turn more important than power. This means that even an extremely good value for petrol consumption cannot compensate for a slightly higher price. Then the criterion vectors $(f_1(x), f_2(x), f_3(x))$ are compared lexicographically (see Table 1.2 for a definition of the lexicographic order and Section 5.1 for more on lexicographic optimization) and we would want to solve

$$\operatorname*{lexmin}_{x \in \mathcal{X}} (f_1(x), f_2(x), f_3(x)). \tag{1.7}$$

In Example 1.1 we should choose the Ford because for this ranking of objectives it is the unique optimal solution (the cheapest).

Let us assume that in Example 1.3 the objective functions measure some negative impacts of a decision (environmental pollution, etc.) to be minimized. We might not want to accept a high value of one criterion for a low value of the other. It is then appropriate to minimize the worst of both objectives. Accordingly we would solve

$$\min_{x \geq 0} \max_{i=1,2} f_i(x). \tag{1.8}$$

This problem is illustrated in Figure 1.5, where the solid line shows the maximum of f_1 and f_2. The optimal solution of the problem is obtained for $x \approx 1.285$, see Figure 1.5.

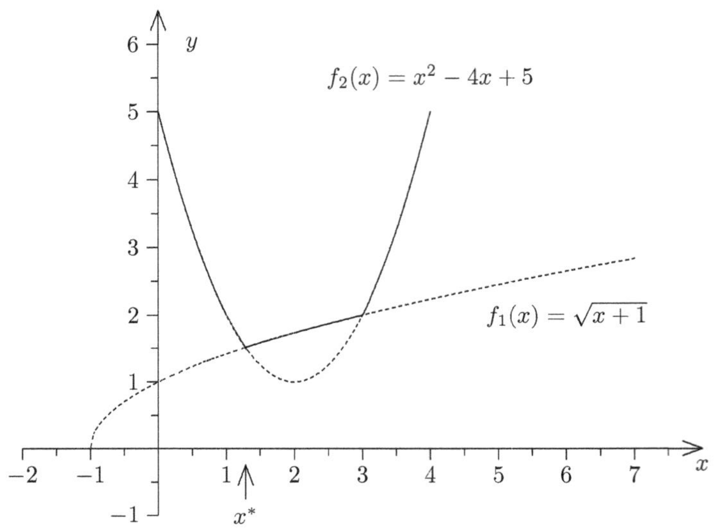

Fig. 1.5. Min-max solution of Example 1.3.

In both examples, we got unique optimal solutions, and there are no incomparable values. And indeed, in the min-max example one could think of this problem as a single objective optimization problem. However, both have to be considered as multicriteria problems, because the multiple objectives are in the formulation of the problems. Thus, in order to define the meaning of "min," we have to define how objective function vectors $(f_1(x), \dots, f_p(x))$ have to be compared for different alternatives $x \in \mathcal{X}$. The different possibilities to do that arise from the fact that for $p \geq 2$ there is no canonical order on $\mathbb{R}^p$ as there is on $\mathbb{R}$. Therefore weaker definitions of orders have to be used.

1.4 Orders and Cones

In this section we will first introduce binary relations and some of their properties to define several classes of orders. The second main topic is cones, defining sets of nonnegative elements of $\mathbb{R}^p$. We will prove the equivalence of properties of orders and geometrical properties of cones. An indication of the relationship between orders and cones has already been shown in Figure 1.4, where we used a cone (the negative orthant of $\mathbb{R}^2$) to confirm that $\hat{y}$ is nondominated.

Let $\mathcal{S}$ be any set. A *binary relation* on $\mathcal{S}$ is a subset $\mathcal{R}$ of $\mathcal{S} \times \mathcal{S}$. We introduce some properties of binary relations.

Definition 1.4. *A binary relation $\mathcal{R}$ on $\mathcal{S}$ is called*

- reflexive *if $(s,s) \in \mathcal{R}$ for all $s \in \mathcal{S}$,*
- irreflexive *if $(s,s) \notin \mathcal{R}$ for all $s \in \mathcal{S}$,*
- symmetric *if $(s^1,s^2) \in \mathcal{R} \Longrightarrow (s^2,s^1) \in \mathcal{R}$ for all $s^1,s^2 \in \mathcal{S}$,*
- asymmetric *if $(s^1,s^2) \in \mathcal{R} \Longrightarrow (s^2,s^1) \notin \mathcal{R}$ for all $s^1,s^2 \in \mathcal{S}$,*
- antisymmetric *if $(s^1,s^2) \in \mathcal{R}$ and $(s^2,s^1) \in \mathcal{R} \Longrightarrow s^1 = s^2$ for all $s^1,s^2 \in \mathcal{S}$,*
- transitive *if $(s^1,s^2) \in \mathcal{R}$ and $(s^2,s^3) \in \mathcal{R} \Longrightarrow (s^1,s^3) \in \mathcal{R}$ for all $s^1,s^2,s^3 \in \mathcal{S}$,*
- negatively transitive *if $(s^1,s^2) \notin \mathcal{R}$ and $(s^2,s^3) \notin \mathcal{R} \Longrightarrow (s^1,s^3) \notin \mathcal{R}$ for all $s^1,s^2,s^3 \in \mathcal{S}$,*
- connected *if $(s^1,s^2) \in \mathcal{R}$ or $(s^2,s^1) \in \mathcal{R}$ for all $s^1,s^2 \in \mathcal{S}$ with $s^1 \neq s^2$,*
- strongly connected *(or* total*) if $(s^1,s^2) \in \mathcal{R}$ or $(s^2,s^1) \in \mathcal{R}$ for all $s^1,s^2 \in \mathcal{S}$.*

Definition 1.5. *A binary relation $\mathcal{R}$ on a set $\mathcal{S}$ is*

- *an* equivalence relation *if it is reflexive, symmetric, and transitive,*
- *a* preorder (quasi-order) *if it is reflexive and transitive.*

Instead of $(s^1,s^2) \in \mathcal{R}$ we shall also write $s^1 \mathcal{R} s^2$. In the case of $\mathcal{R}$ being a preorder the pair $(\mathcal{S},\mathcal{R})$ is called a *preordered set.* In the context of (pre)orders yet another notation for the relation $\mathcal{R}$ is convenient. We shall write $s^1 \preceq s^2$ as shorthand for $(s^1,s^2) \in \mathcal{R}$ and $s^1 \npreceq s^2$ for $(s^1,s^2) \notin \mathcal{R}$ and indiscriminately refer to the relation $\mathcal{R}$ or the relation $\preceq$. This notation can be read as "preferred to."

Given any preorder $\preceq$, two other relations are closely associated with $\preceq$. We define them as follows:

$$s^1 \prec s^2 :\Longleftrightarrow s^1 \preceq s^2 \text{ and } s^2 \npreceq s^1, \tag{1.9}$$

$$s^1 \sim s^2 :\Longleftrightarrow s^1 \preceq s_2 \text{ and } s^2 \preceq s_1. \tag{1.10}$$

Actually, $\prec$ and $\sim$ can be seen as the strict preference and equivalence (or indifference) relation, respectively, associated with the preference defined by preorder $\preceq$.

Proposition 1.6. *Let $\preceq$ be a preorder on $\mathcal{S}$. Then relation $\prec$ defined in (1.9) is irreflexive and transitive and relation $\sim$ defined in (1.10) is an equivalence relation.*

Proof. We consider $\sim$ first. This relation is reflexive because $\preceq$ is. Furthermore $\sim$ is symmetric by definition. Now let $s^1, s^2, s^3 \in \mathcal{S}$ be such that $s^1 \sim s^2$ and $s^2 \sim s^3$. Then using transitivity of $\preceq$

$$\left.\begin{array}{l} s^1 \preceq s^2 \preceq s^3 \implies s^1 \preceq s^3 \\ s^3 \preceq s^2 \preceq s^1 \implies s^3 \preceq s^1 \end{array}\right\} \implies s^1 \sim s^3. \tag{1.11}$$

For $\prec$, note that $\prec$ is irreflexive by definition. Suppose there are $s^1, s^2, s^3 \in \mathcal{S}$ such that $s^1 \prec s^2$ and $s^2 \prec s^3$. Then $s^1 \preceq s^2 \preceq s^3$ and from transitivity of $\preceq$, $s^1 \preceq s^3$. To show that $s^1 \prec s^3$, assume $s^3 \preceq s^1$. But since $s^1 \preceq s^2$ we get $s^3 \preceq s^2$ from transitivity of $\preceq$. This contradiction implies $s^3 \not\preceq s^1$, i.e., $s^1 \prec s^3$. □

Another easily seen result concerns asymmetry and irreflexivity of binary relations.

Proposition 1.7. *An asymmetric binary relation is irreflexive. A transitive, irreflexive binary relation is asymmetric.*

Proof. The proof is left to the reader, see Exercise 1.4 □

Definition 1.8. *A binary relation $\preceq$ on $\mathcal{S}$ is*

- *a* total preorder *if it is reflexive, transitive and connected,*
- *a* total order *if it is an antisymmetric total preorder,*
- *a* strict weak order *if it is asymmetric and negatively transitive.*

From total preorders, strict weak orders can be obtained and vice versa, as Proposition 1.9 shows.

Proposition 1.9. *If $\preceq$ is a total preorder on $\mathcal{S}$, then the associated relation $\prec$ is a strict weak order. If $\prec$ is a strict weak order on $\mathcal{S}$, then $\preceq$ defined by*

$$s^1 \preceq s^2 \iff \text{either } s^1 \prec s^2 \text{ or } (s^1 \not\prec s^2 \text{ and } s^2 \not\prec s^1) \tag{1.12}$$

is a total preorder.

Proof. Let $\preceq$ be a total preorder on $\mathcal{S}$. Then $\prec$ is irreflexive and transitive by Proposition 1.6 and hence asymmetric by Proposition 1.7. For negative transitivity we show that $s^1 \not\prec s^2, s^2 \not\prec s^3$ implies $s^1 \not\prec s^3$ for all $s^1, s^2, s^3 \in \mathcal{S}$. So let $s^1, s^2, s^3 \in \mathcal{S}$ such that $s^1 \not\prec s^2$ and $s^2 \not\prec s^3$ and assume $s^1 \prec s^3$. From

$s^1 \not\prec s^2$ we have $s^2 \prec s^1$ or $s^2 \preceq s^1$ because $\preceq$ is conected. In both cases it follows that $s^2 \prec s^3$, contradicting the assumption.

Let $\prec$ be a strict weak order on $\mathcal{S}$. The relation $\preceq$ is reflexive by definition. For transitivity consider the following cases for $s^1, s^2, s^3 \in \mathcal{S}$ with $s^1 \preceq s^2$ and $s^2 \preceq s^3$:

1. $s^1 \prec s^2$, $s^2 \not\prec s^3$ and $s^3 \not\prec s^2$. Then $s^1 \prec s^3$, because otherwise $s^1 \not\prec s^3$ and $s^3 \not\prec s^2$ imply $s^1 \not\prec s^2$, a contradiction.
2. $s^1 \not\prec s^2$, $s^2 \not\prec s^1$ and $s^2 \prec s^3$. Then $s^1 \prec s^3$ because otherwise $s^1 \not\prec s^3$ and $s^2 \not\prec s^1$ imply $s^2 \not\prec s^3$, again a contradiction.
3. $s^1 \not\prec s^2$, $s^2 \not\prec s^1$, $s^2 \not\prec s^3$, $s^3 \not\prec s^2$. Then $s^1 \not\prec s^3$ and $s^3 \not\prec s^1$ (from negative transitivity) imply $s^1 \preceq s^3$.
4. $s^1 \prec s^2$ and $s^2 \prec s^3$. We get $s^2 \not\prec s^1$ from asymmetry and from $s^1 \prec s^2$. Thus, if $s^1 \not\prec s^3$, negative transitivity implies $s^2 \not\prec s^3$, a contradiction.

In all cases we can conclude $s^1 \preceq s^3$, as desired. Finally, for connectedness let $s^1, s^2 \in \mathcal{S}$, $s^1 \neq s^2$. Then $s^1 \prec s^2$ or $s^2 \prec s^1$ or ($s^1 \not\prec s^2$ and $s^2 \not\prec s^1$) and therefore $s^1 \preceq s^2$ or $s^2 \preceq s^1$. □

The most important classes of relations in multicriteria optimization – partial orders and strict partial orders – are introduced now.

Definition 1.10. *A binary relation $\preceq$ is called*

- partial order *if it is reflexive, transitive and antisymmetric,*
- strict partial order *if it is asymmetric and transitive (or, equivalently, if it is irreflexive and transitive).*

Throughout this book, we use several orders on the Euclidian space $\mathbb{R}^p$ which we define now. Please note that these notations are not unique in multicriteria optimization literature and always check definitions when consulting another source. Let $y^1, y^2 \in \mathbb{R}^p$, and if $y^1 \neq y^2$ let $k^* := \min\{k : y^1_k \neq y^2_k\}$. We shall use the notations and names given in Table 1.2 for the most common ((strict) partial) orders on $\mathbb{R}^p$ appearing in this text.

With the (weak, strict) componentwise orders, we define subsets of $\mathbb{R}^p$ as follows:

- $\mathbb{R}^p_{\geqq} := \{y \in \mathbb{R}^p : y \geqq 0\}$, the nonnegative orthant of $\mathbb{R}^p$;
- $\mathbb{R}^p_{\geq} := \{y \in \mathbb{R}^p : y \geq 0\} = \mathbb{R}^p_{\geqq} \setminus \{0\}$;
- $\mathbb{R}^p_{>} := \{y \in \mathbb{R}^p : y > 0\} = \operatorname{int} \mathbb{R}^p_{\geqq}$, the positive orthant of $\mathbb{R}^p$.

Note that for $p = 1$ we have $\mathbb{R}_{\geq} = \mathbb{R}_{>}$.

We can now proceed to show how the definition of a set of nonnegative elements in $\mathbb{R}^p$ ($\mathbb{R}^2$ for purposes of illustration) can be used to derive a geometric interpretation of properties of orders. These equivalent views on orders

Table 1.2. Some orders on $\mathbb{R}^p$.

Notation	Definition	Name
$y^1 \leqq y^2$	$y_k^1 \leq y_k^2 \quad k = 1, \ldots, p$	weak componentwise order
$y^1 \leq y^2$	$y_k^1 \leq y_k^2 \quad k = 1, \ldots, p;\ y^1 \neq y^2$	componentwise order
$y^1 < y^2$	$y_k^1 < y_k^2 \quad k = 1, \ldots, p$	strict componentwise order
$y^1 \leq_{\text{lex}} y^2$	$y_{k^*}^1 < y_{k^*}^2$ or $y^1 = y^2$	lexicographic order
$y^1 \leq_{MO} y$	$\max_{k=1,\ldots,p} y_k^1 \leq \max_{k=1,\ldots,n} y_k^2$	max-order

will be extremely useful in multicriteria optimization. But first we need the definition of a cone.

Definition 1.11. *A subset $\mathcal{C} \subseteq \mathbb{R}^p$ is called a* cone, *if $\alpha d \in \mathcal{C}$ for all $d \in \mathcal{C}$ and for all $\alpha \in \mathbb{R}, \alpha > 0$.*

Example 1.12. The left drawing in Figure 1.6 shows the cone $\mathcal{C} = \{d \in \mathbb{R}^2 : d_k \geq 0, k = 1, 2\} = \mathbb{R}^2_{\geqq}$. This is the cone of nonnegative elements of the weak componentwise order. The right drawing shows a smaller cone $\mathcal{C} \subset \mathbb{R}^2_{\geqq}$.

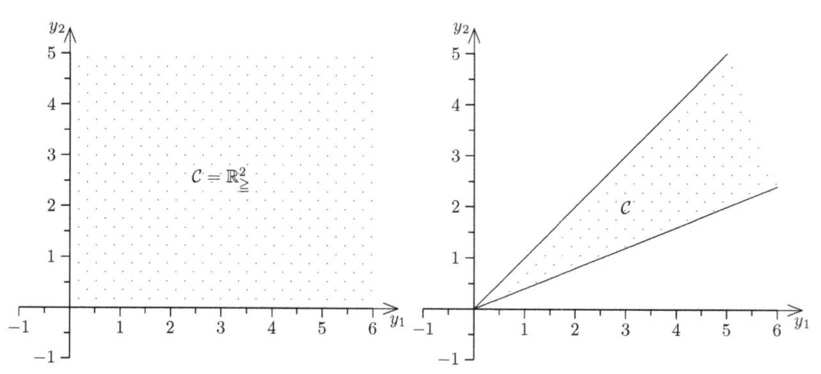

Fig. 1.6. Illustration of two cones.

□

For the following discussion it will be useful to have the operations of the multiplication of a set with a scalar and the sum of two sets. Let $\mathcal{S}, \mathcal{S}_1, \mathcal{S}_2 \subset \mathbb{R}^p$ and $\alpha \in \mathbb{R}$. We denote by

$$\alpha\mathcal{S} := \{\alpha s : s \in \mathcal{S}\}, \tag{1.13}$$

especially $-\mathcal{S} = \{-s : s \in \mathcal{S}\}$. Furthermore, the (algebraic, Minkowski) sum of $\mathcal{S}_1$ and $\mathcal{S}_2$ is

$$\mathcal{S}_1 + \mathcal{S}_2 := \{s^1 + s^2 : s^1 \in \mathcal{S}_1, s^2 \in \mathcal{S}_2\}. \tag{1.14}$$

If $\mathcal{S}_1 = \{s\}$ is a singleton, we also write $s + \mathcal{S}_2$ instead of $\{s\} + \mathcal{S}_2$. Note that these are just simplified notations that do not involve any set arithmetic, e.g. $2\mathcal{S} \neq \mathcal{S} + \mathcal{S}$ in general.

This is also the appropriate place to introduce some further notation used throughout the book. For $\mathcal{S} \subseteq \mathbb{R}^n$ or $\mathcal{S} \subseteq \mathbb{R}^p$

- $\text{int}(\mathcal{S})$ is the interior of $\mathcal{S}$,
- $\text{ri}(\mathcal{S})$ is the relative interior of $\mathcal{S}$,
- $\text{bd}(\mathcal{S})$ is the boundary of $\mathcal{S}$,
- $\text{cl}(\mathcal{S}) = \text{int}(\mathcal{S}) \cup \text{bd}(\mathcal{S})$ is the closure of $\mathcal{S}$,
- $\text{conv}(\mathcal{S})$ is the convex hull of $\mathcal{S}$.

The parentheses might be omitted for simplification of expressions when the argument is clear.

Definition 1.13. *A cone $\mathcal{C}$ in $\mathbb{R}^p$ is called*

- nontrivial *or* proper *if $\mathcal{C} \neq \emptyset$ and $\mathcal{C} \neq \mathbb{R}^n$,*
- convex *if $\alpha d^1 + (1-\alpha)d^2 \in \mathcal{C}$ for all $d^1, d^2 \in \mathcal{C}$ and for all $0 < \alpha < 1$,*
- pointed *if for $d \in \mathcal{C}, d \neq 0$, $-d \notin \mathcal{C}$, i.e., $\mathcal{C} \cap (-\mathcal{C}) \subseteq \{0\}$.*

Due to the definition of a cone, $\mathcal{C}$ is convex if for all $d^1, d^2 \in \mathcal{C}$ we have $d^1 + d^2 \in \mathcal{C}$, too: $\alpha d^1 \in \mathcal{C}$ and $(1-\alpha)d^2 \in \mathcal{C}$ because $\mathcal{C}$ is a cone. Therefore, closedness of $\mathcal{C}$ under addition is sufficient for convexity. Then, using the algebraic sum, we can say that $\mathcal{C} \subset \mathbb{R}^p$ is a convex cone if $\alpha\mathcal{C} \subseteq \mathcal{C}$ for all $\alpha > 0$ and $\mathcal{C} + \mathcal{C} \subseteq \mathcal{C}$. We will only consider nontrivial cones throughout the book.

Given an order relation $\mathcal{R}$ on $\mathbb{R}^p$, we can define a set

$$\mathcal{C}_{\mathcal{R}} := \{y^2 - y^1 : y^1 \mathcal{R} y^2\}, \tag{1.15}$$

which we would like to interpret as the set of nonnegative elements of $\mathbb{R}^p$ according to $\mathcal{R}$. We will now prove some relationships between the properties of $\mathcal{R}$ and $\mathcal{C}_{\mathcal{R}}$.

Proposition 1.14. *Let $\mathcal{R}$ be compatible with scalar multiplication, i.e., for all $(y^1, y^2) \in \mathcal{R}$ and all $\alpha \in \mathbb{R}_>$ it holds that $(\alpha y^1, \alpha y^2) \in \mathcal{R}$. Then $\mathcal{C}_{\mathcal{R}}$ defined in (1.15) is a cone.*

Proof. Let $d \in \mathcal{C}_\mathcal{R}$. Then $d = y^2 - y^1$ for some $y^1, y^2 \in \mathbb{R}^p$ with $(y^1, y^2) \in \mathcal{R}$. Thus $(\alpha y^1, \alpha y^2) \in \mathcal{R}$ for all $\alpha > 0$. Hence $\alpha d = \alpha(y^2 - y^1) = \alpha y^2 - \alpha y^1 \in \mathcal{C}_\mathcal{R}$ for all $\alpha > 0$. □

Example 1.15. Let us consider the weak componentwise order on $\mathbb{R}^p$. Here $y^1 \leqq y^2$ if and only if $y^1_k \leq y^2_k$ for all $k = 1, \ldots, p$ or $y^2_k - y^1_k \geq 0$ for all $k = 1, \ldots, p$. Therefore $\mathcal{C}_\leqq = \{d \in \mathbb{R}^p : d_k \geq 0, k = 1, \ldots, p\} = \mathbb{R}^p_\geqq$. □

It is interesting to consider the definition (1.15) with $y^1 \in \mathbb{R}^p$ fixed, i.e., $\mathcal{C}_\mathcal{R}(y^1) = \{y^2 - y^1 : y^1 \mathcal{R} y^2\}$. If $\mathcal{R}$ is an order relation, $y^1 + \mathcal{C}_\mathcal{R}(y^1)$ is the set of elements of $\mathbb{R}^p$ that y^1 is preferred to or that are dominated by y^1.

A natural question to ask is: Under what conditions is $\mathcal{C}_\mathcal{R}(y)$ the same for all $y \in \mathbb{R}^p$? In order to answer that question, we need another assumption on order relation $\mathcal{R}$. $\mathcal{R}$ is said to be compatible with addition if $(y^1+z, y^2+z) \in \mathcal{R}$ for all $z \in \mathbb{R}^p$ and all $(y^1, y^2) \in \mathcal{R}$.

Lemma 1.16. *If $\mathcal{R}$ is compatible with addition and $d \in \mathcal{C}_\mathcal{R}$ then $0 \mathcal{R} d$.*

Proof. Let $d \in \mathcal{C}_\mathcal{R}$. Then there are $y^1, y^2 \in \mathbb{R}^p$ with $y^1 \mathcal{R} y^2$ such that $d = y^2 - y^1$. Using $z = -y^1$, compatibility with addition implies $(y^1 + z)\mathcal{R}(y^2 + z)$ or $0 \mathcal{R} d$. □

Lemma 1.16 means that if $\mathcal{R}$ is compatible with addition, the sets $\mathcal{C}_\mathcal{R}(y)$, $y \in \mathbb{R}^p$, do not depend on y. In this book, we will be mainly concerned with this case. For relations that are compatible with addition, we obtain further results.

Theorem 1.17. *Let $\mathcal{R}$ be a binary relation on $\mathbb{R}^p$ which is compatible with scalar multiplication and addition. Then the following statements hold.*

1. *$0 \in \mathcal{C}_\mathcal{R}$ if and only if $\mathcal{R}$ is reflexive.*
2. *$\mathcal{C}_\mathcal{R}$ is pointed if and only if $\mathcal{R}$ is antisymmetric.*
3. *$\mathcal{C}_\mathcal{R}$ is convex if and only if $\mathcal{R}$ is transitive.*

Proof.
1. Let $\mathcal{R}$ be reflexive and let $y \in \mathbb{R}^p$. Then $y \mathcal{R} y$ and $y - y = 0 \in \mathcal{C}_\mathcal{R}$. Let $0 \in \mathcal{C}_\mathcal{R}$. Then there is some $y \in \mathbb{R}^p$ with $y \mathcal{R} y$. Now let $y' \in \mathbb{R}^p$. Then $y' = y + z$ for some $z \in \mathbb{R}^p$. Since $y \mathcal{R} y$ and $\mathcal{R}$ is compatible with addition we get $y' \mathcal{R} y'$.
2. Let $\mathcal{R}$ be antisymmetric and let $d \in \mathcal{C}_\mathcal{R}$ such that $-d \in \mathcal{C}_\mathcal{R}$, too. Then there are $y^1, y^2 \in \mathbb{R}^p$ such that $y^1 \mathcal{R} y^2$ and $d = y^1 - y^1$ as well as $y^3, y^4 \in \mathbb{R}^p$ such that $y^3 \mathcal{R} y^4$ and $-d = y^4 - y^3$. Thus, $y^2 - y^1 = y^3 - y^4$ and there must be $y \in \mathbb{R}^p$ such that $y^2 = y^3 + y$ and $y^1 = y^4 + y$. Therefore compatibility with addition implies $y^2 \mathcal{R} y^1$. Antisymmetry of $\mathcal{R}$ now yields $y^2 = y^1$ and therefore $d = 0$, i.e., $\mathcal{C}_\mathcal{R}$ is pointed.

Let $y^1, y^2 \in \mathbb{R}^p$ with $y^1 \mathcal{R} y^2$ and $y^2 \mathcal{R} y^1$. Thus, $d = y^2 - y^1 \in \mathcal{C}_\mathcal{R}$ and $-d = y^1 - y^2 \in \mathcal{C}_\mathcal{R}$. If $\mathcal{C}_\mathcal{R}$ is pointed we know that $\{d, -d\} \subset \mathcal{C}$ implies $d = 0$ and therefore $y^1 = y^2$, i.e., $\mathcal{R}$ is antisymmetric.

3. Let $\mathcal{R}$ be transitive and let $d^1, d^2 \in \mathcal{C}_\mathcal{R}$. Since $\mathcal{R}$ is compatible with scalar multiplication, $\mathcal{C}_\mathcal{R}$ is a cone and we only need to show $d^1 + d^2 \in \mathcal{C}_\mathcal{R}$. By Lemma 1.16 we have $0 \mathcal{R} d^1$ and $0 \mathcal{R} d^2$. Compatibility with addition implies $d^1 \mathcal{R} (d^1 + d^2)$, transitivity yields $0 \mathcal{R} (d^1 + d^2)$, from which $d^1 + d^2 \in \mathcal{C}_\mathcal{R}$.
   Let $\mathcal{C}_\mathcal{R}$ be convex and let $y^1, y^2, y^3 \in \mathbb{R}^p$ be such that $y^1 \mathcal{R} y^2$ and $y^2 \mathcal{R} y^3$. Then $d^1 = y^2 - y^1 \in \mathcal{C}_\mathcal{R}$ and $d^2 = y^3 - y^2 \in \mathcal{C}_\mathcal{R}$. Because $\mathcal{C}_\mathcal{R}$ is convex, $d^1 + d^2 = y^3 - y^1 \in \mathcal{C}_\mathcal{R}$. By Lemma 1.16 we get $0 \mathcal{R} (y^3 - y^1)$ and by compatibility with addition $y^1 \mathcal{R} y^3$. □

Example 1.18. 1. The weak componentwise order $\leqq$ is compatible with addition and scalar multiplication. $\mathcal{C}_\leqq = \mathbb{R}^p_\geqq$ contains 0, is pointed, and convex.

2. The max-order $\leq_{MO}$ is compatible with scalar multiplication, but not with addition (e.g. (-3,2) $\leq_{MO}$ (3,1), but this relation is reversed when adding (0,3)). Furthermore, $\leq_{MO}$ is reflexive, transitive, but not antisymmetric (e.g. (1,0) $\leq_{MO}$ (1,1) and (1,1) $\leq_{MO}$ (1,0)).

□

We have defined cone $\mathcal{C}_\mathcal{R}$ given relation $\mathcal{R}$. We can also use a cone to define an order relation. Let $\mathcal{C}$ be a cone. Define $\mathcal{R}_\mathcal{C}$ by

$$y^1 \mathcal{R}_\mathcal{C} y^2 \iff y^2 - y^1 \in \mathcal{C}. \tag{1.16}$$

Proposition 1.19. *Let $\mathcal{C}$ be a cone. Then $\mathcal{R}_\mathcal{C}$ defined in (1.16) is compatible with scalar multiplication and addition in $\mathbb{R}^p$.*

Proof. Let $y^1, y^2 \in \mathbb{R}^p$ be such that $y^1 \mathcal{R}_\mathcal{C} y^2$. Then $d = y^2 - y^1 \in \mathcal{C}$. Because $\mathcal{C}$ is a cone $\alpha d = \alpha(y^2 - y^1) = \alpha y^2 - \alpha y^1 \in \mathcal{C}$. Thus $\alpha y^1 \mathcal{R}_\mathcal{C} \alpha y^2$ for all $\alpha > 0$. Furthermore, $(y^2 + z) - (y^1 + z) \in \mathcal{C}$ and $(y^1 + z) \mathcal{R}_\mathcal{C} (y^2 + z)$ for all $z \in \mathbb{R}^p$. □

Theorem 1.20. *Let $\mathcal{C}$ be a cone and let $\mathcal{R}_\mathcal{C}$ be as defined in (1.16). Then the following statements hold.*

1. *$\mathcal{R}_\mathcal{C}$ is reflexive if and only if $0 \in \mathcal{C}$.*
2. *$\mathcal{R}_\mathcal{C}$ is antisymmetric if and only if $\mathcal{C}$ is pointed.*
3. *$\mathcal{R}_\mathcal{C}$ is transitive if and only if $\mathcal{C}$ is convex.*

Proof. 1. Let $0 \in \mathcal{C}$ and $y \in \mathbb{R}^p$. Thus, $y - y \in \mathcal{C}$ and $y \mathcal{R}_\mathcal{C} y$ for all $y \in \mathbb{R}^p$.
Let $\mathcal{R}_\mathcal{C}$ be reflexive. Then we have $y \mathcal{R}_\mathcal{C} y$ for all $y \in \mathbb{R}^p$, i.e., $y - y = 0 \in \mathcal{C}$.

2. Let $d \in \mathcal{C}$ and $-d \in \mathcal{C}$. Thus $0\mathcal{R}_{\mathcal{C}}d$ and $0\mathcal{R}_{\mathcal{C}} - d$. Adding d to the latter relation, compatibility with addition yields $d\mathcal{R}_{\mathcal{C}}0$. Then asymmetry implies $d = 0$.
   Let $y^1, y^2 \in \mathbb{R}^p$ be such that $y^1\mathcal{R}_{\mathcal{C}}y^2$ and $y^2\mathcal{R}_{\mathcal{C}}y^1$. Thus, $d = y^2 - y^1$ and $-d = y^1 - y^2 \in \mathcal{C}$. Since $\mathcal{C}$ is pointed, $d = 0$, i.e. $y^1 = y^2$.
3. Let $y^1, y^2, y^3 \in \mathbb{R}^p$ such that $y^1\mathcal{R}_{\mathcal{C}}y^2$ and $y^2\mathcal{R}_{\mathcal{C}}y^3$. Therefore $d^1 = y^2 - y^1 \in \mathcal{C}$ and $d^2 = y^3 - y^2 \in \mathcal{C}$. Because $\mathcal{C}$ is convex, $d^1 + d^2 = y^3 - y^1 \in \mathcal{C}$ and $y^1\mathcal{R}_{\mathcal{C}}y^3$.
   If $d^1, d^2 \in \mathcal{C}$ we have $0\mathcal{R}_{\mathcal{C}}d^1$ and $0\mathcal{R}_{\mathcal{C}}d^2$. Because $\mathcal{R}_{\mathcal{C}}$ is compatible with addition, we get $d^1\mathcal{R}_{\mathcal{C}}(d^1 + d^2)$. By transitivity $0\mathcal{R}_{\mathcal{C}}(d^1 + d^2)$ and $d^1 + d^2 \in \mathcal{C}$. □

Note that Theorem 1.20 does not need the assumption of compatibility with addition since it is a consequence of the definition of $\mathcal{R}_{\mathcal{C}}$. The relationships between cones and binary relations are further investigated in the exercises.

With Theorems 1.17 and 1.20 we have shown equivalence of some partial orders and pointed convex cones containing 0. Since (partial) orders can be used to define "minimization," these results make it possible to analyze multicriteria optimization problems geometrically.

1.5 Classification of Multicriteria Optimization Problems

By the choice of an order $\preceq$ on $\mathbb{R}^p$, we can finally define the meaning of " min" in the problem formulation

$$\text{“}\min_{x \in \mathcal{X}}\text{”} f(x) = \text{“}\min_{x \in \mathcal{X}}\text{”}(f_1(x), \ldots, f_p(x)). \tag{1.17}$$

The different interpretations of "min" pertaining to different orders are the foundation of a classification of multicriteria optimization problems. We only briefly mention it here. A more detailed development can be found in Ehrgott (1997) and Ehrgott (1998).

With the multiple objective functions we can evaluate objective value vectors $(f_1(x), \ldots, f_p(x))$. However, we have seen that these vectors $y = f(x)$, $x \in \mathcal{X}$, are not always compared in objective space, i.e., $\mathbb{R}^p$, directly.

In Example 1.3 we have formulated the optimization problem

$$\min_{x \in \mathcal{X}} \max_{i=1,2} f_i(x). \tag{1.18}$$

That is, we have used a mapping $\theta : \mathbb{R}^2 \to \mathbb{R}$ from objective space $\mathbb{R}^2$ to $\mathbb{R}$, where the min in (1.18) is actually defined by the canonical order on $\mathbb{R}$.

In general, the objective function vectors are mapped from $\mathbb{R}^p$ to an ordered space, e.g. $(\mathbb{R}^P, \preceq)$, where comparisons are made using the order relation $\preceq$. This mapping is called the *model map.*

With the model map, we can now summarize the elements of a multicriteria optimization problem (MOP). These are

- the feasible set $\mathcal{X}$,
- the objective function vector $f = (f_1, \ldots, f_p) : \mathcal{X} \longrightarrow \mathbb{R}^p$,
- the objective space $\mathbb{R}^p$,
- the ordered set $(\mathbb{R}^P, \preceq)$,
- the model map θ.

Feasible set, objective function vector f, and objective space are the *data* of the MOP. The model map provides the link between objective space and ordered set, in which, finally, the meaning of the minimization is defined. Thus with the three main aspects data, model map, and ordered set the classification $(\mathcal{X}, f, \mathbb{R}^p)/\theta/(\mathbb{R}^P, \preceq)$ completely describes a multicriteria optimization problem.

Example 1.21. Let us look at a problem of finding efficient solutions,

$$\min_{x \geq 0}(\sqrt{x+1},\ x^2 - 4x + 1). \tag{1.19}$$

Here $\mathcal{X} = \{x : x \geq 0\} = \mathbb{R}_{\geqq}$ is the feasible set, $f = (f_1, f_2) = (\sqrt{x+1}, x^2 - 4x + 1)$ is the objective function vector, and $\mathbb{R}^p = \mathbb{R}^2$ is the objective space defining the data. Because we compare objective function vectors componentwise, the model map is given by $\theta(y) = y$ and denoted id, the *identity mapping*, henceforth. The ordered set is then $(\mathbb{R}^P, \preceq) = (\mathbb{R}^2, \leq)$. The problem (1.19) is classified as

$$(\mathbb{R}_{\geqq}, f, \mathbb{R}^2)/\text{id}/(\mathbb{R}^2, \leq). \tag{1.20}$$

□

Example 1.22. If we have a ranking of objectives as described in the second example in Section 1.3, we compare objective vectors lexicographically. Let $y^1, y^2 \in \mathbb{R}^p$. Then $y^1 \leq_{\text{lex}} y^2$ if there is some $k^*, 1 \leq k^* \leq p$ such that $y^1_k = y^2_k \;\; k = 1, \ldots, k^* - 1$ and $y^1_{k^*} < y^2_{k^*}$ or $y^1 = y^2$. In the car selection Example 1.1, $\mathcal{X} = \{\text{VW, Opel, Ford, Toyota}\}$ is the set of alternatives (feasible set), f_1 is price, f_2 is petrol consumption, and f_3 is power. We define $\theta(y) = (y_1, y_2, -y_3)$ (note that more power is preferred to less). The problem is then classified as

$$(\mathcal{X}, f, \mathbb{R}^3)/\theta/(\mathbb{R}^3, \leq_{\text{lex}}) \tag{1.21}$$

□

Lexicographic optimality is one of the concepts we cover in Chapter 5.

At the end of this chapter, we formally define optimal solutions and optimal values of multicriteria optimization problems.

Definition 1.23. *A feasible solution* $x^* \in \mathcal{X}$ *is called an* optimal solution *of a multicriteria optimization problem* $(\mathcal{X}, f, \mathbb{R}^p)/\theta/(\mathbb{R}^P, \preceq)$ *if there is no* $x \in \mathcal{X}$, $x \neq x^*$ *such that*

$$\theta(f(x)) \preceq \theta(f(x^*)). \tag{1.22}$$

For an optimal solution $x^*, \theta(f(x^*))$ *is called an* optimal value *of the MOP. The* set of optimal solutions *is denoted by* $\mathrm{Opt}((\mathcal{X}, f, \mathbb{R}^p)/\theta/(\mathbb{R}^P, \preceq))$. *The* set of optimal values *is* $\mathrm{Val}((\mathcal{X}, f, \mathbb{R}^p)/\theta/(\mathbb{R}^P, \preceq))$.

Some comments on this definition are necessary. First, since we are often dealing with orders which are not total, a positive definition of optimality, like $\theta(f(x^*)) \preceq \theta(f(x))$ for all $x \in \mathcal{X}$, is not possible in general. Second, for specific choices of θ and $(\mathbb{R}^P, \preceq)$, specific names for optimal solutions and values are commonly used, such as efficient solutions or lexicographically optimal solutions.

In the following chapters we will introduce shorthand notations for optimal sets, usually $\mathcal{X}$ with an index identifying the problem class, such as $\mathcal{X}_E := \{x \in \mathcal{X} : \text{ there is no } x' \in \mathcal{X} \text{ with } f(x') \leq f(x)\}$ for the set of efficient solutions.

We now check the definition 1.23 with Examples 1.21 and 1.22.

Example 1.24. With the problem $(\mathbb{R}_{\geqq}, f, \mathbb{R}^2)/\mathrm{id}/(\mathbb{R}^2, \leq)$ the optimality definition reads: There is no $x \in \mathcal{X}, x \neq x^*$, such that $f(x) \leq f(x^*)$, i.e., $f_k(x) \leq f_k(x^*)$ for all $k = 1, \ldots, p$, and $f(x) \neq f(x^*)$. This is indeed efficiency as we know it. □

Example 1.25. For $(\mathcal{X}, f, \mathbb{R}^3)/\theta/(\mathbb{R}^3, \leq_{\mathrm{lex}})$ with $\theta(y) = (y_1, y_2, -y_3)$, $x^* \in \mathcal{X}$ is an optimal solution if there is no $x \in \mathcal{X}, x \neq x^*$, such that

$$(f_1(x), f_2(x), -f_3(x)) \leq_{\mathrm{lex}} (f_1(x^*), f_2(x^*), -f_3(x^*)). \tag{1.23}$$

□

Quite often, we will discuss multicriteria optimization problems in the sense of efficiency or lexicographic optimality in general, not referring to specific problem data, and derive results which are independent of problem data. For this purpose it is convenient to introduce classes of multicriteria optimization problems.

Definition 1.26. *A* multicriteria optimization class (MCO class) *is the set of all MOPs with the same model map and ordered set and is denoted by*

$$\bullet/\theta/(\mathbb{R}^P, \preceq). \tag{1.24}$$

For instance, $\bullet/\text{id}/(\mathbb{R}^p, \leq)$ will denote the class of all MOPs, where optimality is understood in the sense of efficiency.

1.6 Notes

Roy (1990) portrays multicriteria decision making and multicriteria decision aid as complementary fundamental attitudes for addressing decision making problems. Multicriteria decision making includes areas such as multiattribute utility theory (Keeney and Raiffa, 1993) and multicriteria optimization (Ehrgott and Gandibleux, 2002b). Multicriteria decision aid, on the other hand, includes research on the elicitation of preferences from decision makers, structuring the decision process, and other more "subjective" aspects. The reader is referred to Figueira *et al.* (2005) for a collection of up-to-date surveys on both multicriteria decision making and aid.

Yu (1974) calls $\{\mathcal{C}_{\mathcal{R}}(y^1) : y^1 \in \mathcal{Y}\}$ a *structure of domination.* Results on structures of domination can also be found in Sawaragi *et al.* (1985). If $\mathcal{C}_{\mathcal{R}}(y^1)$ is independent of y^1, the domination structure is called constant. A cone therefore implies a constant domination structure.

In terms of the relationships between orders and cones, Noghin (1997) performs a similar analysis to Theorems 1.17 and 1.20. He calls a relation $\mathcal{R}$ a cone order, if there *exists* a cone $\mathcal{C}$ such that $y^1\mathcal{R}y^2$ if and only if $y^2 - y^1 \in \mathcal{C}$. He proves that $\mathcal{R}$ is irreflexive, transitive, compatible with addition and scalar multiplication if and only if $\mathcal{R}$ is a cone relation with a pointed convex cone $\mathcal{C}$ not containing 0.

Exercises

1.1. Consider the problem

$$\text{“min”}(f_1(x), f_2(x)) \text{ subject to } x \in [-1, 1],$$

where

$$f_1(x) = \sqrt{5 - x^2}, \quad f_2(x) = \frac{x}{2}.$$

Illustrate the problem in decision and objective space and determine the the nondominated set $\mathcal{Y}_N := \{y \in \mathcal{Y} : \text{there is no } y' \in \mathcal{Y} \text{ with } y' \leq y\}$ and the efficient set $\mathcal{X}_E := \{x \in \mathcal{X} : f(x) \in \mathcal{Y}_N\}$.

1.2. Consider the following binary relations on $\mathbb{R}^p$ (see Table 1.2):

$$\begin{aligned} y^1 \leqq y^2 &\iff y^1_k \leq y^2_k \; k = 1, \ldots, p; \\ y^1 \leq y^2 &\iff y^1 \leqq y^2 \text{ and } y^1 \neq y^2; \\ y^1 < y^2 &\iff y^1_k < y^2_k \; k = 1, \ldots, p. \end{aligned}$$

Which of the properties listed in Definition 1.4 do these relations have?

1.3. Solve the problem of Exercise 1.1 as max-ordering and lexicographic problems:

$$\begin{aligned} \min_{x \in [-1,1]} \; &\max_{i=1,2} f_i(x), \\ \text{lexmin}_{x \in [-1,1]} \; &(f_1(x), f_2(x)), \\ \text{lexmin}_{x \in [-1,1]} \; &(f_2(x), f_1(x)). \end{aligned}$$

Compare the optimal solutions with efficient solutions. What do you observe?

1.4. Prove the following statements.

1. An asymmetric relation is irreflexive.
2. A transitive and irreflexive relation is asymmetric.
3. A negatively transitive and asymmetric relation is transitive.
4. A transitive and connected relation is negatively transitive.

1.5. This exercise is about cones and orders.

1. Determine the cones related to the (strict and weak) componentwise order and the lexicographic order on $\mathbb{R}^2$.
2. Find and illustrate $\mathcal{C}_{\leq_{MO}}(y)$ for $y = 0, y = (2, 1)$ and $y = (-1, 3)$.
3. Give an example of a non-convex cone $\mathcal{C}$ and list the properties of the related order $\mathcal{R}_\mathcal{C}$.

1.6. A cone $\mathcal{C}$ is called acute, if there exists an open halfspace $H_a = \{x \in \mathbb{R}^p : \langle x, a\rangle > 0\}$ such that $\text{cl}(\mathcal{C}) \subset H_a \cup \{0\}$. Is a pointed cone always acute? What about a convex cone?

1.7. Consider the order relations $\leqq, \leq, <, \leq_{lex}$, and $\leq_{MO}$ on $\mathbb{R}^p$ and determine their relationships, i.e., statements of the form

$$y^1 \mathcal{R}_a y^2 \Longrightarrow y^1 \mathcal{R}_b y^2,$$

where $\mathcal{R}_a, \mathcal{R}_b \in \{\leqq, \leq, <, \leq_{lex}, \leq_{MO}\}$. What do these statements imply for the related cones $\mathcal{C}_\mathcal{R}$?

1.8. Let $|| \; || : \mathbb{R}^p \longrightarrow \mathbb{R}_\geqq$ be a norm. Define $y^1 \leq_{|| \; ||} y^2 \Longleftrightarrow ||y^1|| \leq ||y^2||$. Is $\leq_{|| \; ||}$ a partial order? Is it connected? Determine $\mathcal{C}_{\leq_{|| \; ||}}$ for some norm $|| \; ||$ of your choice.

1.9. A cone $\mathcal{C}$ in some vector space $\mathcal{V}$ is called *generating* if $\mathcal{V} = \mathcal{C} - \mathcal{C}$ (loosely speaking, every $v \in \mathcal{V}$ can be written as the difference of two nonnegative elements).

Consider $\mathcal{V} = C[0,1]$, the vector space of all continuous functions $f : [0,1] \longrightarrow \mathbb{R}$. Show that

$$\mathcal{C} := \{f \in C[0,1] : f(x) \geq 0 \text{ for all } x \in [0,1]\}$$

is a cone that defines a partial order $\mathcal{R}_\mathcal{C}$, and that $C[0,1] = \mathcal{C} - \mathcal{C}$, i.e., for all $f \in C[0,1]$ there are $f^1, f^2 \in \mathcal{C}$ such that $f = f^1 - f^2$. Can you give an example of a cone $\mathcal{C} \subset \mathbb{R}^p$ with $\mathcal{C} - \mathcal{C} \neq \mathbb{R}^p$ and find a relationship between the cone property "generating" and a property of the order $\mathcal{R}_\mathcal{C}$?

1.10. In this exercise, the relationships between cones and relations are further developed.

1. Let $\mathcal{R}$ be a relation. Define $\mathcal{C}_\mathcal{R}$ as in (1.15). Define $\mathcal{R}_{\mathcal{C}_\mathcal{R}}$ as in (1.16) with $\mathcal{C} = \mathcal{C}_\mathcal{R}$. Under what conditions is $\mathcal{R}_{\mathcal{C}_\mathcal{R}} = \mathcal{R}$, i.e., $y^1 \mathcal{R}_{\mathcal{C}_\mathcal{R}} y^2 \Longleftrightarrow y^1 \mathcal{R} y^2$?
2. Let $\mathcal{C}$ be a cone. Define $\mathcal{R}_\mathcal{C}$ as in (1.16). Define $\mathcal{C}_{\mathcal{R}_\mathcal{C}}$ as in (1.15) with $\mathcal{R} = \mathcal{R}_\mathcal{C}$. Is $\mathcal{C}_{\mathcal{R}_\mathcal{C}} = \mathcal{C}$ always, i.e., $d \in \mathcal{C}_{\mathcal{R}_\mathcal{C}} \Longleftrightarrow d \in \mathcal{C}$?

1.11. Generalize the definition of $\mathcal{R}_\mathcal{C}$ for the case where $\mathcal{C}$ is an arbitrary set. Derive relationships between properties of $\mathcal{C}$ and $\mathcal{R}_\mathcal{C}$.

2

Efficiency and Nondominance

This chapter covers the fundamental concepts of efficiency and nondominance. We first present some fundamental properties of nondominated points and several existence results for nondominated points and efficient solutions in Section 2.1. Section 2.2 introduces ideal and nadir points as bounds on the set of nondominated solutions. Then we briefly review weakly and strictly efficient solutions in Section 2.3. The same section also includes a geometric characterization of the three optimality concepts, with some extensions for the case of weakly efficient solutions. Finally, in Section 2.4 we introduce several definitions of properly efficient solutions, important subsets of efficient solutions from a computational point of view and in applications, and investigate their relationships.

Most of the material in this chapter can be found in the two books Göpfert and Nehse (1990) and Sawaragi *et al.* (1985), where the results are presented in more generality. We will also refer to the original publications for the main results.

2.1 Efficient Solutions and Nondominated Points

In this chapter we consider multicriteria optimization problems of the class $\bullet/\mathrm{id}/(\mathbb{R}^p, \leqq)$:

$$\begin{aligned} \min\ & (f_1(x), \ldots, f_p(x)) \\ & \text{subject to } x \in \mathcal{X}. \end{aligned} \tag{2.1}$$

The image of the feasible set $\mathcal{X}$ under the objective function mapping f is denoted as $\mathcal{Y} := f(\mathcal{X})$. Let us formally repeat the definition of efficient solutions and nondominated points. Definition 2.1 also introduces the notion of dominance.

Definition 2.1. *A feasible solution* $\hat{x} \in \mathcal{X}$ *is called* efficient *or* Pareto optimal, *if there is no other* $x \in \mathcal{X}$ *such that* $f(x) \leq f(\hat{x})$. *If* $\hat{x}$ *is efficient,* $f(\hat{x})$ *is called* nondominated point. *If* $x^1, x^2 \in \mathcal{X}$ *and* $f(x^1) \leq f(x^2)$ *we say* x^1 dominates x^2 *and* $f(x^1)$ dominates $f(x^2)$. *The set of all efficient solutions* $\hat{x} \in \mathcal{X}$ *is denoted* $\mathcal{X}_E$ *and called the* efficient set. *The set of all nondominated points* $\hat{y} = f(\hat{x}) \in \mathcal{Y}$, *where* $\hat{x} \in \mathcal{X}_E$, *is denoted* $\mathcal{Y}_N$ *and called the* nondominated set—.

We have to remark that these notations are not unique in literature, unfortunately. Some authors use Pareto optimal for what we call efficient and efficient for what we call nondominated (e.g. this notation was used in the first edition of this book). The term noninferior solution has also been used. We will use the terms of Definition 2.1, but whenever consulting literature, the reader should check the definitions the respective author adopts.

Several other, equivalent, definitions of efficiency are frequently used, and we shall often refer to the one which is best suited in a given context. In particular, $\hat{x}$ is efficient if

1. there is no $x \in \mathcal{X}$ such that $f_k(x) \leq f_k(\hat{x})$ for $k = 1, \ldots, p$ and $f_i(x) < f_i(\hat{x})$ for some $i \in \{1, \ldots, k\}$;
2. there is no $x \in X$ such that $f(x) - f(\hat{x}) \in -\mathbb{R}^p_{\geqq} \setminus \{0\}$;
3. $f(x) - f(\hat{x}) \in \mathbb{R}^p \setminus \left\{-\mathbb{R}^p_{\geqq} \setminus \{0\}\right\}$ for all $x \in \mathcal{X}$;
4. $f(\mathcal{X}) \cap \left(f(\hat{x}) - \mathbb{R}^p_{\geqq}\right) = \{f(\hat{x})\}$;
5. there is no $f(x) \in f(\mathcal{X}) \setminus \{f(\hat{x})\}$ with $f(x) \in f(\hat{x}) - \mathbb{R}^p_{\geqq}$;
6. $f(x) \leqq f(\hat{x})$ for some $x \in \mathcal{X}$ implies $f(x) = f(\hat{x})$.

With the exception of the last, these definitions can be illustrated graphically. Definition 2.1 and equivalent definitions 1., 4., and 5. consider $f(\hat{x})$ and check for images of feasible solutions to the left and below (in direction of $-\mathbb{R}^p_{\geqq}$) of that point. See the left part of Figure 2.1. In equivalent definitions 2. and 3., through $f(x) - f(\hat{x})$, the set $\mathcal{Y} = f(\mathcal{X})$ is translated so that the origin coincides with $f(\hat{x})$, and the intersection of the translated set $\mathcal{Y}$ with the negative orthant is checked. This intersection contains only $f(\hat{x})$ if $\hat{x}$ is efficient. See the right part of Figure 2.1.

The first questions we discuss are the existence and the properties of the efficient set $\mathcal{X}_E$ and the nondominated set $\mathcal{Y}_N$. It is convenient to consider $\mathcal{Y}_N$ first, and then use properties of f to derive results on $\mathcal{X}_E$. So let $\mathcal{Y} \subset \mathbb{R}^p$ be a set. According to our definitions, $\hat{y} \in \mathcal{Y}$ is nondominated, if there is no $y \in \mathcal{Y}$ such that $y \leq \hat{y}$.

First we show by means of an example that even for convex sets $\mathcal{X}$ and $\mathcal{Y}$ the efficient set $\mathcal{X}_E$ and the nondominated set set $\mathcal{Y}_N$ might be empty or

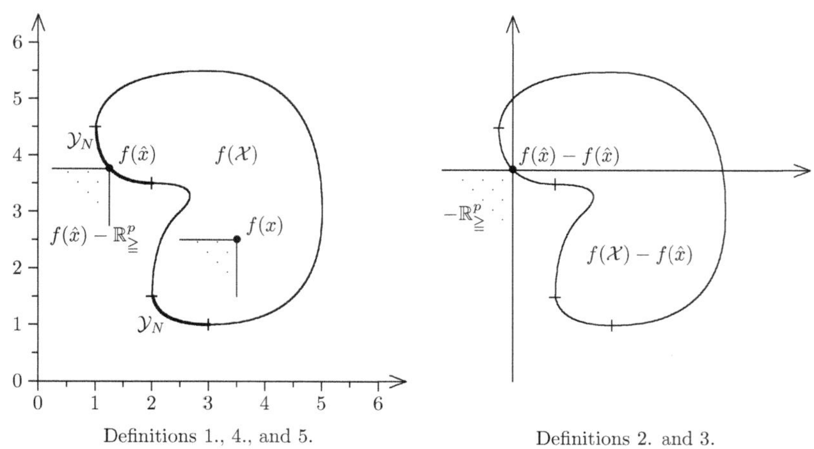

Fig. 2.1. Illustration of definitions of efficient solutions.

consist of isolated points. We will then proceed to prove some basic properties of nondominated sets, before we present several existence theorems for efficient solutions/nondominated points. Results on connectedness of $\mathcal{Y}_N$ and $\mathcal{X}_E$ will be given in Chapter 3.

Example 2.2 (Göpfert and Nehse (1990)). Consider a bicriterion optimization problem with feasible set

$$\mathcal{X} = \left\{ (x_1, x_2) \in \mathbb{R}^2 \;\middle|\; \begin{array}{r} -1 \leq x_1 \leq 1, \\ -\sqrt{-x_1^2+1} < x_2 \leq 0 \quad \text{if} \quad -1 \leq x_1 \leq 0, \\ -\sqrt{-x_1^2+1} \leq x_2 \leq 0 \quad \text{if} \quad 0 < x_1 \leq 1 \end{array} \right\} \tag{2.2}$$

and objective function

$$f(x_1, x_2) = (x_1, x_2). \tag{2.3}$$

The feasible sets $\mathcal{X}$ in decision space and $\mathcal{Y}$ in criterion space (the latter coincides with $\mathcal{X}$ in this example) are depicted in Figure 2.2.

Clearly, there are no nondominated points, and therefore the bicriterion problem given by (2.2) and (2.3)does not have any efficient solutions: $\mathcal{Y}_N = \mathcal{X}_E = \emptyset$, even though $\mathcal{X}$ and $\mathcal{Y}$ are convex and f is continuous.

If we modify the problem slightly by letting

$$\mathcal{X} = \left\{ (x_1, x_2) \in \mathbb{R}^2 \;\middle|\; \begin{array}{r} -1 \leq x_1 \leq 1, \\ x_2 = 0 \quad \text{if} \quad x_1 = -1, \\ -\sqrt{-x_1^2+1} < x_2 \leq 0 \quad \text{if} \quad -1 < x_1 < 0, \\ -\sqrt{-x_1^2+1} \leq x_2 \leq 0 \quad \text{if} \quad 0 \leq x_1 \leq 1 \end{array} \right\} \tag{2.4}$$

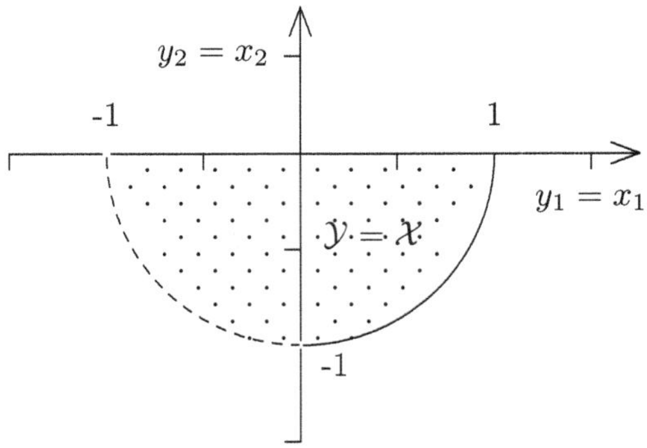

Fig. 2.2. Feasible set of the original problem in Example 2.2.

$\mathcal{Y}_N = \{(-1, 0), (0, -1)\}$ is no longer empty (Figure 2.3), but consists of only two disconnected points, which are "far apart" from one another in $\mathcal{Y}_N$.

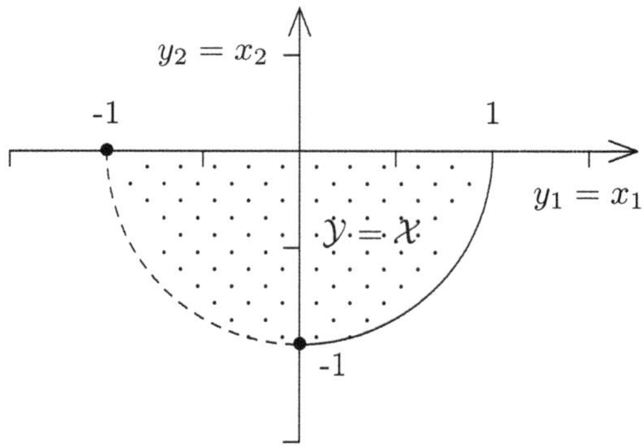

Fig. 2.3. Feasible set of the modified problem in Example 2.2.

□

Example 2.2 shows that conditions for existence of efficient solutions and nondominated points must be our first concern in the study of multicriteria optimization. In multicriteria optimization, the "trick" of Example 2.2, to use $y = f(x) = x$ is quite useful, as it allows to identify decision and criterion space and enables the study of both $\mathcal{X}_E$ and $\mathcal{Y}_N$ at the same time. We will often apply it in the examples to come.

The following properties of nondominated sets are mainly proved as tools for the proofs of theorems later in the text. However, they may well enhance an intuitive understanding of the concept of nondominance. First we show that nondominated points are located in the "lower left part" of $\mathcal{Y}$: Adding $\mathbb{R}^p_{\geqq}$ to $\mathcal{Y}$ does not change the nondominated set.

So let $\mathcal{Y} \subset \mathbb{R}^p$. Let $\mathcal{Y}_N = \{y \in \mathcal{Y} : \text{ there is no } y' \in \mathcal{Y} \text{ such that } y' \leq y\}$. In particular $\mathcal{Y}_N \subset \mathcal{Y}$.

Proposition 2.3. $\mathcal{Y}_N = \left(\mathcal{Y} + \mathbb{R}^p_{\geqq}\right)_N$.

Proof. The result is trivial if $\mathcal{Y} = \emptyset$, because $\mathcal{Y} + \mathbb{R}^p_{\geqq} = \emptyset$ and the nondominated subsets of both are empty, too.

So let $\mathcal{Y} \neq \emptyset$. First, assume $y \in (\mathcal{Y} + \mathbb{R}^p_{\geqq})_N$, but $y \notin \mathcal{Y}_N$. There are two possibilities. If $y \notin \mathcal{Y}$ there is $y' \in \mathcal{Y}$ and $0 \neq d \in \mathbb{R}^p_{\geqq}$ such that $y = y' + d$. Since $y' = y' + 0 \in \mathcal{Y} + \mathbb{R}^p_{\geqq}$ we get $y \notin (\mathcal{Y} + \mathbb{R}^p_{\geqq})_N$, a contradiction. If $y \in \mathcal{Y}$ there is $y' \in \mathcal{Y}$ such that $y' \leq y$. Let $d = y - y'$, which is in $\mathbb{R}^p_{\geqq} \setminus \{0\}$. Therefore $y = y' + d$ and $y \notin (\mathcal{Y} + \mathbb{R}^p_{\geqq})_N$, again contradicting the assumption. Hence in either case $y \in \mathcal{Y}_N$.

Second, assume $y \in \mathcal{Y}_N$ but $y \notin (\mathcal{Y}+\mathbb{R}^p_{\geqq})_N$. Then there is some $y' \in \mathcal{Y}+\mathbb{R}^p_{\geqq}$ with $y - y' = d' \in \mathbb{R}^p_{\geqq} \setminus \{0\}$. I.e. $y' = y'' + d''$ with $y'' \in \mathcal{Y}$, $d'' \in \mathbb{R}^p_{\geqq}$ and therefore $y = y' + d' = y'' + (d' + d'') = y'' + d$ with $d = d' + d'' \in \mathbb{R}^p_{\geqq} \setminus \{0\}$. This implies $y \notin \mathcal{Y}_N$, contradicting the assumption. Hence, $y \in (\mathcal{Y} + \mathbb{R}^p_{\geqq})_N$. □

Proposition 2.3 is illustrated in Figure 2.4.

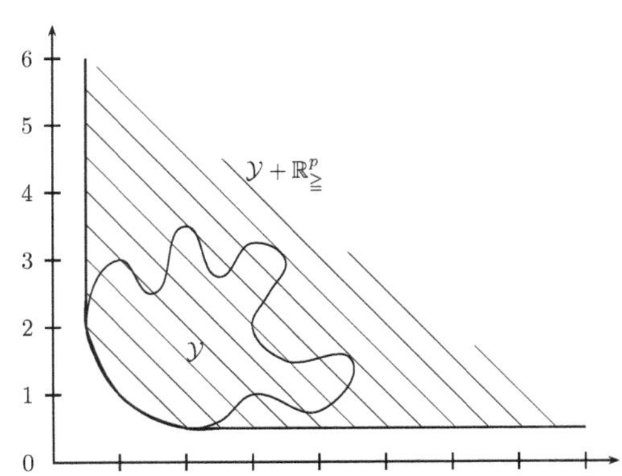

Fig. 2.4. Nondominated points of $\mathcal{Y}$ and $\mathcal{Y} + \mathbb{R}^p_{\geqq}$ are the same.

A second result, which is intuitively clear, is that efficient points must belong to the boundary of $\mathcal{Y}$.

Proposition 2.4. $\mathcal{Y}_N \subset \text{bd}(\mathcal{Y})$.

Proof. Let $y \in \mathcal{Y}_N$ and suppose $y \notin \text{bd}(\mathcal{Y})$. Therefore $y \in \text{int}\,\mathcal{Y}$ and there exists an ε-neighbourhood $B(y, \varepsilon)$ of y (with $B(y, \varepsilon) := y + B(0, \varepsilon) \subset \mathcal{Y}$, $B(0, \varepsilon)$ is an open ball with radius ε centered at the origin). Let $d \neq 0$, $d \in \mathbb{R}^p_{\geqq}$. Then we can choose some $\alpha \in \mathbb{R}$, $0 < \alpha < \varepsilon$ such that $\alpha d \in B(0, \varepsilon)$. Now, $y - \alpha d \in \mathcal{Y}$ with $\lambda d \in \mathbb{R}^p_{\geqq} \setminus \{0\}$, i.e. $y \notin \mathcal{Y}_N$. □

From Propositions 2.3 and 2.4 we immediately get conditions for $\mathcal{Y}_N$ being empty.

Corollary 2.5. *If* $\mathcal{Y}$ *is open or if* $\mathcal{Y} + \mathbb{R}^p_{\geqq}$ *is open* $\mathcal{Y}_N = \emptyset$.

The next results concern the nondominated set of the Minkowski sum of two sets and of a set multiplied by a positive scalar.

Proposition 2.6. $(\mathcal{Y}_1 + \mathcal{Y}_2)_N \subset (\mathcal{Y}_1)_N + (\mathcal{Y}_2)_N$.

Proof. Let $y \in (\mathcal{Y}_1 + \mathcal{Y}_2)_N$. Then $y = y^1 + y^2$ for some $y^1 \in \mathcal{Y}_1, y^2 \in \mathcal{Y}_2$. Assuming $y^1 \notin (\mathcal{Y}_1)_N$ it follows that there must be some $y' \in \mathcal{Y}_1$ and $d \in \mathbb{R}^p_{\geq}$ such that $y^1 = y' + d$ and thus $y = y' + y^2 + d$ with $y' + y^2 \in \mathcal{Y}_1 + \mathcal{Y}_2$ whence $y \notin (\mathcal{Y}_1 + \mathcal{Y}_2)_N$, contradicting the assumption.

Analogously, $y^2 \in (\mathcal{Y}_2)_N$, i.e. $y^1 + y^2 \in (\mathcal{Y}_1)_N + (\mathcal{Y}_\in)_\mathcal{N}$. □

The inclusion $(\mathcal{Y}_1)_N + (\mathcal{Y}_2)_N \subset (\mathcal{Y}_1 + \mathcal{Y}_2)_N$ is not satisfied in general, Exercise 2.1 asks for a counterexample.

Proposition 2.7. $(\alpha\mathcal{Y})_N = \alpha(\mathcal{Y}_N)$, *for* $\alpha \in \mathbb{R}$, $\alpha > 0$.

Proof. The easy proof is left to the reader, see Exercise 2.4. □

With these propositions we have some tools to facilitate working with nondominated sets. In order to prove existence results for nondominated points we have to introduce another fundamental statement, Zorn's Lemma.

Definition 2.8. *Let* $(\mathcal{S}, \preceq)$ *be a preordered set, i.e.* $\preceq$ *is reflexive and transitive.* $(\mathcal{S}, \preceq)$ *is* inductively ordered, *if every totally ordered subset of* $(\mathcal{S}, \preceq)$ *has a lower bound. A totally ordered subset of* $(\mathcal{S}, \preceq)$ *is also called a* chain.

Theorem 2.9 (Zorn's lemma). *Let the preordered set* $(\mathcal{S}, \preceq)$ *be inductively ordered. Then* $\mathcal{S}$ *contains a minimal element, i.e. there is* $\hat{s} \in \mathcal{S}$ *such that* $s \preceq \hat{s}$ *implies* $\hat{s} \preceq s$.

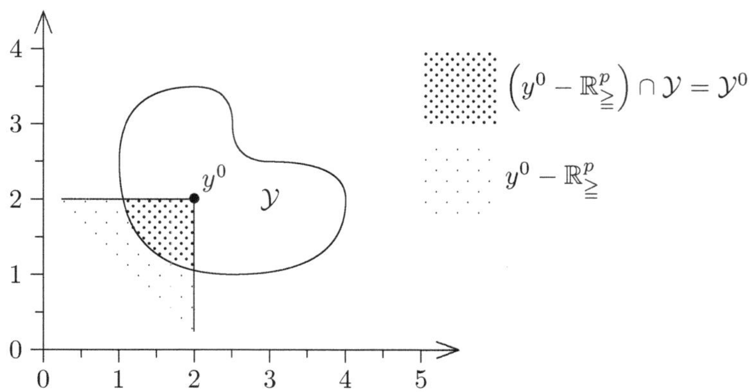

Fig. 2.5. $\mathcal{Y}^0$ is a compact section of $\mathcal{Y}$.

Theorem 2.10 (Borwein (1983)). *Let $\mathcal{Y}$ be a nonempty set and suppose there is some $y^0 \in \mathcal{Y}$ such that the section $\mathcal{Y}^0 = \{y \in \mathcal{Y} : y \leqq y^0\} = (y^0 - \mathbb{R}^p_{\geqq}) \cap \mathcal{Y}$ is compact (we say "$\mathcal{Y}$ contains a compact section"). Then $\mathcal{Y}_N$ is nonempty.*

Proof. The idea of the proof is as follows. We use the compactness of $\mathcal{Y}^0$ to show that every chain in $\mathcal{Y}^0$ has a lower bound. Thus $\mathcal{Y}^0$ is inductively ordered, and by Zorn's Lemma contains a minimal element $\hat{y}$. Showing that $\hat{y}$ is efficient in $\mathcal{Y}$ completes the proof.

Let $\mathcal{Y}^0$ be the compact section that exists by assumption and let $\mathcal{Y}^{\mathcal{I}} = \{y^i : i \in \mathcal{I}\}$, where $\mathcal{I}$ is some index set, be a chain in $\mathcal{Y}^0$. We prove that $\{y^i : i \in \mathcal{I}\}$ has a lower bound. To that end let $\mathcal{J} := \{J \subset \mathcal{I} : |J| < \infty\}$ be the set of all finite subsets of index set $\mathcal{I}$. For all $J \in \mathcal{J}$ finiteness of J and $\mathcal{Y}^{\mathcal{I}}$ being a chain in $\mathcal{Y}^0$ imply that $y^J := \inf\{y^i : i \in J\}$ exists and $y^J \in \mathcal{Y}^0$. Consider all sets $\mathcal{Y}^i := (y^i - \mathbb{R}^p_{\geqq}) \cap \mathcal{Y}^0$, where $i \in \mathcal{I}$. Obviously $\mathcal{Y}^i \subset \mathcal{Y}^0$ and $\mathcal{Y}^i$ is compact as a closed subset of the compact set $\mathcal{Y}^0$. Furthermore, if $J \in \mathcal{J}$, i.e. J is finite, $\cap_{i \in J} \mathcal{Y}^i \neq \emptyset$ because it contains y^J. Finally, by compactness of $\mathcal{Y}^0$ it follows that $\cap_{i \in \mathcal{I}} \mathcal{Y}^i \neq \emptyset$, which means there is some

$$y' \in \bigcap_{i \in \mathcal{I}} \left(y^i - \mathbb{R}^p_{\geqq}\right) \cap \mathcal{Y}^0. \tag{2.5}$$

In terms of the componentwise order this means $y' \leqq y^i$ for all $i \in \mathcal{I}$, or, in other words, $y' \in \mathcal{Y}^0$ is a lower bound of $\{y^i : i \in \mathcal{I}\}$, which is therefore inductively ordered.

We can now apply Zorn's Lemma (Theorem 2.9) to conclude that $\mathcal{Y}^0$ contains a minimal element $\hat{y}$. It remains to be shown that $\hat{y} \in \mathcal{Y}_N$. Assume the contrary. Then there would be some $y'' \in \mathcal{Y}$ with $y'' \leq \hat{y}$. For y'' we have

$$\begin{aligned} y'' \in \left(\hat{y} - \mathbb{R}^p_{\geqq}\right) \cap \mathcal{Y} &\subset \left(\left(y^0 - \mathbb{R}^p_{\geqq}\right) \cap \mathcal{Y} - \mathbb{R}^p_{\geqq}\right) \cap \mathcal{Y} \\ &\subset \left(y^0 - \mathbb{R}^p_{\geqq}\right) \cap \mathcal{Y} - \mathbb{R}^p_{\geqq} = \mathcal{Y}^0 - \mathbb{R}^p_{\geqq}. \end{aligned} \tag{2.6}$$

The first inclusion holds because $\hat{y} \in \mathcal{Y}^0$, the second is clear. Since $y'' \in \mathcal{Y}$ this implies $y'' \in \mathcal{Y}^0$, so that $y'' \leq \hat{y}$ contradicts minimality of $\hat{y}$ in $\mathcal{Y}^0$. □

Note that we have used the following fact about compact sets: If $\mathcal{Y}$ is compact and $(\mathcal{Y}^i)$, $i \in \mathcal{I}$ is a family of closed subsets of $\mathcal{Y}$ for some index set $\mathcal{I}$ such that $\cap_{k=1}^n \mathcal{Y}_{i_k} \neq \emptyset$ for all finite subsets of $\{i_1, \ldots, i_n\}$ of $\mathcal{I}$ then $\cap_{i \in \mathcal{I}} \mathcal{Y}_i \neq \emptyset$.

Another existence result does not use a compact section but a condition on $\mathcal{Y}$ which is similar to the finite subcover property of compact sets: the $\mathbb{R}^p_{\geqq}$-semicompactness condition, which considers open covers with special sets.

Definition 2.11. *A set $\mathcal{Y} \subset \mathbb{R}^p$ is called* $\mathbb{R}^p_{\geqq}$-semicompact *if every open cover of $\mathcal{Y}$ of the form $\left\{(y^i - \mathbb{R}^p_{\geqq})^c : y^i \in \mathcal{Y}, i \in \mathcal{I}\right\}$ has a finite subcover. This means that whenever $\mathcal{Y} \subset \cup_{i \in \mathcal{I}} (y^i - \mathbb{R}^p_{\geqq})^c$ there exist $m \in \mathbb{N}$ and $\{i_1, \ldots, i_m\} \subset \mathcal{I}$ such that*

$$\mathcal{Y} \subset \bigcup_{k=1}^{m} \left(y^{i_k} - \mathbb{R}^p_{\geqq}\right)^c. \tag{2.7}$$

Here $(y^i - \mathbb{R}^p_{\geqq})^c$ denotes the complement $\mathbb{R}^p \setminus (y^i - \mathbb{R}^p_{\geqq})$ of $y^i - \mathbb{R}^p_{\geqq}$. Note that these sets are always open.

Based on Zorn's Lemma again, we can prove that $\mathbb{R}^p_{\geqq}$-semicompactness guarantees existence of efficient points.

Theorem 2.12 (Corley (1980)). *If $\mathcal{Y} \neq \emptyset$ is $\mathbb{R}^p_{\geqq}$-semicompact then $\mathcal{Y}_N \neq \emptyset$.*

Proof. The main steps of the proof are the same as for Theorem 2.10. We show that $\mathcal{Y}$ is inductively ordered and apply Zorn's Lemma. First, we construct an open cover of $\mathcal{Y}$ as in Definition 2.11 and derive a contradiction when we assume that $\mathcal{Y}$ is not inductively ordered.

So assume $\mathcal{Y}$ is not inductively ordered. Then there is a totally ordered subset (a chain) of $\mathcal{Y}$, say $\mathcal{Y}' = \{y^i : i \in \mathcal{I}\}$ which has no lower bound. Therefore

$$\bigcap_{i \in \mathcal{I}} \left(\left(y^i - \mathbb{R}^p_{\geqq}\right) \cap \mathcal{Y}\right) = \emptyset. \tag{2.8}$$

As seen in the proof of Theorem 2.10, any element in this intersection would be a lower bound of $\mathcal{Y}'$. Then for each $y \in \mathcal{Y}$ there is some $y^i \in \mathcal{Y}'$ such that $y \notin y^i - \mathbb{R}^p_{\geqq}$.

Since $y^i - \mathbb{R}^p_{\geqq}$ is closed, $\{(y^i - \mathbb{R}^p_{\geqq})^c : i \in \mathcal{I}\}$ defines an open cover of $\mathcal{Y}$. Moreover, $y^i - \mathbb{R}^p_{\geqq} \subset y^{i'} - \mathbb{R}^p_{\geqq}$ if and only if $y^i \leqq y^{i'}$ and the sets of the cover are totally ordered by inclusion because $\mathcal{Y}'$ is a chain. Also, $\mathcal{Y}$ is $\mathbb{R}^p_{\geqq}$-semicompact and there is a finite subcover of $\{(y^i - \mathbb{R}^p_{\geqq})^c : i \in \mathcal{I}\}$.

Combining the last two observations, it follows that there is a minimal set (with respect to inclusion) in the finite subcover and hence there exists a single $y^* \in \mathcal{Y}'$ such that $\mathcal{Y} \subset (y^* - \mathbb{R}^p_{\geqq})^c$. This implies $y^* \leqq y^i$ for all $i \in \mathcal{I}$ and $y^* \notin \mathcal{Y}$, which is not possible. Therefore $\mathcal{Y}$ is inductively ordered.

Knowing that, we proceed as in the proof of Theorem 2.10 to conclude $\mathcal{Y}_N \neq \emptyset$. □

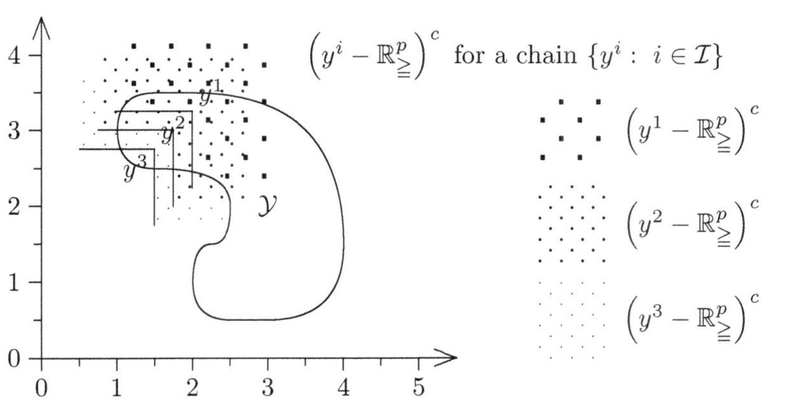

Fig. 2.6. Constructing a cover from a chain.

Although theoretically interesting, Theorem 2.12 gives a condition which is usually not easy to check: $\mathbb{R}^p_{\geqq}$-semicompactness. A weaker result is obtained if we use the stronger assumption of $\mathbb{R}^p_{\geqq}$-compactness.

Definition 2.13. *A set $\mathcal{Y} \subset \mathbb{R}^p$ is called* $\mathbb{R}^p_{\geqq}$-compact, *if for all $y \in \mathcal{Y}$ the section $(y - \mathbb{R}^p_{\geqq}) \cap \mathcal{Y}$ is compact.*

Proposition 2.14. *If $\mathcal{Y}$ is $\mathbb{R}^p_{\geqq}$-compact then $\mathcal{Y}$ is $\mathbb{R}^p_{\geqq}$-semicompact.*

Proof. Let $\{(y^i - \mathbb{R}^p_{\geqq})^c : y^i \in \mathcal{Y}, i \in \mathcal{I}\}$ be an open cover of $\mathcal{Y}$. For arbitrary $y^{i'} \in \mathcal{Y}$ take

$$\left\{\left(y^i - \mathbb{R}^p_{\geqq}\right)^c : y^i \in \mathcal{Y}, i \in \mathcal{I}, i \neq i'\right\}. \tag{2.9}$$

(2.9) defines an open cover of $(y^{i'} - \mathbb{R}^p_{\geqq}) \cap \mathcal{Y}$, a compact set, since $\mathcal{Y}$ is $\mathbb{R}^p_{\geqq}$-compact. But compactness implies that the cover in (2.9) contains a finite subcover of $(y^{i'} - \mathbb{R}^p_{\geqq}) \cap \mathcal{Y}$. This finite subcover together with $(y^{i'} - \mathbb{R}^p_{\geqq})^c$ yields a finite cover of $\mathcal{Y}$, of the structure required for $\mathbb{R}^p_{\geqq}$-semicompactness. □

Corollary 2.15 (Hartley (1978)). *If $\mathcal{Y} \subset \mathbb{R}^p$ is nonempty and $\mathbb{R}^p_{\geqq}$-compact, then $\mathcal{Y}_N \neq \emptyset$.*

Proof. The result follows immediately from Theorem 2.12 and Proposition 2.14. □

So far, we focused on existence of nondominated points. Let us now consider existence of efficient solutions, i.e. conditions that guarantee $\mathcal{X}_E \neq \emptyset$, which is an important issue when practical problems are considered. We can use Theorem 2.12 and properties of f to get an existence result for $\mathcal{X}_E$. Theorem 2.19 below is a multicriteria analogon to the well known result that a lower semicontinuous function attains its minimum over a compact set.

Definition 2.16. *A function $f : \mathbb{R}^n \to \mathbb{R}^p$ is* $\mathbb{R}^p_{\geqq}$-semicontinuous *if*

$$f^{-1}\left(y - \mathbb{R}^p_{\geqq}\right) = \left\{x \in \mathbb{R}^n : y - f(x) \in \mathbb{R}^p_{\geqq}\right\} \tag{2.10}$$

is closed for all $y \in \mathbb{R}^p$, i.e. the preimage of the translated negative orthant is always closed.

Lemma 2.17 below establishes $\mathbb{R}^p_{\geqq}$-semicontinuity as a proper generalization of lower semicontinuity of scalar valued functions.

Lemma 2.17. *A function $f : \mathbb{R}^n \to \mathbb{R}^p$ is $\mathbb{R}^p_{\geqq}$-semicontinuous if and only if the component functions $f_k : \mathbb{R}^n \to \mathbb{R}$ are lower semicontinuous for all $k = 1, \ldots, p$.*

The proof is left to the reader.

Proposition 2.18. *Let $\mathcal{X} \subset \mathbb{R}^n$ be nonempty and compact, $f : \mathbb{R}^n \to \mathbb{R}^p$ be $\mathbb{R}^p_{\geqq}$-semicontinuous. Then $\mathcal{Y} = f(\mathcal{X})$ is $\mathbb{R}^p_{\geqq}$-semicompact.*

Proof. Let $\{(y^i - \mathbb{R}^p_{\geqq})^c : y^i \in \mathcal{Y}, i \in \mathcal{I}\}$ be an open cover of $\mathcal{Y}$. By $\mathbb{R}^p_{\geqq}$-semicontinuity of f, $\{f^{-1}((y^i - \mathbb{R}^p_{\geqq})^c) : y^i \in \mathcal{Y}, i \in \mathcal{I}\}$ is an open cover of $\mathcal{X}$. Because $\mathcal{X}$ is compact there is a finite subcover in this open cover. The image of this subcover is a finite subcover of $\mathcal{Y}$ whence $\mathcal{Y}$ is $\mathbb{R}^p_{\geqq}$ semicompact. □

Theorem 2.19. *Let $\mathcal{X} \subset \mathbb{R}^n$ be a nonempty and compact set. Let f be $\mathbb{R}^p_{\geqq}$-semicontinuous. Then $\mathcal{X}_E \neq \emptyset$.*

Proof. The result follows directly from Theorem 2.12 and Proposition 2.18. □

Given a set $\mathcal{Y} \subset \mathbb{R}^p$ with nonempty nondominated set $\mathcal{Y}_N \neq \emptyset$, it is clear that for any $y \in \mathcal{Y} \setminus \mathcal{Y}_N$ there is some $\hat{y} \in \mathcal{Y}$ such that $\hat{y} \leq y$. But is it always guaranteed that a nondominated $\hat{y}$ dominating y exists? It turns out that under existence conditions for nondominated points this is true.

Definition 2.20. *The nondominated set $\mathcal{Y}_N$ is said to be* externally stable, *if for each $y \in \mathcal{Y} \setminus \mathcal{Y}_N$ there is $\hat{y} \in \mathcal{Y}_N$ such that $y \in \hat{y} + \mathbb{R}^p_{\geqq}$.*

Theorem 2.21. *Let $\mathcal{Y} \subset \mathbb{R}^p_{\geqq}$ be nonempty and $\mathbb{R}^p_{\geqq}$-compact. Then $\mathcal{Y}_N$ is externally stable, i.e.*

$$\mathcal{Y} \subset \mathcal{Y}_N + \mathbb{R}^p_{\geqq}.$$

Proof. Let $y \in \mathcal{Y}$. Define

$$\mathcal{Y}' := \left(y - \mathbb{R}^p_{\geqq}\right) \cap \mathcal{Y},$$

i.e. all points in $\mathcal{Y}$ dominating y. We need to show that $\mathcal{Y}' \cap \mathcal{Y}_N \neq \emptyset$. To do so it is enough to show that $\mathcal{Y}'_N \neq \emptyset$ and that $\mathcal{Y}'_N \subset \mathcal{Y}_N$.

$\mathcal{Y}'$ is $\mathbb{R}^p_{\geqq}$-compact since $\mathcal{Y}$ is (see Definition 2.13). Therefore $\mathcal{Y}'_N \neq \emptyset$ according to Corollary 2.15.

Assume that y' is not in $\mathcal{Y}_N$, but $y' \in \mathcal{Y}'$ (otherwise y' is certainly not contained in $\mathcal{Y}'_N$). Thus $y' \in \mathcal{Y}$ and there is some $y'' \in \mathcal{Y}$ such that $y'' \leq y'$. Therefore $y'' \leq y' \leq y$ and $y'' \in \mathcal{Y}'$. This implies $y' \notin \mathcal{Y}'_N$. □

2.2 Bounds on the Nondominated Set

In this section, we define the ideal and nadir points as lower and upper bounds on nondominated points. These points give an indication of the range of the values which nondominated points can attain. They are often used as reference points in compromise programming (see Section 4.5) or in interactive methods the aim of which is to find a most preferred solution for a decision maker.

We assume that $\mathcal{X}_E$ and $\mathcal{Y}_N$ are nonempty, and want to find real numbers $\underline{y}_k, \overline{y}_k,\ k = 1, \ldots, p$ with $\underline{y}_k \leq y_i \leq \overline{y}_i$ for all $y \in \mathcal{Y}_N$, as shown in Figure 2.7. An obvious possibility is to choose

$$\underline{y}_k := \min_{y \in \mathcal{Y}} y_i, \tag{2.11}$$

$$\overline{y}_k := \max_{y \in \mathcal{Y}} y_i. \tag{2.12}$$

While the lower bound (2.11) is tight (there is always an efficient point $y \in \mathcal{Y}_N$ with $y_k = \underline{y}_k$), the upper bound (2.12) tends to be far away from actual nondominated points. For this reason, the upper bound is defined as the maximum over nondominated points only.

Definition 2.22. *1. The point $y^I = (y^I_1, \ldots, y^I_p)$ given by*

$$y^I_k := \min_{x \in \mathcal{X}} f_k(x) = \min_{y \in \mathcal{Y}} y_k \tag{2.13}$$

is called the ideal point *of the multicriteria optimization problem* $\min_{x\in\mathcal{X}} (f_1(x), \ldots, f_p(x))$.

2. The point $y^N = (y^N_1, \ldots, y^N_p)$ given by

$$y^N_k := \max_{x \in \mathcal{X}_E} f_k(x) = \max_{y \in \mathcal{Y}_N} y_k \tag{2.14}$$

is called the nadir point *of the multicriteria optimization problem.*

The ideal and nadir points for a nonconvex problem are shown in Figure 2.7.

Obviously, we have $y^I_k \leq y_k$ and $y_k \leq y^N_k$ for any $y \in \mathcal{Y}_N$. Furthermore y^I and y^N are tight lower and upper bounds on the efficient set. Since the ideal point is found by solving p single objective optimization problems its computation can be considered easy (from a multicriteria point of view). On the other hand, computation of y^N involves optimization over the efficient set, a very difficult problem. No efficient method to determine y^N for a general MOP is known.

Due to the difficulty of computing y^N, heuristics are often used. A basic estimation of the nadir point uses pay-off tables. We describe the approach now.

First, we solve p single objective problems $\min_{x\in\mathcal{X}} f_k(x)$. Let the optimal solutions be $x^k,\ k = 1, \ldots, p$, i.e. $f_k(x^k) = \min_{x\in X} f_k(x)$. Using these optimal solutions compute the pay-off table shown in Table 2.1.

Finally, from the pay-off table, clearly $y^I_k = f_k(x^k), k = 1, \ldots, p$. We define

$$\tilde{y}^N_i := \max_{k=1,\ldots,p} f_i(x^k), \tag{2.15}$$

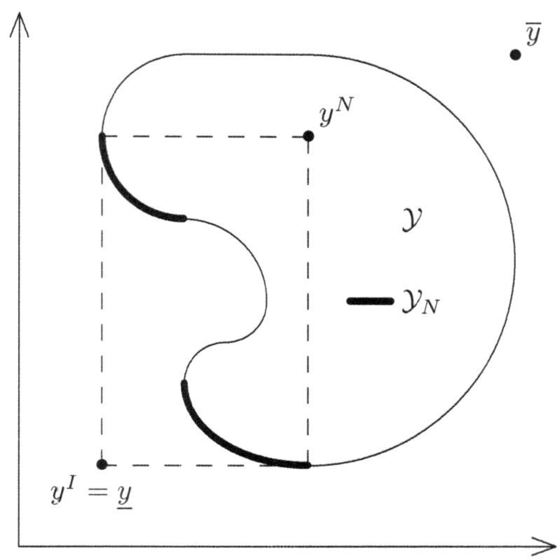

Fig. 2.7. Efficient set, ideal, and nadir point.

Table 2.1. Pay-off table and ideal point.

	x^1	x^2	$\cdots$	x^{p-1}	x^p
f_1	y_1^I	$f_1(x^2)$	$\cdots$	$f_1(x^{p-1})$	$f_1(x^p)$
f_2	$f_2(x^1)$	$\ddots$	$\cdots$	$\cdots$	$f_2(x^p)$
$\vdots$	$\vdots$	$\vdots$	$\ddots$	$\vdots$	$\vdots$
f_{p-1}	$f_{p-1}(x^1)$	$\cdots$	$\cdots$	$\ddots$	$f_{p-1}(x^p)$
f_p	$f_p(x^1)$	$f_p(x^2)$	$\cdots$	$f_p(x^{p-1})$	y_p^I

the largest element in row i, as an estimate for y_i^N.

Although appealing at first glance, the problem with pay-off tables is that $\tilde{y}^N$ may over- or under-estimate y^N, when more than two objectives are present, and when there are multiple optimal solutions of the single objective problems $\min_{x \in \mathcal{X}} f_k(x)$. The example below illustrates the phenomenon.

Example 2.23 (Korhonen et al. *(1997)).* Consider the multicriteria linear programming problem

$$\begin{array}{llllllllll}
\min & & -11\,x_2 & -11\,x_3 & -12\,x_4 & -9\,x_5 & -9\,x_6 & +9\,x_7 & \\
\min & -11\,x_1 & & -11\,x_3 & -9\,x_4 & -12\,x_5 & -9\,x_6 & +9\,x_7 & \\
\min & -11\,x_1 & -11\,x_2 & & -9\,x_4 & -9\,x_5 & -12\,x_6 & -12\,x_7 & \\
\text{subject to} & x_1 & +\,x_2 & +\,x_3 & +\,x_4 & +\,x_5 & +\,x_6 & +\,x_7 & = 1 \\
 & & & & & & & x & \geqq 0.
\end{array}$$

The image of the feasible set $\mathcal{Y} = f(\mathcal{X})$ is illustrated in Figure 2.8.

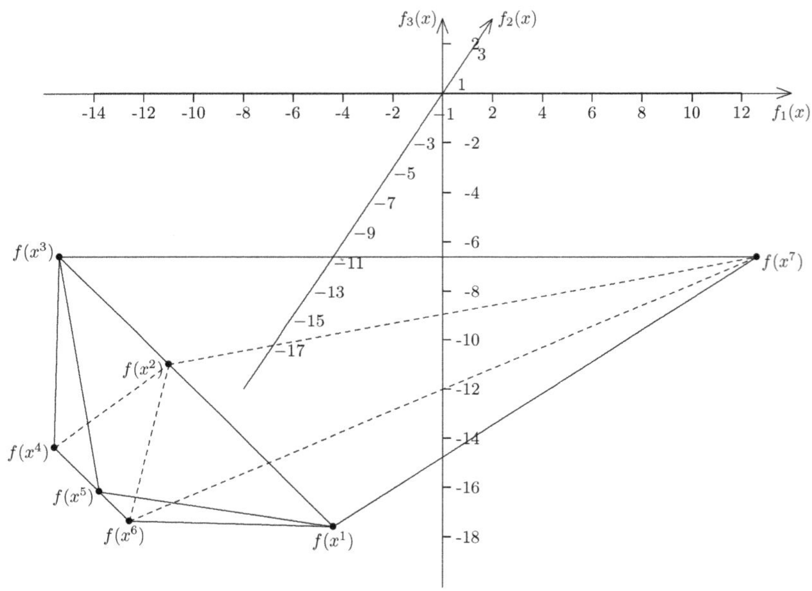

Fig. 2.8. Feasible set in Example 2.23.

To check the pay-off table approach, we proceed as follows. Solving the single objective problems, we get the solutions shown in Table 2.2, where e^i denotes the i-th unit vector.

The pay-off table is shown in Table 2.3, with two different choices of the optimal solution of the third problem, namely $x = e^6$ and $x = e^7$.

We shall now show that the nadir point cannot be obtained from the pay-off table. By solving appropriate weighted sum problems with positive weights, it can be seen that $x^i = e^i, i \in \{1, \ldots, 6\}$ are (properly) efficient (cf. Chapter 3) The feasible solution $x^7 = e^7$ is obviously weakly efficient, as a minimizer of one objective, but not efficient since x^6 dominates x^7.

For $x = e^1 \in \mathcal{X}_E$ we have $f(x) = (0, -11, -11)$. For $x = e^2 \in \mathcal{X}_E$ we have $f(x) = (-11, 0, -11)$ and for $x = e^3 \in \mathcal{X}_E$ we have $f(x) = (-11, -11, 0)$.

Table 2.2. Single objectives and minimizers in Example 2.23.

Problem	All optimal solutions
$\min_{x\in\mathcal{X}} f_1(x)$	$x_4 = 1,\ x_i = 0,\ i \neq 4$, i.e. $x = e^4$
$\min_{x\in\mathcal{X}} f_2(x)$	$x_5 = 1,\ x_i = 0,\ i \neq 5$, i.e. $x = e^5$
$\min_{x\in\mathcal{X}} f_3(x)$	$x_6 = \alpha,\ x_7 = 1-\alpha,\ x_i = 0,\ i \neq 6,7$,
	where $\alpha \in [0,1]$, i.e. $x = \alpha e^6 + (1-\alpha)e^7$

Table 2.3. Pay-off table in Example 2.23.

	e^4	e^5	e^6	e^7
f_1	-12	-9	-9	9
f_2	-9	-12	-9	9
f_3	-9	-9	-12	-12

Therefore $y_i^N \geq 0,\ i = 1,2,3$. But because no efficient solution can have positive objective values in this example, the Nadir point is $y^N = (0,0,0)$.

For the values in the pay-off table, we observe that

- with $x = e^7$ we overestimate y_1^N (arbitrarily far: replace +9 by $M > 0$ arbitrarily large), whereas
- with $x = e^6$ we underestimate y_1^N severely (arbitrarily far, if we modify the cost coefficients appropriately).

□

The reason for overestimation in Example 2.23 is, that $\overline{x}^3$ is only weakly efficient. If we choose efficient solutions to determine x^i, overestimation is of course impossible. The presence of weakly efficient solutions is caused by the multiple optimal solutions of $\min_{x\in\mathcal{X}} f_3(x)$. In general, it is difficult to be sure that the single objective optimizers are efficient.

The only case where y^N can be determined is for $p = 2$. Here the worst value for y_2 is attained when y_1 is minimal and vice versa, and by a two step optimization process, we can eliminate weakly efficient choices in the pay-off table.

Algorithm 2.1 (Nadir point for $p = 2$.)

Input: *Feasible set* $\mathcal{X}$ *and objective function* f *of an MOP.*

Solve the single objective problems $\min_{x \in \mathcal{X}} f_1(x)$ *and* $\min_{x \in \mathcal{X}} f_2(x)$. *Denote the optimal objective values by* y_1^I, y_2^I.

Solve $\min_{x \in \mathcal{X}} f_2(x)$ *with the additional constraint* $f_1(x) \leq y_1^I$.

Solve $\min_{x \in \mathcal{X}} f_1(x)$ *with the additional constraint* $f_2(x) \leq y_2^I$.

Denote the optimal objective values by y_2^N, y_1^N, *respectively.*

Output: $y^N = (y_1^N, y_2^N)$ *is the nadir point,* $y^I = (y_1^I, y_2^I)$ *is the ideal point.*

It is easy to see from the definition of y^N that the procedure indeed finds y^N. The optimal solutions of the constrained problems in the second step are efficient. Unfortunately, this approach cannot be generalized to more than two objectives, because if $p > 2$ we do not know, which objectives to fix in the second step. Indeed, the reader can check, that in Example 8.5 of Section 8.1, the Nadir point is $y^N = (11, 9, 11, 8)$, where y_2^N and y_4^N are determined by efficient solutions, which are not optimal for any of the single objectives.

2.3 Weakly and Strictly Efficient Solutions

Nondominated points are defined by the componentwise order on $\mathbb{R}^p$. When we use the the weak and strict componentwise order instead, we obtain definitions of strictly and weakly nondominated points, respectively. In this section, we prove an existence result for weakly nondominated points and weakly efficient solutions. We then give a geometric characterization of all three types of efficiency and some further results on the structure of weakly efficient solutions of convex multicriteria optimization problems.

Definition 2.24. *A feasible solution* $\hat{x} \in \mathcal{X}$ *is called* weakly efficient *(weakly Pareto optimal) if there is no* $x \in \mathcal{X}$ *such that* $f(x) < f(\hat{x})$, *i.e.* $f_k(x) < f_k(\hat{x})$ *for all* $k = 1, \ldots, p$. *The point* $\hat{y} = f(\hat{x})$ *is then called* weakly nondominated.

A feasible solution $\hat{x} \in \mathcal{X}$ *is called* strictly efficient *(strictly Pareto optimal) if there is no* $x \in \mathcal{X}$, $x \neq \hat{x}$ *such that* $f(x) \leqq f(\hat{x})$. *The weakly (strictly) efficient and nondominated sets are denoted* $\mathcal{X}_{wE}(\mathcal{X}_{sE})$ *and* $\mathcal{Y}_{wE}$, *respectively.*

Some authors say that a weakly nondominated point is a nondominated point with respect to $\operatorname{int} \mathbb{R}^p_{\geqq} = \mathbb{R}^p_{>}$, a notation that is quite convenient in the context of cone-efficiency and cone-nondominance. Because in this text we focus on the case of the nonnegative orthant we shall distinguish between efficiency/nondominance and their weak counterparts.

From the definitions it is obvious that

$$\mathcal{Y}_N \subset \mathcal{Y}_{wN} \tag{2.16}$$

and

$$\mathcal{X}_{sE} \subset \mathcal{X}_E \subset \mathcal{X}_{wE}. \tag{2.17}$$

As in the case of efficiency, weak efficiency has several equivalent definitions. We mention only two. A feasible solution $\hat{x} \in \mathcal{X}$ is weakly efficient if and only if

1. there is no $x \in \mathcal{X}$ such that $f(\hat{x}) - f(x) \in \operatorname{int} \mathbb{R}^p_{\geqq} = \mathbb{R}^p_{>}$
2. $(f(\hat{x}) - \mathbb{R}^p_{>}) \cap \mathcal{Y} = \emptyset$.

It is also of interest that there is no such concept as strict nondominance for sets $\mathcal{Y} \subset \mathbb{R}^p$. By definition, strict efficiency prohibits solutions x^1, x^2 with $f(x^1) = f(x^2)$, i.e. strict efficiency is the multicriteria analogon of unique optimal solutions in scalar optimization:

$$\hat{x} \in \mathcal{X}_{sE} \Longleftrightarrow \hat{x} \in \mathcal{X}_E \text{ and } |\{x : f(x) = f(\hat{x})\}| = 1. \tag{2.18}$$

It is obvious that all existence results for $\mathcal{Y}_N$ imply existence of $\mathcal{Y}_{wN}$ as well. However, we shall see that $\mathcal{Y}_{wN}$ can be nonempty, even if $\mathcal{Y}_N$ is empty. Therefore, independent conditions for $\mathcal{Y}_{wN}$ to be nonempty are interesting. We give a rather weak one here, another one is in Exercise 2.6. Note that the proof does not require Zorn's Lemma.

Theorem 2.25. *Let $\mathcal{Y} \subset \mathbb{R}^p$ be nonempty and compact. Then $\mathcal{Y}_{wN} \neq \emptyset$.*

Proof. Suppose $\mathcal{Y}_{wN} = \emptyset$. Then for all $y \in \mathcal{Y}$ there is some $y' \in \mathcal{Y}$ such that $y \in y' + \mathbb{R}^p_{>}$. Taking the union over all $y \in \mathcal{Y}$ we obtain

$$\mathcal{Y} \subset \bigcup_{y'} (y' + \mathbb{R}^p_{>}). \tag{2.19}$$

Because $\mathbb{R}^p_{>}$ is open, (2.19) defines an open cover of $\mathcal{Y}$. By compactness of $\mathcal{Y}$ there exists a finite subcover, i.e.

$$\mathcal{Y} \subset \bigcup_{i=1}^{k} (y^i + \mathbb{R}^p_{>}). \tag{2.20}$$

Choosing y^i on the left hand side, this yields that for all $i = 1, \ldots, k$ there is some $1 \leq j \leq k$ with $y^i \in y^j + \mathbb{R}^p_{>}$. In other words, for all i there is some j such that $y^j < y^i$. By transitivity of the strict componentwise order $<$ and because there are only finitely many y^i there exist i^*, m, and a chain of inequalities s.t. $y^{i^*} < y^{i_1} < \ldots < y^{i_m} < y^{i^*}$, which is impossible. □

The essential difference as compared to the proofs of Theorems 2.10 and 2.12 is that in those theorems we deal with sets $y - \mathbb{R}^p_{\geqq}$ which are closed. Here, we have sets $y + \mathbb{R}^p_{>}$ which are open. Note that $y \notin y + \mathbb{R}^p_{>}$.

Theorem 2.25 and continuity of f can now be used to prove existence of weakly efficient solutions.

Corollary 2.26. *Let $\mathcal{X} \subset \mathbb{R}^n$ be nonempty and compact. Assume that $f : \mathbb{R}^n \to \mathbb{R}^p$ is continuous. Then $\mathcal{X}_{wE} \neq \emptyset$.*

Proof. The result follows from Theorem 2.19 and $\mathcal{X}_E \subset \mathcal{X}_{wE}$ or from Theorem 2.25 and the fact that $f(\mathcal{X})$ is compact for compact $\mathcal{X}$ and continuous f. □

As indicated earlier, the inclusion $\mathcal{Y}_N \subset \mathcal{Y}_{wN}$ is in general strict. The following example shows that $\mathcal{Y}_{wN}$ can be nonempty, even if $\mathcal{Y}_N$ is empty, and also, of course, if $\mathcal{Y}$ is not compact. It also illustrates that $\mathcal{Y}_{wN} \setminus \mathcal{Y}_N$ might be a rather large set.

Example 2.27. Consider the set

$$\mathcal{Y} = \left\{(y_1, y_2) \in \mathbb{R}^2 : \ 0 < y_1 < 1, \ 0 \leq y_2 \leq 1\right\}. \tag{2.21}$$

Then $\mathcal{Y}_N = \emptyset$ but $\mathcal{Y}_{wN} = (0,1) \times \{0\} = \{y \in \mathcal{Y} : \ 0 < y_1 < y_2, y_2 = 0\}$ (Figure 2.9).

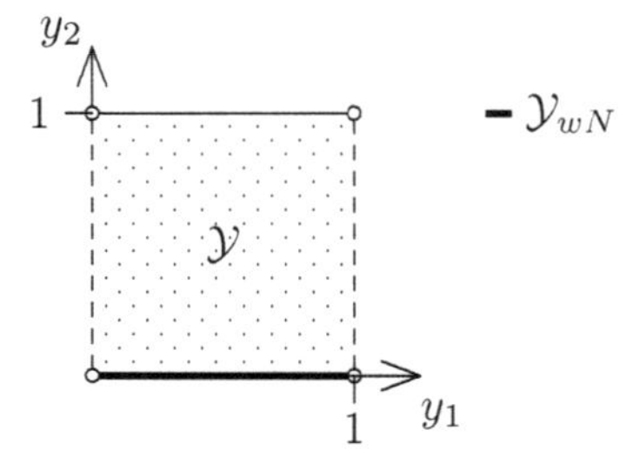

Fig. 2.9. $\mathcal{Y}_N$ is empty, $\mathcal{Y}_{wN}$ is not.

Let us now look at the closed square, i.e.

$$\mathcal{Y} = \{(y_1, y_2) \in \mathbb{R}^2 : \ 0 \leq y_i \leq 1\}. \tag{2.22}$$

We have $\mathcal{Y}_N = \{0\}$ and $\mathcal{Y}_{wN} = \{(y_1, y_2) \in \mathcal{Y} : \ y_1 = 0 \text{ or } y_2 = 0\}$. (Figure 2.10)

□

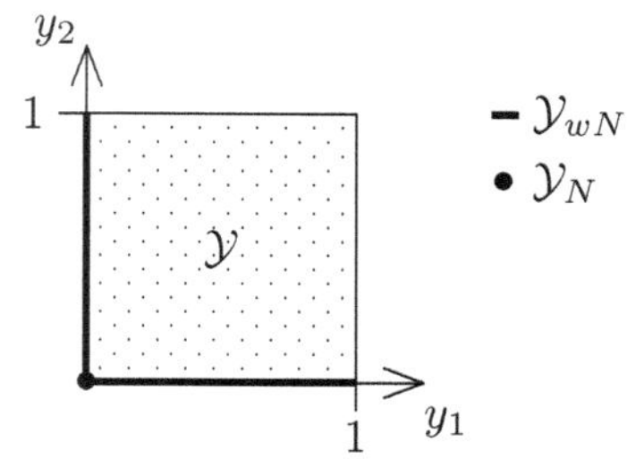

Fig. 2.10. Nondominated and weakly nondominated points.

$\mathcal{X}_E, \mathcal{X}_{sE}$ and $\mathcal{X}_{wE}$ can be characterized geometrically. To derive this characterization, we introduce *level sets* and *level curves* of functions.

Definition 2.28. *Let* $\mathcal{X} \subset \mathbb{R}^n$*,* $f : \mathcal{X} \to \mathbb{R}$*, and* $\hat{x} \in \mathcal{X}$*.*

$$\mathcal{L}_{\leq}(f(\hat{x})) = \{x \in \mathcal{X} : \ f(x) \leq f(\hat{x})\} \tag{2.23}$$

is called the level set *of* f *at* $\hat{x}$*.*

$$\mathcal{L}_{=}(f(\hat{x})) = \{x \in \mathcal{X} : \ f(x) = f(\hat{x})\} \tag{2.24}$$

is called the level curve *of* f *at* $\hat{x}$*.*

$$\begin{aligned} \mathcal{L}_{<}(f(\hat{x})) &= \mathcal{L}_{\leq}(f(\hat{x})) \setminus \mathcal{L}_{=}(f(\hat{x})) \\ &= \{x \in \mathcal{X} : \ f(x) < f(\hat{x})\} \end{aligned} \tag{2.25}$$

is called the strict level set *of* f *at* $\hat{x}$*.*

Obviously $\mathcal{L}_{=}(f(\hat{x})) \subset \mathcal{L}_{\leq}(f(\hat{x}))$ and $x \in \mathcal{L}_{=}(f(\hat{x}))$.

Example 2.29. We use an example with $\mathcal{X} = \mathbb{R}^2$ for illustration purposes. Let $f(x_1, x_2) = x_1^2 + x_2^2$. Let $\hat{x} = (3, 4)$. Hence

$$\mathcal{L}_{\leq}(f(\hat{x})) = \left\{(x_1, x_2) \in \mathbb{R}^2 : \ x_1^2 + x_2^2 \leq 25\right\}, \tag{2.26}$$

$$\mathcal{L}_{=}(f(\hat{x})) = \left\{(x_1, x_2) \in \mathbb{R}^2 : \ x_1^2 + x_2^2 = 25\right\}. \tag{2.27}$$

The level set and level curve are illustrated in Figure 2.11, as disk and circle in the x_1-x_2-plane, respectively.

□

For a multicriteria optimization problem we consider the level sets and level curves of all objectives $f_1, \ldots, f_p$ at $\hat{x}$. The following observation shows how level sets can be used to decide efficiency of $\hat{x}$.

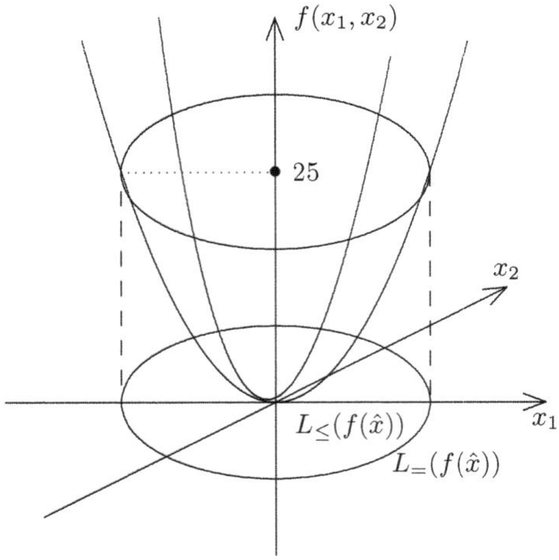

Fig. 2.11. Level set and level curve in Example 2.29.

Let us consider a bicriterion problem, and assume that we have determined $\mathcal{L}_{\leq}(f_1(\hat{x}))$ and $\mathcal{L}_{\leq}(f_2(\hat{x}))$ for feasible solution $\hat{x}$, as shown in Figure 2.12. We shall assume that the level curves are the boundaries of the level sets and the strict level sets are the interiors of the level sets.

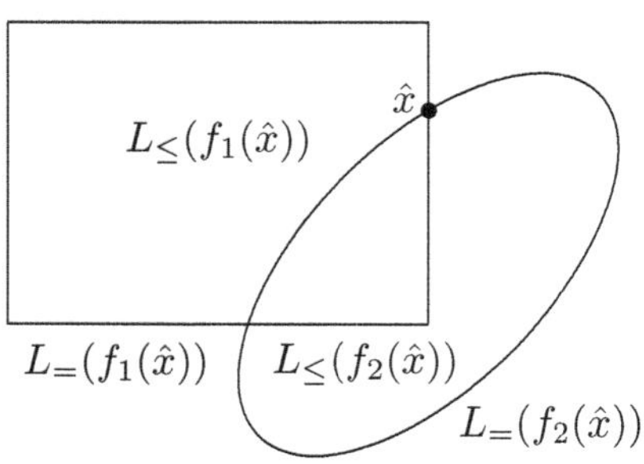

Fig. 2.12. Level sets and efficiency.

Can $\hat{x}$ be efficient? The answer is no: It is possible to move into the interior of the intersection of both level sets and thus find feasible solutions, which are better with respect to both f_1 and f_2. In fact, $\hat{x}$ is not even weakly efficient. Thus, $\hat{x}$ can only be (weakly) efficient if the intersection of strict level sets is empty or level sets intersect in level curves, respectively. We can now formulate the characterization of (strict, weak) efficiency using level sets.

Theorem 2.30 (Ehrgott *et al.* (1997)). *Let $\hat{x} \in \mathcal{X}$ be a feasible solution and define $\hat{y}_k := f_k(\hat{x})$, $k = 1, \ldots, p$. Then*

1. $\hat{x}$ is strictly efficient *if and only if*

$$\bigcap_{k=1}^{p} \mathcal{L}_{\leq}(\hat{y}_k) = \{\hat{x}\}. \tag{2.28}$$

2. $\hat{x}$ is efficient *if and only if*

$$\bigcap_{k=1}^{p} \mathcal{L}_{\leq}(\hat{y}_k) = \bigcap_{k=1}^{p} \mathcal{L}_{=}(\hat{y}_k). \tag{2.29}$$

3. $\hat{x}$ is weakly efficient *if and only if*

$$\bigcap_{k=1}^{p} \mathcal{L}_{<}(\hat{y}_k) = \emptyset. \tag{2.30}$$

Proof. 1. $\hat{x}$ is strictly efficient
$\iff$ there is no $x \in \mathcal{X}, x \neq \hat{x}$ such that $f(x) \leqq f(\hat{x})$
$\iff$ there is no $x \in \mathcal{X}, x \neq \hat{x}$ such that $f_k(x) \leq f_k(\hat{x})$ for all $k = 1, \ldots, p$
$\iff$ there is no $x \in \mathcal{X}, x \neq \hat{x}$ such that $x \in \cap_{k=1}^{p} \mathcal{L}_{\leq}(\hat{y}_k)$
$\iff \cap_{k=1}^{p} \mathcal{L}_{\leq}(\hat{y}_k) = \{\hat{x}\}$

2. $\hat{x}$ is efficient
$\iff$ there is no $x \in \mathcal{X}$ such that both $f_k(x) \leq f_k(\hat{x})$ for all $k = 1, \ldots, p$ and $f_j(x) < f_j(\hat{x})$ for some j
$\iff$ there is no $x \in \mathcal{X}$ such that both $x \in \cap_{k=1}^{p} \mathcal{L}_{\leq}(\hat{y}_q)$ and $x \in \mathcal{L}_{<}(\hat{y}_j)$ for some j
$\iff \cap_{k=1}^{p} \mathcal{L}_{\leq}(\hat{y}_k) = \cap_{k=1}^{p} \mathcal{L}_{=}(y_k)$

3. $\hat{x}$ is weakly efficient
$\iff$ there is no $x \in \mathcal{X}$ such that $f_k(x) < f_k(\hat{x})$ for all $k = 1, \ldots, p$
$\iff$ there is no $x \in \mathcal{X}$ such that $x \in \cap_{k=1}^{p} \mathcal{L}_{<}(\hat{y}_q)$
$\iff \cap_{k=1}^{p} \mathcal{L}_{<}(\hat{y}_k) = \emptyset.$ □

Clearly, Theorem 2.30 is most useful when the level sets are available graphically, i.e. when $n \leq 3$. We illustrate the use of the geometric characterization by means of an example with two variables. Exercises 2.8 – 2.11 show how the(strictly, weakly) efficient solutions can be described explicitly for problems with one variable.

Example 2.31. Consider three points in the Euclidean plane, $x^1 = (1,1), x^2 = (1,4)$, and $x^3 = (4,4)$. The l_2^2-location problem is to find a point $x = (x_1, x_2) \in \mathbb{R}^2$ such that the sum of weighted squared distances from x to the three points $x^i, i = 1,2,3$ is minimal. We consider a bicriterion l_2^2-location problem, i.e. two weights for each of the points x^i are given through two weight vectors $w^1 = (1,1,1)$ and $w^2 = (2,1,4)$.

The two objectives measuring weighted distances are given by

$$f_k(x) = \sum_{i=1}^{3} w_i^k((x_1^i - x_1)^2 + (x_2^i - x_2)^2). \tag{2.31}$$

Evaluating these functions we obtain

$$\begin{aligned} f_1(x) &= 2(1-x_1)^2 + (4-x_1)^2 + (1-x_2)^2 + 2(4-x_2)^2 \\ &= (x_1^2 - 4x_1 + x_2^2 - 6x_2) + 51 \\ f_2(x) &= 3(1-x_1)^2 + 4(4-x_1)^2 + 2(1-x_2)^2 + 5(4-x_2)^2 \\ &= 7\left(x_1^2 - \frac{38}{7}x_1 + x_2^2 - \frac{44}{7}x_2\right) + 149. \end{aligned}$$

We want to know if $x = (2,2)$ is efficient. So we check the level sets and level curves of f_1 and f_2 at $(2,2)$. The objective values are $f_1(2,2) = 15$ and $f_2(2,2) = 41$.

The level set $\mathcal{L}_=(f_1(2,2)) = \{x \in \mathbb{R}^2 : \; f_1(x) = 15\}$ is given by

$$\begin{aligned} f_1(x) = 15 &\iff 3(x_1^2 - 4x_1 + x_2^2 - 6x_2) + 51 = 15 \\ &\iff (x_1 - 2)^2 + (x_2 - 3)^2 = 1, \end{aligned}$$

i.e. $\mathcal{L}_=(f_1(2,2)) = \{x \in \mathbb{R}^2 \; : \; (x_1-2)^2 + (x_2-3)^2 = 1\}$, a circle with center $(2,3)$ and radius 1. Analogously, for f_2 we have

$$\begin{aligned} f_2(x) = 41 &\iff 7\left(x_1^2 - \frac{38}{7}x_1 + x_2^2 - \frac{44}{7}x_2\right) + 149 = 41 \\ &\iff \left(x_1 - \frac{19}{7}\right)^2 + \left(x_2 - \frac{22}{7}\right)^2 = \frac{89}{49}, \end{aligned}$$

and $\mathcal{L}_=(f_2(2,2)) = \left\{x \in \mathbb{R}^2 \; : \; (x_1 - 19/7)^2 + (x_2 - 22/7)^2 = 89/49\right\}$, a circle around $(19/7, 22/7)$ with radius $\sqrt{89}/7$.

In Figure 2.13 we see that $\cap_{i=1}^2 \mathcal{L}_\leq(f_i(2,2)) \neq \cap_{i=1}^2 \mathcal{L}_=(f_i(2,2))$ because the intersection of the discs has nonempty interior. Therefore, from Theorem 2.30 $x = (2,2)$ is not efficient. Note that in this case the level sets are simply the whole discs, the level curves are the circles and the strict level sets are the interiors of the discs.

Let us now check $x = (2,3)$. We have $f_1(2,3) = 12$ and $f_2(2,3) = 32$. Repeating the computations from above, we obtain

$$f_1(x) = 12 \Longleftrightarrow (x_1 - 2)^2 + (x_2 - 3)^2 = 0,$$

whence $\mathcal{L}_=(f_1(2,3)) = \{x \in \mathbb{R}^2 : (x_1-2)^2 + (x_2-3)^2 = 0\} = \{(2,3)\}$. For f_2

$$f_2(x) = 32 \Longleftrightarrow \left(x_1 - \frac{19}{7}\right)^2 + \left(x_2 - \frac{22}{7}\right)^2 = \frac{26}{49}$$

and $\mathcal{L}_=(f_2(2,3)) = \left\{x \in \mathbb{R}^2 : (x_1 - 19/7)^2 + (x_2 - 22/7)^2 = 26/49\right\}$, a circle around $(19/7, 22/7)$ with radius $\sqrt{26}/7$.

We have to check if $\mathcal{L}_=(f_1(2,3)) \cap \mathcal{L}_=(f_2(2,3))$ is the same as $\mathcal{L}_\leq(f_1(2,3)) \cap \mathcal{L}_\leq(f_2(2,3))$. But for $x = (2,3)$ $\mathcal{L}_=(f_1(2,3)) = \{(2,3)\}$, i.e. the level set consists of only one point, which is on the boundary of $\mathcal{L}_\leq(f_2(2,3))$. Thus $(2,3)$ is efficient. In fact, it is even strictly efficient.

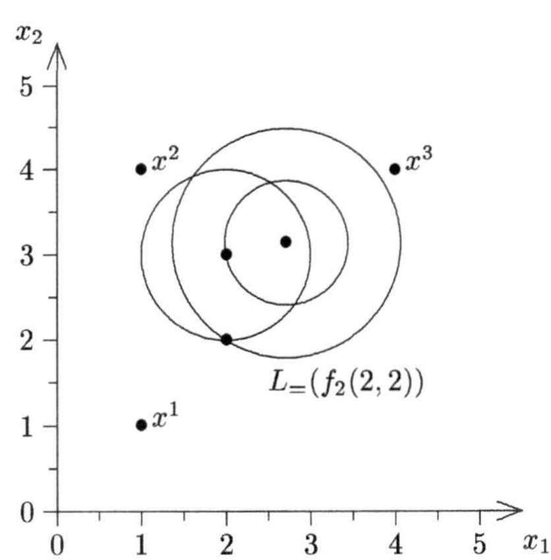

$\mathcal{L}_=(f_1(2,2))$ is the circle around $(2,3)$ with radius 1.

$\mathcal{L}_=(f_1(2,3))$ is $\{(2,3)\}$.

$\mathcal{L}_=(f_2(2,2))$ is the circle around $(19/7, 22/7)$ with radius $\sqrt{89}/7$.

$\mathcal{L}_=(f_2(2,3))$ is the circle around $(19/7, 22/7)$ with radius $\sqrt{26}/7$.

Fig. 2.13. Location problem of Example 2.31.

□

Theorem 2.30 shows that sometimes not all the criteria are needed to see if a feasible solution $\hat{x}$ is weakly or strictly efficient: Once the intersection of some level sets contains only $\hat{x}$, or the intersection of some strict level sets is empty, it will remain so when intersected with more (strict) level sets. This observation leads us to investigating the question of how many objectives are actually needed to determine if a feasible solution $\hat{x}$ is (strictly, weakly) efficient or not.

Let $\mathcal{P} \subset \{1, \ldots, p\}$ and denote by $f^{\mathcal{P}} := (f_j : j \in \mathcal{P})$ the objective function vector that only contains $f_j, j \in \mathcal{P}$.

Corollary 2.32. *Let $\mathcal{P} \subset \{1, \ldots, p\}$ be nonempty and let $\hat{x} \in \mathcal{X}$. Then the following statements hold.*

1. *If $\hat{x}$ is a weakly efficient solution of $(\mathcal{X}, f^{\mathcal{P}}, \mathbb{R}^{|\mathcal{P}|})/\mathrm{id}/(\mathbb{R}^{|\mathcal{P}|}, <)$ it is also a weakly efficient solution of $(\mathcal{X}, f, \mathbb{R}^p)/\mathrm{id}/(\mathbb{R}^p, <)$.*
2. *If $\hat{x}$ is a strictly efficient solution of $(\mathcal{X}, f^{\mathcal{P}}, \mathbb{R}^{|\mathcal{P}|})/\mathrm{id}/(\mathbb{R}^{|\mathcal{P}|}, \leqq)$ it is also a strictly efficient solution of $(\mathcal{X}, f, \mathbb{R}^p)/\mathrm{id}/(\mathbb{R}^p, \leqq)$.*

Corollary 2.32 says that weak or strict efficiency of some solution $\hat{x}$ for a problem with a subset of the p objectives implies weak (strict) efficiency for the problem with all objectives. Let us now investigate whether it is possible to find *all* weakly (strictly) efficient solutions by solving only problems with less than p objectives. For weakly efficient solutions this is possible for convex functions..

For the rest of this section we suppose that $\mathcal{X} \subset \mathbb{R}^n$ is a convex set and that $f_k : \mathbb{R}^n \to \mathbb{R}$ are convex functions. This implies that all level sets are convex. Theorem 2.30 is then about intersections of convex sets. A fundamental theorem on such intersections is known in convex analysis: Helly's Theorem.

Theorem 2.33 (Helly (1923)). *Let $p > n$ and let $C_1, \ldots, C_p \subset \mathbb{R}^n$ be convex sets. Then*

$$\bigcap_{i=1}^{p} C_i \neq \emptyset$$

if and only if for all collections of $n+1$ sets $C_{i_1}, \ldots, C_{i_{n+1}}$

$$\bigcap_{j=1}^{n+1} C_{i_j} \neq \emptyset.$$

Equivalently stated, we can say that

$$\bigcap_{i=1}^{p} C_i = \emptyset$$

if and only if there is a subset of $n+1$ sets C_i, $\{C_{i_1}, \ldots, C_{i_{n+1}}\}$ such that

$$\bigcap_{j=1}^{n+1} C_{i_j} = \emptyset.$$

In the multicriteria optimization context we will choose strict level sets as C_i. Combining Theorem 2.30, Corollary 2.32 and Helly's Theorem we immediately obtain the following "reduction result" for weakly efficient solutions of convex multicriteria optimization problems.

Proposition 2.34. *Consider the multicriteria optimization problem* $(\mathcal{X}, f, \mathbb{R}^p)/\text{id}/(\mathbb{R}^p, <)$, *where* $\mathcal{X} \subset \mathbb{R}^n$ *is convex,* $f_k : \mathbb{R}^n \to \mathbb{R}, k = 1 \ldots, p$ *are convex and* $p > n$. *Then* $\hat{x} \in \mathcal{X}$ *is weakly efficient if and only if there is a subset* $\mathcal{P} \subset \{1, \ldots, p\}$, $0 < |\mathcal{P}| \leq n + 1$ *such that* $\hat{x}$ *is a weakly efficient solution of* $(\mathcal{X}, f^{\mathcal{P}}, \mathbb{R}^{|\mathcal{P}|})/\text{id}/(\mathbb{R}^{|\mathcal{P}|}, <)$.

We shall adopt the notation $\mathcal{X}_{wE}(f)$, $\mathcal{X}_{wE}(f^{\mathcal{P}})$, and $\mathcal{X}_E(f)$, $\mathcal{X}_E(f^{\mathcal{P}})$ here to refer to the (weakly) efficient sets of the problems with f and $f^{\mathcal{P}}$, to avoid confusion. Proposition 2.34 is called a "reduction result", because it shows that the p-criteria problem $(\mathcal{X}, f, \mathbb{R}^p)/\text{id}/(\mathbb{R}^p, <)$ can be solved by solving problems with at most $n + 1$ criteria $(\mathcal{X}, f^{\mathcal{P}}, \mathbb{R}^{|\mathcal{P}|})/\text{id}/(\mathbb{R}^{|\mathcal{P}|}, <)$ at a time. Indeed, we observe that the structure of $\mathcal{X}_{wE}(f)$ is described by

$$\mathcal{X}_{wE}(f) = \bigcup_{\substack{\mathcal{P} \subset \{1,\ldots,p\} \\ |\mathcal{P}| \leq n+1}} \mathcal{X}_{wE}\left(f^{\mathcal{P}}\right). \tag{2.32}$$

Investing some more effort, it is even possible to describe $\mathcal{X}_{wE}(f)$ in terms of efficient solutions of subproblems with at most $n+1$ objectives. The following results show that on the right hand side of (2.32), $\mathcal{X}_{wE}$ can be replaced by $\mathcal{X}_E$.

Proposition 2.35 (Malivert and Boissard (1994)). *When the objective functions* f_k *are convex functions and the feasible set* $\mathcal{X}$ *is convex we have*

$$\mathcal{X}_{wE}(f) = \bigcup_{\substack{\mathcal{P} \subset \{1,\ldots,p\} \\ \mathcal{P} \neq \emptyset}} \mathcal{X}_E(f^{\mathcal{P}}). \tag{2.33}$$

Proof. We prove both set inclusions by showing the contrapositive.

"$\supset$" Choose $x \in \mathcal{X}$ with $x \notin \mathcal{X}_{wE}(f)$. Consequently, there is some $x' \in \mathcal{X}$ with $f_k(x') < f_k(x)$ for all $k = 1, \ldots, p$, which implies that x cannot be in $\mathcal{X}_E(f^{\mathcal{P}})$ for any choice of $\mathcal{P} \subset \{1, \ldots, p\}$.

"$\subset$" We prove, by induction, that for each $l = 1, \ldots, p$ there is a subset $\mathcal{P}_l$ of $\{1, \ldots, p\}$ of cardinality $p - l$ and a feasible solution x^l such that

$f_i(x^l) \leq f_i(x)$ whenever $i \in \mathcal{P}_l$ and $f_i(x^l) < f_i(x)$ otherwise. For $p = l$ this implies that x is not weakly efficient.

Choose $x \in \mathcal{X}$ with $x \notin \cup_{\mathcal{P} \subset \{1,\ldots,p\}} \mathcal{X}_E(f^{\mathcal{P}})$. In particular, $x \notin \mathcal{X}_E(f)$. Thus letting $\mathcal{P} = \{1, \ldots, p\}$, there is some $i_1 \in \mathcal{P}$ and some $x^1 \in \mathcal{X}$ such that $f_{i_1}(x^1) < f_{i_1}(x)$ and $f_i(x^1) \leq f_i(x)$, $i \neq i_1$. We now define $\mathcal{P}_1 := \mathcal{P} \setminus \{i_1\}$.

Now for $l \geq 1$ suppose we found $\mathcal{P}_l = \{1, \ldots, p\} \setminus \{i_1, \ldots, i_l\}$ and $x^l \in \mathcal{X}$ such that $f_i(x^l) < f_i(x)$ for all $i \in \{i_1, \ldots, i_l\}$ and $f_i(x^l) \leq f_i(x)$ for all $i \in \mathcal{P}_l$. Since $x \notin \mathcal{X}_E(f^{\mathcal{P}_l})$ by assumption, there is some $i_{l+1} \in \mathcal{P}_l$ and $\tilde{x}^{l+1} \in \mathcal{X}$ such that $f_{i_{l+1}}(\tilde{x}^{l+1}) < f_{i_{l+1}}(x)$ and $f_i(\tilde{x}^{l+1}) \leq f_i(x)$ for all $i \in P_l$. However, $\tilde{x}^{l+1}$ itself does not suffice to prove the condition for objectives f_i, $i \in \{i_1, \ldots, i_l\}$. We exploit convexity here. Let $x^{l+1} = \alpha x^l + (1-\alpha)\tilde{x}^{l+1}$, where $\lambda \in (0,1)$. Then

$$f_i(x^{l+1}) < f_i(x) \text{ for all } i \in \{i_1, \ldots, i_l\}, \tag{2.34}$$

whenever $(1-\alpha)$ is sufficiently small, due to the continuity of f_i and applying $f_i(x^l) < f_i(x)$ from the induction hypothesis. Furthermore,

$$\begin{aligned} f_{i_{l+1}}(x^{l+1}) &\leq \alpha f_{i_{l+1}}(x^l) + (1-\alpha) f_{i_{l+1}}(\tilde{x}^{l+1}) \\ &< \alpha f_{i_{l+1}}(x) + (1-\alpha) f_{i_{l+1}}(x) \\ &= f_{i_{l+1}}(x) \end{aligned} \tag{2.35}$$

by applying convexity for the first inequality and the induction hypothesis as well as the choice of $\tilde{x}^{l+1}$ for the second. Finally,

$$f_i(x^{l+1}) \leq f_i(x) \text{ for all } i \in P_{l+1} = \{1, \ldots, p\} \setminus \{i_1, \ldots, i_{l+1}\} \tag{2.36}$$

follows from convexity and the choice of $\tilde{x}^{l+1}$.

After p applications of this construction we have found x^p such that $f_i(x^p) < f_i(x)$ for $i = 1, \ldots, p$, i.e. $x \notin \mathcal{X}_{wE}(f)$. □

The preliminary result of Proposition 2.35 can now be combined with Helly's Theorem to obtain the structure result for weakly efficient solutions of convex multicriteria problems.

Theorem 2.36 (Malivert and Boissard (1994)). *Assume that $\mathcal{X}$ is a nonempty convex set and that the objective functions $f_k, k = 1, \ldots, p$ are convex functions. Then*

$$\mathcal{X}_{wE}(f) = \bigcup_{\substack{\mathcal{P} \subset \{1,\ldots,p\} \\ 1 \leq |\mathcal{P}| \leq n+1}} \mathcal{X}_E(f^{\mathcal{P}}). \tag{2.37}$$

Proof. Of course we need only consider the case $p > n+1$ and we only have to prove "$\subset$", because the other inclusion is an immediate consequence of Proposition 2.35 and the fact that $\mathcal{X}_E(f^{\mathcal{P}}) \subset \mathcal{X}_{wE}(f^{\mathcal{P}})$.

So, again, choose $x \in \mathcal{X}$, where $x \notin \cup_{1\leq|\mathcal{P}|\leq n+1}\mathcal{X}_E(f^{\mathcal{P}})$ and let $\mathcal{J} \subset \{1,\ldots,p\}$, $\mathcal{J} \neq \emptyset$, $|\mathcal{J}| \leq n+1$ be any nonempty subset of at most $n+1$ indices. By the assumption on x we know that $x \notin \cup_{\mathcal{I}\subset\mathcal{J}}\mathcal{X}_E(f^{\mathcal{I}})$. Then by Proposition 2.35 $x \notin \mathcal{X}_{wE}(f^{\mathcal{J}})$ and there is some $x^{\mathcal{J}} \in \mathcal{X}$ such that

$$f_j(x^{\mathcal{J}}) < f_j(x) \text{ for all } j \in \mathcal{J}. \tag{2.38}$$

For all indices $i \in \{1,\ldots,p\}$ we define

$$C_i = \operatorname{conv}\left\{x^{\mathcal{J}} : \mathcal{J} \subset \{1,\ldots,p\},\ \mathcal{J} \neq \emptyset,\ |\mathcal{J}| \leq n+1,\ i \in \mathcal{J}\right\}. \tag{2.39}$$

By (2.38) it follows that $f_i(x^{\mathcal{J}}) < f_i(x)$ for each $\mathcal{J} \subset \{1,\ldots,p\}$, $1 \leq |\mathcal{J}| \leq n+1$ and each $i \in \mathcal{J}$. Furthermore by convexity

$$f_i(x') < f_i(x) \text{ for all } x' \in C_i. \tag{2.40}$$

When we look at some $\mathcal{J}$, fixed for the moment, we know that $\cap_{i\in\mathcal{J}}C_i \supset \{x^{\mathcal{J}}\}$, i.e. $\cap_{i\in\mathcal{J}}C_i \neq \emptyset$. Therefore we can apply Helly's Theorem to conclude that there is at least one $\hat{x} \in \cap_{i=1}^{p}C_i$ and (2.40) tells us $f_i(\hat{x}) < f_i(x)$, thus $x \notin \mathcal{X}_{wE}(f)$. □

With a reduction result like (2.32) and a structure result like Theorem 2.36 for weakly efficient solutions, we may ask if similar results are possible for (strictly) efficient solutions. We give a counterexample to see that

$$\mathcal{X}_{sE}(f) = \bigcup_{\substack{\mathcal{P}\subset\{1,\ldots,p\}\\ |\mathcal{P}|\leq n+1}} \mathcal{X}_{sE}\left(f^{\mathcal{P}}\right)$$

does not hold for strictly efficient solutions.

Example 2.37 (Ehrgott and Nickel (2002)). Consider the MOP

$$\begin{aligned} \min \quad & (x_1,\ldots,x_n,-x_1,\ldots-x_n) \\ \text{subject to } & x \in [-1,1]^n. \end{aligned}$$

$\hat{x} = 0$ is a strictly efficient solution. Consider subsets $\mathcal{P} \subset \{1,\ldots,p\}$ with $|\mathcal{P}| < p = 2n$. If $\mathcal{P}$ is such that $|\mathcal{P}| = 2k$ and $i \in \mathcal{P} \Leftrightarrow 2i \in \mathcal{P}$ then $\hat{x} \in \mathcal{X}_E(f^{\mathcal{P}})$. However, for such subsets $\hat{x}$ is not strictly efficient, because all vectors $e^j, j \notin \mathcal{P}$ have the same objective function values for objectives in $\mathcal{P}$ (e^j denotes the i-th unit vector in $\mathbb{R}^n$).

In any other case there is some $i \leq n$ such that either $i \in \mathcal{P}, 2i \notin \mathcal{P}$ or $2i \in \mathcal{P}, i \notin \mathcal{P}$. Thus either $-e^i$ or e^i dominate $\hat{x}$. □

This shows that strict efficiency of $\hat{x}$ can only be confirmed using all p objectives. Thus for a problem with n variables, a lower bound on the maximal number of criteria needed to decide strict efficiency is $2n$.

2.4 Proper Efficiency and Proper Nondominance

According to Definition 2.1, an efficient solution does not allow improvement of one objective function while retaining the same values on the others. Improvement of some criterion can only be obtained at the expense of the deterioration of at least one other criterion. These trade-offs among criteria can be measured by computing the increase in objective f_i, say, per unit decrease in objective f_j. In some situations such trade-offs can be unbounded. We give an example below and introduce Geoffrion's definition of efficient solutions with bounded trade-offs, so called *properly efficient solutions*. Then some further definitions of proper efficiency by Borwein, Benson, and Kuhn and Tucker are presented. The results proved thereafter give an overview about the relationships between the various types of proper efficiency.

Example 2.38. Let the feasible set in decision and objective space be given by

$$\mathcal{X} = \{(x_1, x_2) \in \mathbb{R}^2 : \ (x_1 - 1)^2 + (x_2 - 1)^2 \leq 1, \ 0 \leq x_1, x_2 \leq 1\},$$

and $\mathcal{Y} = \mathcal{X}$ as shown in Figure 2.14.

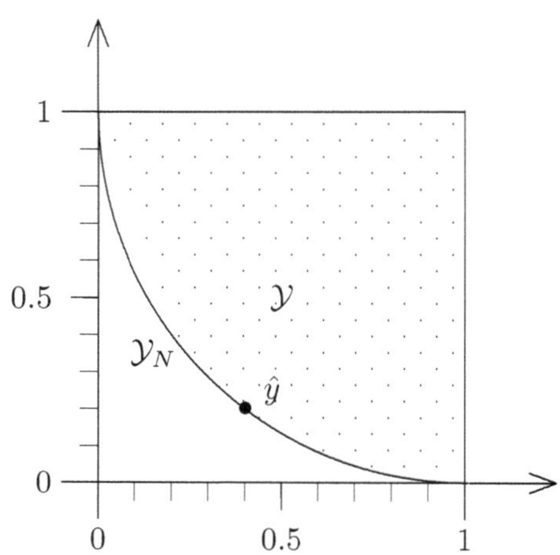

Fig. 2.14. Properly nondominated point $\hat{y}$.

Clearly, $\mathcal{Y}_N = \{(y_1, y_2) \in \mathcal{Y} : (y_1 - 1)^2 + (y_2 - 1)^2 = 1\}$. We observe that the closer $\hat{y}$ is moved towards $(1, 0)$, the larger an increase of y^1 is necessary to achieve a unit decrease in y_2. In the limit, an infinite increase of y_1 is needed to obtain a unit decrease in y_2. □

Definition 2.39 (Geoffrion (1968)). *A feasible solution $\hat{x} \in \mathcal{X}$ is called* properly efficient, *if it is efficient and if there is a real number $M > 0$ such that for all i and $x \in \mathcal{X}$ satisfying $f_i(x) < f_i(\hat{x})$ there exists an index j such that $f_j(\hat{x}) < f_j(x)$ such that*

$$\frac{f_i(\hat{x}) - f_i(x)}{f_j(x) - f_j(\hat{x})} \leq M. \tag{2.41}$$

The corresponding point $\hat{y} = f(\hat{x})$ is called properly nondominated.

According to Definition 2.39 properly efficient solutions therefore are those efficient solutions that have bounded trade-offs between the objectives.

Example 2.40. In Example 2.38 consider the solution $\hat{x} = (1, 0)$. We show that $\hat{x}$ is not properly efficient. To do so, we have to prove that for all $M > 0$ there is an index $i \in \{1, 2\}$ and some $x \in \mathcal{X}$ with $f_i(x) < f_i(\hat{x})$ such that

$$\frac{f_i(\hat{x}) - f_i(x)}{f_j(x) - f_j(\hat{x})} > M$$

for all $j \in \{1, 2\}$ with $f_j(x) > f_j(\hat{x})$.

Let $i = 1$ and choose x^ε with $x_1^\varepsilon = 1 - \varepsilon$, $0 < \varepsilon < 1$ and $x_2^\varepsilon = 1 - \sqrt{1 - \varepsilon^2}$, i.e. x^ε is efficient because $(x_1^\varepsilon - 1)^2 + (x_2^\varepsilon - 1)^2 = 1$. Since $x^\varepsilon \in \mathcal{X}$, $x_1^\varepsilon < \hat{x}_1$ and $x_2^\varepsilon > \hat{x}_2$ we have $i = 1, j = 2$. Thus

$$\frac{f_i(\hat{x}) - f_i(x^\varepsilon)}{f_j(x^\varepsilon) - f_j(\hat{x})} = \frac{1 - (1 - \varepsilon)}{1 - \sqrt{1 - \varepsilon^2}} = \frac{\varepsilon}{1 - \sqrt{1 - \varepsilon^2}} \overset{\varepsilon \to 0}{\longrightarrow} \infty. \tag{2.42}$$

□

The main results about properly efficient solutions show that they can be obtained by minimizing a weighted sum of the objective functions where all weights are positive. For convex problems optimality for the weighted sum scalarization is a necessary and sufficient condition for proper efficiency. We will prove these results in Section 3.2.

In the previous section we have given conditions for the existence of nondominated points/efficient solutions. These imply, of course, existence of weakly nondominated points/weakly efficient solutions. They do not guarantee existence of properly nondominated points. This can be seen from the following example.

Example 2.41. Let $\mathcal{Y} = \{y \in \mathbb{R}^2 : y_1 < 0, y_2 = 1/y_1\}$. Then $\mathcal{Y}_N = \mathcal{Y}$, but $\mathcal{Y}_{pN} = empty$. To see this, take any $\hat{y} \in \mathcal{Y}_N$ and a sequence y^k with $y_2^k > \hat{y}_2$ and $y_1^k \to -\infty$ or $y_1^k > \hat{y}_1$ and $y_2^k \to -\infty$. □

As mentioned in the introduction of this section, Geoffrion is not the only one to introduce properly efficient solutions. Before we can present the definitions of Borwein and Benson, we have to introduce two cones related to sets $\mathcal{Y} \subset \mathbb{R}^p$.

Definition 2.42. *Let $\mathcal{Y} \subset \mathbb{R}^p$ and $y \in \mathcal{Y}$.*

1. *The* tangent cone *of $\mathcal{Y}$ at $y \in \mathcal{Y}$ is*

$$T_{\mathcal{Y}}(y) := \left\{d \in \mathbb{R}^p : \exists \{t_k\} \subset \mathbb{R}, \{y^k\} \subset \mathcal{Y} \text{ s.t. } y^k \to y, t_k(y^k - y) \to d\right\}. \tag{2.43}$$

2. *The* conical hull *of $\mathcal{Y}$ is*

$$\text{cone}(\mathcal{Y}) = \{\alpha y : \alpha \geq 0,\ y \in \mathcal{Y}\} = \bigcup_{\alpha \geq 0} \alpha \mathcal{Y}. \tag{2.44}$$

Note that the conditions $y^k \to y$ and $t_k(y^k - y) \to d$ in the definition of the tangent cone imply that $t_k \to \infty$. One could equivalently require $y^k \to y$ and $(1/(t_k))(y^k - y) \to d$, whence $t_k \to 0$. Both definitions can be found in the literature. Examples of the conical hull of a set $\mathcal{Y}$ and the tangent cone of $\mathcal{Y}$ at a point y are shown in Figure 2.15. The tangent cone is translated from the origin to the point y to illustrate where its name comes from: It is the cone of all directions tangential to $\mathcal{Y}$ in y.

Proposition 2.43 on properties of tangent cones and conical hulls will be helpful later.

Proposition 2.43. *1. The tangent cone $T_{\mathcal{Y}}(y)$ is a closed cone.*
2. If $\mathcal{Y}$ is convex then $T_{\mathcal{Y}}(y) = \text{cl}(\text{cone}(\mathcal{Y} - y))$, which is a closed convex cone.

Proof. 1. Note first that $0 \in T_{\mathcal{Y}}(y)$ (take $y^k = y$ for all k) and $T_{\mathcal{Y}}(y)$ is indeed a cone: For $\alpha > 0$, $d \in T_{\mathcal{Y}}(y)$ we have $\alpha d \in T_{\mathcal{Y}}(y)$. To see this, just take αt_k instead of t_k when constructing the sequence t_k.
To see that it is closed take a sequence $\{d^l\} \subset T_{\mathcal{Y}}(y)$, $y \in Y$, with $d^l \to d$, for some $d \in \mathbb{R}^p$. Since $d_l \in T_{\mathcal{Y}}(y)$, for all l there are sequences $\{y^{l,k}\}$, $\{t_{l,k}\}$ as in the Definition 2.42. From the convergence we get that for fixed l there is some k_l s.t.

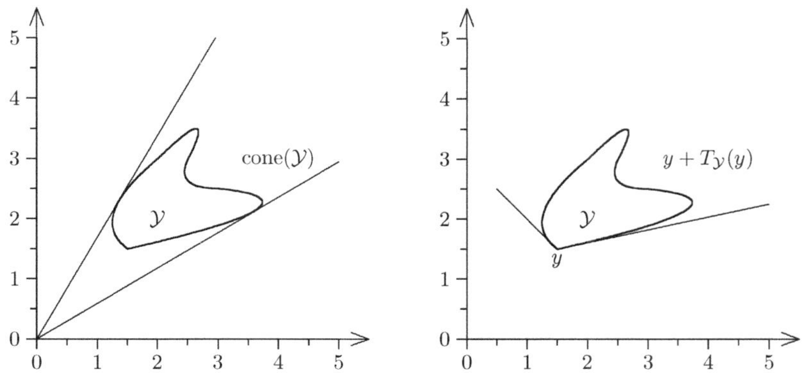

Fig. 2.15. Conical hull and tangent cone.

$$\left\|t_{l,k_l}(y^{l,k_l} - y) - d^l\right\| \leq \frac{1}{l} \tag{2.45}$$

for all $k \geq k_l$. We fix the k_l and observe that because of (2.45) if $l \to \infty$ the sequence $t_{l,k_l}(y^{l,k_l} - y) \to d$, i.e. $d \in T_{\mathcal{Y}}(y)$.

2. Let $\mathcal{Y}$ be convex, $y \in \mathcal{Y}$. By definition of closure and conical hull, it is obvious that $\mathrm{cl}(\mathrm{cone}(\mathcal{Y} - y))$ is a closed convex cone.
To see that $T_{\mathcal{Y}}(y) \subset \mathrm{cl}(\mathrm{cone}(\mathcal{Y} - y))$ let $d \in T_{\mathcal{Y}}(y)$. Then there are sequences $\{t_k\}, \{y^k\}$ with $t_k(y^k - y) \to d$. Since $t_k(y^k - y) \in \alpha(\mathcal{Y} - y)$ for some $\alpha > 0$closedness implies $d \in \mathrm{cl}(\mathrm{cone}(\mathcal{Y} - y))$.
For $\mathrm{cl}(\mathrm{cone}(\mathcal{Y} - y)) \subset T_{\mathcal{Y}}(y)$ we know that $T_{\mathcal{Y}}(y)$ is closed and only show $\mathrm{cone}(\mathcal{Y} - y) \subset T_{\mathcal{Y}}(y)$. Let $d \in \mathrm{cone}(\mathcal{Y} - y)$, i.e. $d = \alpha(y' - y)$ with $\alpha \geq 0$, $y' \in \mathcal{Y}$. Now define $y^k := (1-1/k)y + (1/k)y' \in \mathcal{Y}$ and $t_k = \alpha k \geq 0$. Hence

$$t_k(y^k - y) = \alpha k\left(\left(\frac{k-1}{k}y + \frac{1}{k}y'\right) - y\right) = \alpha((k-1)y + y' - ky) = \alpha(y' - y).$$

So $y^k \to y$ and $t_k(y^k - y) \to d$ implying $d \in T_{\mathcal{Y}}(y)$. □

Definition 2.44. *1. (Borwein (1977)) A solution* $\hat{x} \in \mathcal{X}$ *is called* properly efficient *(in Borwein's sense) if*

$$T_{\mathcal{Y}+\mathbb{R}^p_{\geqq}}(f(\hat{x})) \cap \left(-\mathbb{R}^p_{\geqq}\right) = \{0\}. \tag{2.46}$$

2. (Benson (1979)) A solution $\hat{x} \in \mathcal{X}$ *is called* properly efficient *if*

$$\mathrm{cl}\left(\mathrm{cone}\left(\mathcal{Y} + \mathbb{R}^p_{\geqq} - f(\hat{x})\right)\right) \cap \left(-\mathbb{R}^p_{\geqq}\right) = \{0\}. \tag{2.47}$$

As we observed in Proposition 2.43 it is immediate from the definitions of conical hulls and tangent cones that

$$T_{\mathcal{Y}+\mathbb{R}^p_{\geqq}}(f(\hat{x})) \subset \text{cl}\left(\text{cone}\left(\mathcal{Y} + \mathbb{R}^p_{\geqq} - f(\hat{x})\right)\right) \tag{2.48}$$

so that Benson's definition is stronger than Borwein's.

Theorem 2.45. *1. If $\hat{x}$ is properly efficient in Benson's sense, it is also properly efficient in Borwein's sense.*
2. If $\mathcal{X}$ is convex and $f_k : \mathbb{R}^n \to \mathbb{R}$ are convex then both definitions coincide.

Example 2.46. Consider $\mathcal{X} = \{(x_1, x_2) : x_1^2 + x_2^2 \leq 1\}$ and, as usual, $f_1(x) = x_1$, $f_2(x) = x_2$. Then $(-1, 0)$ and $(0, -1)$ are efficient, but not properly efficient in the sense of Borwein (and thus not in the sense of Benson).

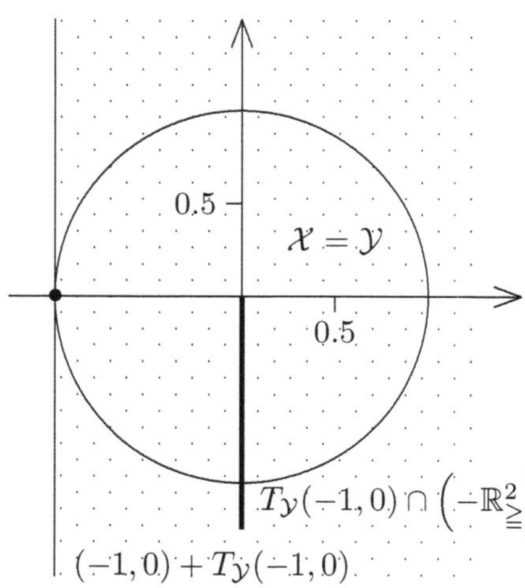

Fig. 2.16. Benson's proper efficiency.

The tangent cone translated to the point $y = (-1, 0)$ contains all directions in which $\mathcal{Y}$ extends from y, including the limits, i.e. the tangents. The tangent to the circle at $(-1, 0)$ is a vertical line, and therefore

$$T_{\mathcal{Y}}(-1, 0) = \{(y_1, y_2) \in \mathbb{R}^2 : \ y_1 \geq 0\}. \tag{2.49}$$

The intersection with the nonpositive orthant is therefore not $\{0\}$:

$$T_{\mathcal{Y}}(-1, 0) \cap (-\mathbb{R}^p_{\geqq}) = \{(y_1, y_2) \in \mathbb{R}^2 \ : \ y_1 = 0, \ y_2 \leq 0\}, \tag{2.50}$$

indicated by the bold line in Figure 2.16. A similar interpretation applies to $(0, -1)$. □

That convexity is indeed needed for Borwein's definition to imply Benson's can be seen in Exercise 2.13. Definition 2.44 does not require $\hat{x}$ to be efficient, as does Definition 2.39. It is therefore legitimate to ask whether properly efficient solutions in Benson's or Borwein's sense are always efficient.

Proposition 2.47. *If $\hat{x}$ is properly efficient in Borwein's sense, then $\hat{x}$ is efficient.*

Proof. The proof is left to the reader as Exercise 2.12. □

Benson's and Borwein's definitions of proper efficiency are not restricted to the componentwise order. In fact, in these definitions $\mathbb{R}^p_{\geqq}$ can be replaced by an arbitrary closed convex cone $\mathcal{C}$. They are therefore applicable in the more general context of orders defined by cones. Geoffrion's definition on the other hand explicitly uses the componentwise order. Our next result shows that in the case of $\mathcal{C} = \mathbb{R}^p_{\geqq}$ the definitions of Geoffrion and Benson actually coincide, so that Benson's proper efficiency is a proper generalization of Geoffrion's.

Theorem 2.48 (Benson (1979)). *Feasible solution $\hat{x} \in \mathcal{X}$ is properly efficient in Geoffrion's sense (Definition 2.39) if and only if it is properly efficient in Benson's sense.*

Proof. "$\Longrightarrow$" Suppose $\hat{x}$ is efficient, but not properly efficient in Benson's sense. Then we know that a nonzero $d \in \text{cl}(\text{cone}(\mathcal{Y} + \mathbb{R}^p_{\geqq} - f(\hat{x}))) \cap (-\mathbb{R}^p_{\geqq})$ exists. Without loss of generality we may assume that $d_1 < -1$, $d_i \leq 0$, $i = 2, \ldots, p$ (otherwise we can reorder the components of f and rescale d). Consequently there are sequences $\{t_k\} \subset \mathbb{R}_{>}$, $\{x^k\} \subset \mathcal{X}$, $\{r^k\} \subset \mathbb{R}^p_{\geqq}$ such that $t_k(f(x^k) + r^k - f(\hat{x})) \to d$.
Choosing subsequences if necessary, we can assume that $\mathcal{Q} := \{i \in \{1, \ldots, p\} : f_i(x^k) > f_i(\hat{x})\}$ is the same for all k and nonempty (since $\hat{x}$ is efficient). Now let $M > 0$. From convergence we get existence of k_0 such that for all $k \geq k_0$

$$f_1(x^k) - f_1(\hat{x}) < -\frac{1}{2 \cdot t_k} \tag{2.51}$$

$$\text{and } f_i(x^k) - f_i(\hat{x}) \leq \frac{1}{2 \cdot M t_k} \quad i = 2, \ldots, p \tag{2.52}$$

because $t_k \to \infty$. In particular, for $i \in \mathcal{Q}$, we have

$$0 < f_i(x^k) - f_i(\hat{x}) \leq \frac{1}{2 \cdot M t_k} \quad \forall\, k \geq k_0 \tag{2.53}$$

and therefore, from (2.51) and (2.53)

$$\frac{f_1(\hat{x}) - f_1(x^k)}{f_i(x^k) - f_i(\hat{x})} > \frac{\frac{1}{2 \cdot t_k}}{\frac{1}{2 \cdot M t_k}} = M. \tag{2.54}$$

Because M was chosen arbitrarily chosen, $\hat{x}$ is not properly efficient in Geoffrion's sense.

"$\Longleftarrow$" Suppose $\hat{x}$ is efficient, but not properly efficient in Geoffrion's sense. Let $M_k > 0$ be an unbounded sequence of positive real numbers. Without loss of generality we assume that for all M_k there is an $x^k \in \mathcal{X}$ such that $f_1(x^k) < f_1(\hat{x})$ and

$$\frac{f_1(\hat{x}) - f_1(x^k)}{f_j(x^k) - f_j(\hat{x})} > M_k \quad \forall\, j \in \{2, \ldots, p\} \text{ with } f_j(x^k) > f_j(\hat{x}). \tag{2.55}$$

Again, choosing a subsequence if necessary, we can assume $\mathcal{Q} = \{i \in \{1, \ldots, p\} : f_i(x^k) > f_i(\hat{x})\}$ is constant for all k and nonempty. We construct appropriate sequences $\{t_k\}$, $\{r^k\}$ such that the limit of $t_k(f(x^k) + r^k - f(\hat{x}))$ converges to $d \in \text{cl}(\text{cone}(f(\mathcal{X}) + \mathbb{R}^p_{\geqq} - f(\hat{x}))) \cap (-\mathbb{R}^p_{\geq})$.

Define $t_k := (f_1(\hat{x}) - f_1(x^k))^{-1}$ which means $t_k > 0$ for all k. Define $r^k \in \mathbb{R}^p_{\geqq}$ through

$$r_i^k := \begin{cases} 0 & i = 1,\ i \in \mathcal{Q} \\ f_i(\hat{x}) - f_i(x^k) & \text{otherwise.} \end{cases} \tag{2.56}$$

With these sequences we compute

$$t_k(f_i(x^k) + r_i^k - f_i(\hat{x})) \begin{cases} = -1 & i = 1 \\ = 0 & i \neq 1,\ i \notin \mathcal{Q} \\ \in (0, M_k^{-1}) & i \in \mathcal{Q}. \end{cases} \tag{2.57}$$

This sequence converges due to the choice of $M_k \to \infty$ to some $d \in \mathbb{R}^p$, where $d_i = \lim_{k \to \infty} t_k(f_i(x^k) + r_i^k - f_i(\hat{x}))$ for $i = 1, \ldots, p$. Thus, from (2.57) $d_1 = -1$, $d_i = 0$, $i \neq 1$, $i \notin \mathcal{Q}$, $d_i = 0$, $i \in \mathcal{Q}$. Because $d = (-1, 0, \ldots, 0) \in \text{cl}(\text{cone}(f(\mathcal{X}) + \mathbb{R}^p_{\geqq} - f(\hat{x}))) \cap (-\mathbb{R}^p_{\geq})$, $\hat{x}$ is not properly efficient in Benson's sense. □

In multicriteria optimization, especially in applications, we will often encounter problems, where $\mathcal{X}$ is given implicitly by constraints, i.e.

$$\mathcal{X} = \{x \in \mathbb{R}^n : (g_1(x), \ldots, g_m(x)) \leq 0\}. \tag{2.58}$$

For such constrained multicriteria optimization problems yet another definition of proper efficiency can be given. Let us assume that the objective functions $f_i,\ i = 1, \ldots, p$ as well as the constraint functions $g_j,\ j = 1, \ldots, m$ are continuously differentiable. We consider the multiobjective programme

$$\begin{aligned} &\min f(x) \\ &\text{subject to } g(x) \leqq 0, \end{aligned} \tag{2.59}$$

where $f : \mathbb{R}^n \to \mathbb{R}^p$ and $g : \mathbb{R}^n \to \mathbb{R}^p$.

Definition 2.49 (Kuhn and Tucker (1951)). *A feasible solution $\hat{x} \in \mathcal{X}$ is called* properly efficient *(in Kuhn and Tucker's sense) if it is efficient and if there is no $d \in \mathbb{R}^n$ satisfying*

$$\langle \nabla f_k(\hat{x}), d \rangle \leq 0 \quad \forall\ k = 1, \ldots, p \tag{2.60}$$

$$\langle \nabla f_i(\hat{x}), d \rangle < 0 \quad \textit{for some } i \in \{1, \ldots, p\} \tag{2.61}$$

$$\langle \nabla g_j(\hat{x}), d \rangle \leq 0 \quad \forall\ j \in \mathcal{J}(\hat{x}) = \{j = 1, \ldots, m : \ g_j(\hat{x}) = 0\} \tag{2.62}$$

The set $\mathcal{J}(\hat{x})$ is called the set of active indices. As for Geoffrion's definition, efficiency according to the componentwise order is implicitly assumed here, and the definition is not applicable to orders derived from closed convex cones. Intuitively, the existence of a vector d satisfying (2.60) – (2.62) means that moving from $\hat{x}$ in direction d no objective function increases (2.60), one strictly decreases (2.61), and the feasible set is not left (2.62). Thus d is a feasible direction of descent. Note that a slight movement in *all* directions is always possible without violating any inactive constraint.

We prove equivalence of Kuhn and Tucker's and Geoffrion's definitions under some constraint qualification. This constraint qualification of Definition 2.50 below means that the feasible set $\mathcal{X}$ has a local description as a differentiable curve (at a feasible solution $\hat{x}$): Every feasible direction d can be written as the gradient of a feasible curve starting at $\hat{x}$.

Definition 2.50. *A differentiable MOP (2.59) satisfies the KT constraint qualification at $\hat{x} \in X$ if for any $d \in \mathbb{R}^n$ with $\langle \nabla g_j(\hat{x}), h \rangle \leq 0$ for all $j \in \mathcal{J}(\hat{x})$ there is a real number $\bar{t} > 0$, a function $\theta : [0, \bar{t}] \to \mathbb{R}^n$, and $\alpha > 0$ such that $\theta(0) = \hat{x}, g(\theta(t)) \leq 0$ for all $t \in [0, \bar{t}]$ and $\theta'(0) = \alpha d$.*

Theorem 2.51 (Geoffrion (1968)). *If a differentiable MOP satisfies the KT constraint qualification at $\hat{x}$ and $\hat{x}$ is properly efficient in Geoffrion's sense, then it is properly efficient in Kuhn and Tucker's sense.*

Proof. Suppose $\hat{x}$ is efficient, but not properly efficient according to Definition 2.49. Then there is some $d \in \mathbb{R}^n$ such that (without loss of generality, after renumbering the objectives)

$$\langle \nabla f_1(\hat{x}), d \rangle < 0 \tag{2.63}$$

$$\langle \nabla f_k(\hat{x}), d \rangle \leq 0 \quad \forall\, k = 2, \ldots, p \tag{2.64}$$

$$\langle \nabla g_j(\hat{x}), d \rangle \leq 0 \quad \forall\, j \in \mathcal{J}(\hat{x}). \tag{2.65}$$

Using the function θ from the constraint qualification we take a sequence $t_k \to 0$, and if necessary a subsequence such that

$$\mathcal{Q} = \{i : \ f_i(\theta(t_k)) > f_i(\hat{x})\} \tag{2.66}$$

is the same for all k. Since for $i \in \mathcal{Q}$ by the Taylor expansion of f_i at $\theta(t_k)$)

$$f_i(\theta(t_k)) - f_i(\hat{x}) = t_k \langle \nabla f_i(\hat{x}), \alpha d \rangle + o(t_k) > 0 \tag{2.67}$$

and $\langle \nabla f_i(\hat{x}), d \rangle \leq 0$ it must be that

$$\langle \nabla f_i(\hat{x}), \alpha d \rangle = 0 \quad \forall\, i \in \mathcal{Q}. \tag{2.68}$$

But since $\langle \nabla f_1(\hat{x}), d \rangle < 0$ the latter implies

$$\Longrightarrow \qquad \frac{f_1(\hat{x}) - f_1(\theta(t_k))}{f_i(\theta(t_k)) - f_i(\hat{x})} = \frac{-\langle \nabla f_1(\hat{x}), \alpha d \rangle + \frac{o(t_k)}{t_k}}{\langle \nabla f_i(\hat{x}), \alpha d \rangle + \frac{o(t_k)}{t_k}} \to \infty \tag{2.69}$$

whenever $i \in \mathcal{Q}$. Hence $\hat{x}$ is not properly efficient according to Geoffrion's definition. □

The converse of Theorem 2.51 holds without the constraint qualification. It turns out that this result is an immediate consequence of Theorem 3.25 (necessary conditions for Kuhn and Tucker's proper efficiency) and Theorem 3.27 (sufficient conditions for Geoffrion's proper efficiency). These results are proved in Section 3.3. In Section 3.3 we shall also see that without constraint qualification, Geoffrion's proper efficiency does not necessarily imply proper efficiency in Kuhn and Tucker's sense.

Theorem 2.52. *Let $f_k, g_j : \mathbb{R}^n \to \mathbb{R}$ be convex, continuously differentiable functions and suppose $\hat{x}$ is properly efficient in Kuhn and Tucker's sense. Then $\hat{x}$ is properly efficient in Geoffrion's sense.*

Let us conclude the section by a summary of the definitions of proper efficiency and their relationships. Figure 2.17 illustrates these (see also Sawaragi *et al.* (1985)). The arrows indicate implications. Corresponding results and the conditions under which the implications hold are mentioned alongside the arrows. On the right of the picture, the orders and problem types for which the respective definition is applicable are given.

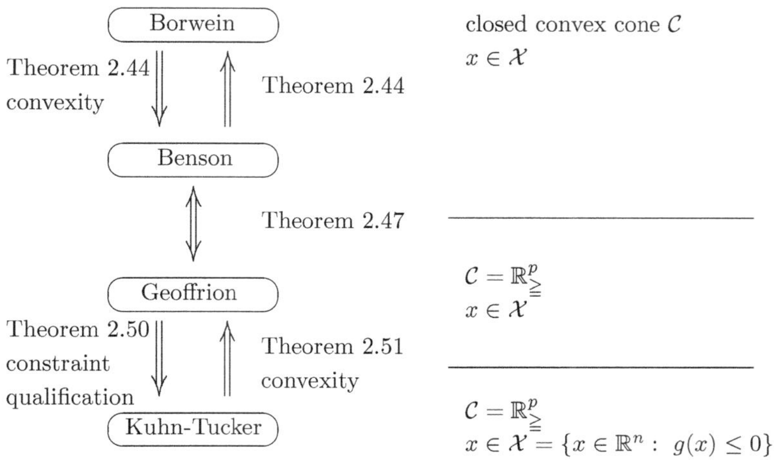

Fig. 2.17. Relationships among definitions of proper efficiency for the case $\mathcal{C} = \mathbb{R}^p_{\geqq}$.

In order to derive further results on proper efficiency and important properties of (weakly) efficient sets we have to investigate weighted sum scalarizations in greater detail, i.e. the relationships between those types of solutions and optimal solutions of single objective optimization problems

$$\min_{x \in \mathcal{X}} \sum_{k=1}^{p} \lambda_k f_k(x),$$

where $\lambda \in \mathbb{R}^p_{\geq}$ is a vector of nonnegative weights of the objective functions. This is the topic of Chapter 3.

2.5 Notes

As we have pointed out after the definition of efficient solutions and nondominated points (Definition 2.1) notation for efficient solutions and nondominated points is not unique in the literature. Table 2.4 below gives an overview of some of the notations used. Another term for efficient point is admissible point (Arrow *et al.*, 1953), but this is rarely used today. Although some authors distinguish between the case that the decision space is $\mathbb{R}^n$ or a more general vector space (Jahn) or the order is defined by $\mathbb{R}^p_{\geqq}$ or a more general cone (Miettinen), most of these definitions use the same terms in decision and

Table 2.4. Terminology for efficiency and nondominance.

Author	Decision space	Objective space
Sawaragi *et al.* (1985)	efficient solution	efficient element
Chankong and Haimes (1983)	noninferior solution	noninferior solution
Yu (1985)	Pareto optimal point	Pareto optimal outcome
	N-point	N-point
Miettinen (1999)	Pareto optimal decision vector	Pareto optimal criterion vector
	efficient decision vector	efficient criterion vector
Deb (2001)	Pareto optimal solution	Pareto optimal solution
Jahn (2004)	Edgeworth-Pareto optimal point	minimal element
	minimal solution	minimal element
Göpfert and Nehse (1990)	Pareto optimal solution	efficient element
Steuer (1985)	efficient point	nondominated criterion vector

criterion space, which might cause confusion and does not help distinguish two very different things.

The condition of $\mathbb{R}^p_{\geqq}$-compactness in Corollary 2.15 can be replaced by $\mathbb{R}^p_{\geqq}$-closedness and $\mathbb{R}^p_{\geqq}$-boundedness, which are generalizations of closedness and boundedness, see Exercises 2.4 and 2.5. For closed convex sets $\mathcal{Y}$ it can be shown that the conditions of Theorem 2.10, Corollary 2.15 and $\mathbb{R}^p_{\geqq}$-closedness and $\mathbb{R}^p_{\geqq}$-boundedness coincide, see for example Sawaragi *et al.* (1985, page 56). Other existence results are known, which often use a more general setting than we adopt in this text. We refer, e.g.xs to Göpfert and Nehse (1990); Sawaragi *et al.* (1985); Hazen and Morin (1983) or Henig (1986). A review of existence results for nondominated and efficient sets is provided by Sonntag and Zălinescu (2000).

We remark that all existence results presented in this Chapter are still valid, if $\mathbb{R}^p_{\geqq}$ is replaced by a convex, pointed, nontrivial, closed cone $\mathcal{C}$ with the proofs unchanged. Furthermore, the closedness assumption for $\mathcal{C}$ is not required if $(y - \operatorname{cl}\mathcal{C})$ is used instead of $(y - \mathcal{C})$ everywhere. In Exercises 2.2 and 2.7 nondominance with respect to a cone is formally defined, and the reader is asked to check some of the results about efficient sets in this more general context.

Similarly, Theorem 2.21 is valid for any nonempty, closed, convex cone $\mathcal{C}$. In fact, $\mathcal{C}$-compactness can be replaced by $\mathcal{C}$-closedness and $\mathcal{C}$-boundedness, see Sawaragi *et al.* (1985) for more details. External stability of $\mathcal{Y}_N$ has been shown for closed convex $\mathcal{Y}$ by Luc (1989). More results can be found in Hirschberger (2002). A counterpart to the external stability is internal stability of a set. A set $\mathcal{Y}$ is called internally stable with respect to $\mathcal{C}$, if $y - y' \notin \mathcal{C}$ for all $y, y' \in \mathcal{Y}$. Obviously, nondominated sets are always internally stable.

The computation of the nadir point is difficult, because it amounts to solving an optimization problem over the efficient set of a multicriteria optimization problem, see Yamamoto (2002) for a survey on that topic. Nevertheless, interactive methods often assume that the ideal and nadir point are known (see Miettinen (1999)[Part II, Chapter 5] for an overview on interactive methods). A discussion of heuristics and exact methods to compute nadir points and implications for interactive methods can be found in Ehrgott and Tenfelde-Podehl (2003).

For reduction results on the number of criteria to determine (strict, weak) efficiency of a feasible solution $\hat{x}$, we remark that the case of efficiency is much more difficult than either strict or weak efficiency. Ehrgott and Nickel (2002) show that a reduction result is true for strictly quasi-convex problems with $n = 2$ variables. For the general case of $n > 2$ neither a proof nor a counterexample is known.

In addition to the definitions of proper efficiency mentioned here, the following authors define properly efficient solutions Klinger (1967), Wierzbicki (1980) and Henig (1982). Borwein and Zhuang (1991, 1993) define super efficient solutions. Henig (1982) gives two definitions that generalize the definitions of Borwein and Benson (Definition 2.44) but that coincide with these in the case $\mathcal{C} = \mathbb{R}^p_{\geqq}$ discussed in this book.

Exercises

2.1. Give a counterexample to the converse inclusion in Proposition 2.6.

2.2. Given a cone $\mathcal{C} \subset \mathbb{R}^p$ and the induced order $\leq_\mathcal{C}$, $\hat{y} \in \mathcal{Y}$ is said to be $\mathcal{C}$-nondominated if there is no $y \in \mathcal{Y}, y \neq \hat{y}$ such that $y \in \hat{y} - \mathcal{C}$. The set of $\mathcal{C}$-nondominated points is denoted $\mathcal{Y}_{\mathcal{C}N}$. Let $\mathcal{C}_1, \mathcal{C}_2$ be two cones in $\mathbb{R}^p$ and assume $\mathcal{C}_1 \subset \mathcal{C}_2$. Prove that if $\hat{y}$ is $\mathcal{C}_2$-nondominated it is also $\mathcal{C}_1$-nondominated. Illustrate this "larger cone – fewer nondominated points" result graphically.

2.3. Prove that $(\alpha\mathcal{Y})_N = \alpha(\mathcal{Y}_E)$ where $\mathcal{Y} \subset \mathbb{R}^p$ is a nonempty set and α is a positive real number.

2.4. Let $\mathcal{Y} \subset \mathbb{R}^p$ be a convex set. The recession cone (or asymptotic cone) $\mathcal{Y}_\infty$ of $\mathcal{Y}$, is defined as

$$\mathcal{Y}_\infty := \{d \in \mathbb{R}^p : \exists\, y \text{ s.t. } y + \alpha d \in \mathcal{Y} \quad \forall\, \alpha > 0\},$$

i.e. the set of directions in which $\mathcal{Y}$ extends infinitely.

1. Show that $\mathcal{Y}$ is bounded if and only if $\mathcal{Y}_\infty = \{0\}$.
2. Let $\mathcal{Y} = \{(y_1, y_2) \in \mathbb{R}^2 : y_2 \geq y_1^2\}$. Determine $\mathcal{Y}_\infty$.

2.5. A set $\mathcal{Y} \subset \mathbb{R}^p$ is called $\mathbb{R}^p_\geqq$-closed, if $\mathcal{Y} + \mathbb{R}^p_+$ is closed and $\mathbb{R}^p_\geqq$-bounded, if $\mathcal{Y}_\infty \cap (-\mathbb{R}^p_\geqq) = \{0\}$. Give examples of sets $\mathcal{Y} \subset \mathbb{R}^2$ that are

1. $\mathbb{R}^2_\geqq$-compact, $\mathbb{R}^2_\geqq$-bounded, but not $\mathbb{R}^2_\geqq$-closed,
2. $\mathbb{R}^2_\geqq$-bounded, $\mathbb{R}^2_\geqq$-closed, but not $\mathbb{R}^2_\geqq$-compact.

2.6. Prove the following existence result for weakly nondominated points. Let $\emptyset \neq \mathcal{Y} \subset \mathbb{R}^p$ be $\mathbb{R}^p_\geqq$-compact. Show that $\mathcal{Y}_{wN} \neq \emptyset$. Do not use Corollary 2.15 nor the fact that $\mathcal{Y}_N \subset \mathcal{Y}_{wN}$.

2.7. Recall the definition of $\mathcal{C}$-nondominance from Exercise 2.2: $\hat{y} \in \mathcal{Y}$ is $\mathcal{C}$-nondominated if there is no $y \in \mathcal{Y}$ such that $\hat{y} \in y + \mathcal{C}$. Verify that Proposition 2.3 is still true if $\mathcal{C}$ is a pointed, convex cone. Give examples that the inclusion $\mathcal{Y}_{\mathcal{C}N} \subset (\mathcal{Y} + \mathcal{C})_{\mathcal{C}N}$ is not true when $\mathcal{C}$ is not pointed and when $\mathcal{C}$ is not convex.

2.8. Let $[a, b] \subset \mathbb{R}$ be a compact interval. Suppose that all $f_k : \mathbb{R} \to \mathbb{R}$ are convex, $k = 1, \ldots, p$. Let

$$x_k^m = \min\left\{x \in [a, b] : f_k(x) = \min_{x \in [a,b]} f_k(x)\right\}$$

and

$$x_k^M = \max\left\{x \in [a,b] : f_k(x) = \min_{x\in[a,b]} f_k(x)\right\}.$$

Using Theorem 2.30 show that

$$\mathcal{X}_E = \left[\min_{k=1,\ldots,p} x_k^M, \quad \max_{k=1,\ldots,p} x_k^m\right] \cup \left[\max_{k=1,\ldots,p} x_k^m, \quad \min_{k=1,\ldots,p} x_k^M\right]$$

$$\mathcal{X}_{wE} = \left[\min_{k=1,\ldots,p} x_k^m, \quad \max_{k=1,\ldots,p} x_k^M\right].$$

2.9. Use the result of Exercise 2.8 to give an example of a multicriteria optimization problem with $\mathcal{X} \subset \mathbb{R}$ where $\mathcal{X}_{sE} \subset \mathcal{X}_E \subset \mathcal{X}_{wE}$, with strict inclusions. Use two or three objective functions.

2.10 (Hirschberger (2002)). Let $\mathcal{Y} = \{y \in \mathbb{R}^2 : y_1 < 0, y_2 = 1/y_1\}$. Show that $\mathcal{Y}_N = \mathcal{Y}$ but $\mathcal{Y}_{pN} = \emptyset$.

2.11. Let $\mathcal{X} = \{x \in \mathbb{R} : x \geq 0\}$ and $f_1(x) = e^x$,

$$f_2(x) = \begin{cases} \frac{1}{x+1} & 0 \leq x \leq 5 \\ (x-5)^2 + \frac{1}{6} & x \geq 5. \end{cases}$$

Using the result of Exercise 2.8, determine $\mathcal{X}_E$. Which of these solutions are strictly efficient? Can you prove a sufficient condition on f for $x \in \mathbb{R}$ to be a strictly efficient solution of $\min_{x\in\mathcal{X}\subset\mathbb{R}} f(x)$? Derive a conjecture from the example and try to prove it.

2.12. Show that if $\hat{x}$ is properly efficient in the sense of Borwein, then $\hat{x}$ is efficient.

2.13 (Benson (1979)). Consider the following example:

$$\begin{aligned}\mathcal{X} = \left\{(x_1,x_2) \in \mathbb{R}^2 : x_1 + x_2 \geq 0\right\} &\cup \left\{(x_1,x_2) \in \mathbb{R}^2 : x_1 \geq 1\right\} \\ &\cup \left\{(x_1,x_2) \in \mathbb{R}^2 : x_2 \geq 1\right\}\end{aligned}$$

with $f_1(x) = x_1, f_2(x) = x_2$. Show that $x = 0$ is properly efficient in the sense of Borwein, but not in the sense of Benson.

2.14. Consider an MOP $\min_{x\in\mathcal{X}} f(x)$ with p objectives. Add a new objective f_{p+1}. Is the efficient set of the new problem bigger or smaller than that of the original problem or does it remain unchanged?

2.15. The following definition of an ideal point was given by Balbás *et al.* (1998). Let $\min_{x\in\mathcal{X}}(f_1(x), \ldots, f_p(x))$ be a multicriteria optimization problem. A point $y \in \mathbb{R}^p$ is called an ideal point if there exists a closed, convex, pointed cone $\mathcal{C} \subseteq \mathbb{R}^p$ such that $\mathcal{Y} \subset y + \mathcal{C}$. If in addition $y \in \mathcal{Y}$, y is a proper ideal point.

1. Show that the ideal point y^I (Section 2.2) is ideal in Balbas' sense. Which cone C is used?
2. For the MOP with

$$\mathcal{X} = \{x \in \mathbb{R}^2 : x_1 + x_2 \geq 1, \ -x_1 + x_2 \geq 0, \ x_1, x_2 \geq 0\}$$

   and $f_1(x) = x_1, \ f_2(x) = x_2$, determine $\mathcal{Y}_N$ and the set of all ideal points.
3. Give an example of a problem where ideal points exist, but at most finitely many of them are proper ideal points. Can you find an example with no proper ideal points?

2.16. Prove formally that Algorithm 2.1 is correct, i.e. that it finds the nadir point for bicriterion problems.

3

The Weighted Sum Method and Related Topics

In this chapter we will investigate to what extent an MOP

$$\min_{x \in \mathcal{X}} (f_1(x), \ldots, f_p(x)) \tag{3.1}$$

of the Pareto class

$$(\mathcal{X}, f, \mathbb{R}^p)/\mathrm{id}/(\mathbb{R}^p, \leq)$$

can be solved (i.e. its efficient solutions be found) by solving single objective problem problems of the type

$$\min_{x \in \mathcal{X}} \sum_{k=1}^{p} \lambda_k f_k(x), \tag{3.2}$$

which in terms of the classification of Section 1.5 is written as

$$(\mathcal{X}, f, \mathbb{R}^p)/\langle \lambda, \cdot \rangle/(\mathbb{R}, \leq), \tag{3.3}$$

where $\langle \lambda, \cdot \rangle$ denotes the scalar product in $\mathbb{R}^p$. We call the single objective (or scalar) optimization problem (3.2) a *weighted sum scalarization* of the MOP (3.1).

As in the previous chapter, we will usually look at the objective space $\mathcal{Y}$ first and prove results on the relationships between (weakly, properly) nondominated points and values $\sum_{k=1}^{p} \lambda_k y_k$. From those, we can derive results on the relationships between $\mathcal{X}_{(w,p)E}$ and optimal solutions of (3.2).

We use these results to prove Fritz-John and Kuhn-Tucker type optimality conditions for (weakly, properly) efficient solutions (Section 3.3). Finally, we investigate conditions that guarantee that nondominated and efficient sets are connected (Section 3.4).

Let $\mathcal{Y} \subset \mathbb{R}^p$. For a fixed $\lambda \in \mathbb{R}^p_{\geqq}$ we denote by

$$\mathcal{S}(\lambda, \mathcal{Y}) := \left\{ \hat{y} \in \mathcal{Y} : \ \langle \lambda, \hat{y} \rangle = \min_{y \in \mathcal{Y}} \langle \lambda, y \rangle \right\} \tag{3.4}$$

the set of optimal points of $\mathcal{Y}$ with respect to λ.

Figure 3.1 gives an example of a set $\mathcal{S}(\lambda, \mathcal{Y})$ consisting of two points y^1 and y^2. These points are the intersection points of a line $\{y \in \mathbb{R}^p : \langle \lambda, y \rangle = \hat{c}$. Obviously, y^1 and y^2 are nondominated. Considering c as a parameter, and the family of lines $\{y \in \mathbb{R}^p : \langle \lambda, y \rangle = c\}$, we see that in Figure 3.1 $\hat{c}$ is chosen as the smallest value of c such that the intersection of the line with $\mathcal{Y}$ is nonempty.

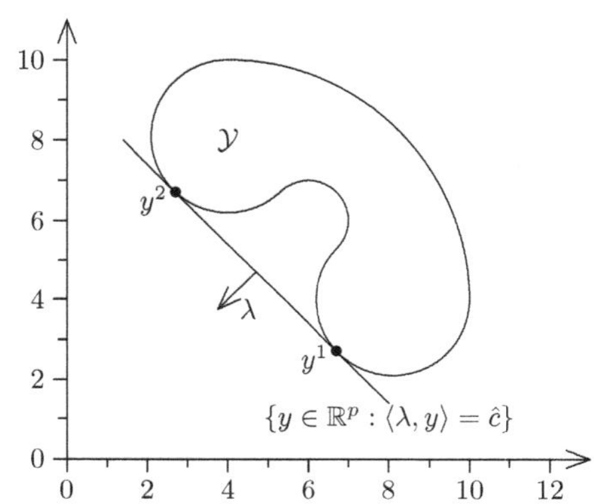

Fig. 3.1. A set $\mathcal{S}(\lambda, \mathcal{Y})$.

Graphically, to find $\hat{c}$ we can start with a large value of the parameter c and translate the line in parallel towards the origin as much as possible while keeping a nonempty intersection with $\mathcal{Y}$. Analytically, this means finding elements of $\mathcal{S}(\lambda, \mathcal{Y})$. The obvious questions are:

1. Does this process always yield nondominated points? (Is $\mathcal{S}(\lambda, \mathcal{Y}) \subset \mathcal{Y}_N$?) and
2. if so, can all nondominated points be detected this way? (Is $\mathcal{Y}_N \subset \bigcup_{\lambda \in \mathbb{R}^p_{\geq}} \mathcal{S}(\lambda, \mathcal{Y})$?)

Note that due to the definition of nondominated points, we have to consider nonnegative weighting vectors $\lambda \in \mathbb{R}^p_{\geq}$ only. However, the distinction between nonnegative and positive weights turns out to be essential. Therefore we distinguish optimal points of $\mathcal{Y}$ with respect to nonnegative and strictly positive weights, and define

$$\mathcal{S}(\mathcal{Y}) := \bigcup_{\lambda \in \mathbb{R}^p_>} \mathcal{S}(\lambda, \mathcal{Y}) = \bigcup_{\{\lambda > 0 : \sum_{k=1}^p \lambda_k = 1\}} \mathcal{S}(\lambda, \mathcal{Y}) \tag{3.5}$$

$$\text{and } \mathcal{S}_0(\mathcal{Y}) := \bigcup_{\lambda \in \mathbb{R}^p_\geq} \mathcal{S}(\lambda, \mathcal{Y}) = \bigcup_{\{\lambda \geq 0 : \sum_{k=1}^p \lambda_k = 1\}} \mathcal{S}(\lambda, \mathcal{Y}). \tag{3.6}$$

Clearly, the assumption $\sum_{k=1}^p \lambda_k = 1$ can always be made. It just normalizes the weights, but does not change $\mathcal{S}(\lambda, \mathcal{Y})$. It will thus be convenient to have the notation

$$\Lambda := \left\{ \lambda \in \mathbb{R}^p_{\geqq} : \sum_{k=1}^p \lambda_k = 1 \right\}$$

$$\Lambda^0 := \operatorname{ri} \Lambda = \left\{ \lambda \in \mathbb{R}^p_{\geq} : \sum_{k=1}^p \lambda_k = 1 \right\}.$$

It is also evident that using $\lambda = 0$ does not make sense, as $\mathcal{S}(0, \mathcal{Y}) = \mathcal{Y}$. We exclude this case henceforth. Finally,

$$\mathcal{S}(\mathcal{Y}) \subset \mathcal{S}_0(\mathcal{Y}) \tag{3.7}$$

follows directly from the definition. The results in the following two sections extend (3.7) by including links with efficient sets.

In many of the results of this chapter we will need some convexity assumptions. However, requiring $\mathcal{Y}$ to be convex is usually too restrictive a requirement. After all, we are looking for nondominated points, which, bearing Proposition 2.3 in mind are located in the "south-west" of $\mathcal{Y}$. Hence, we define $\mathbb{R}^p_{\geqq}$-convexity.

Definition 3.1. *A set $\mathcal{Y} \in \mathbb{R}^p$ is called $\mathbb{R}^p_{\geqq}$-convex, if $\mathcal{Y} + \mathbb{R}^p_{\geqq}$ is convex.*

Every convex set $\mathcal{Y}$ is clearly $\mathbb{R}^p_{\geqq}$-convex. The set $\mathcal{Y}$ of Figure 3.1 is neither convex nor $\mathbb{R}^p_{\geqq}$-convex. Figure 2.4 shows a nonconvex set $\mathcal{Y}$ which is $\mathbb{R}^p_{\geqq}$-convex.

A fundamental result about convex sets is that nonintersecting convex sets can be separated by a hyperplane.

Theorem 3.2. *Let $\mathcal{Y}_1, \mathcal{Y}_2 \subset \mathbb{R}^p$ be nonempty convex sets. There exists some $y^* \in \mathbb{R}^p$ such that*

$$\inf_{y \in \mathcal{Y}_1} \langle y, y^* \rangle \geq \sup_{y \in \mathcal{Y}_2} \langle y, y^* \rangle \tag{3.8}$$

$$\text{and } \sup_{y \in \mathcal{Y}_1} \langle y, y^* \rangle > \inf_{y \in \mathcal{Y}_2} \langle y, y^* \rangle \tag{3.9}$$

if and only if $\text{ri}(\mathcal{Y}_1) \cap \text{ri}(\mathcal{Y}_2) = \emptyset$. *In this case* $\mathcal{Y}_1$ *and* $\mathcal{Y}_2$ *are said to be properly separated by a hyperplane with normal* y^*.

Recall that $\text{ri}(\mathcal{Y}_i)$ is the relative interior of $\mathcal{Y}_i$, i.e. the interior in the space of appropriate dimension $\dim(\mathcal{Y}_i) \leq p$. A proof of Theorem 3.2 can be found in Rockafellar (1970, p. 97).

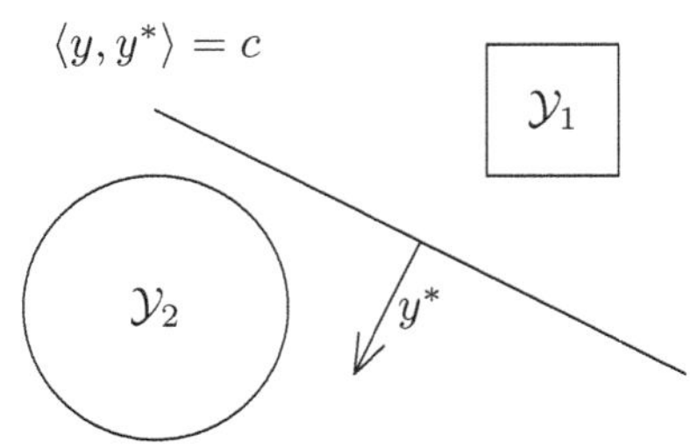

Fig. 3.2. Properly separated sets $\mathcal{Y}_1$ and $\mathcal{Y}_2$.

We will also use the following separation theorem.

Theorem 3.3. *Let* $\mathcal{Y} \subset \mathbb{R}^p$ *be a nonempty, closed, convex set and let* $y^0 \in \mathbb{R}^p \setminus \mathcal{Y}$. *Then there exists a* $y^* \in \mathbb{R}^p \setminus \{0\}$ *and* $\alpha \in \mathbb{R}$ *such that*

$$\langle y^*, y^0 \rangle < \alpha < \langle y^*, y \rangle$$

for all $y \in \mathcal{Y}$.

We will derive results on efficient solutions $x \in \mathcal{X}$ of a multicriteria optimization problem from results on nondominated points $y \in \mathcal{Y}$. This is done as follows. For results that are valid for any set $\mathcal{Y}$ we obtain analogous results simply by invoking the fact that efficient solutions are preimages of nondominated points. For results that are only valid under some conditions on $\mathcal{Y}$ (usually $\mathbb{R}^p_{\geqq}$-convexity) , appropriate assumptions on $\mathcal{X}$ and f are required. To ensure $\mathbb{R}^p_{\geqq}$-convexity of $\mathcal{Y}$ the assumption of convexity of $\mathcal{X}$ and all objective functions f_k.

3.1 Weighted Sum Scalarization and (Weak) Efficiency

In this section, we show that optimal solutions of the weighted sum problem (3.2) with positive (nonnegative) weights are always (weakly) efficient and that under convexity assumptions all (weakly) efficient solutions are optimal solutions of scalarized problems with positive (nonnegative) weights.

Theorem 3.4. *For any set* $\mathcal{Y} \subset \mathbb{R}^p$ *we have* $\mathcal{S}_0(\mathcal{Y}) \subset \mathcal{Y}_{wN}$.

Proof. Let $\lambda \in \mathbb{R}^p_{\geq}$ and $\hat{y} \in \mathcal{S}(\lambda, \mathcal{Y})$. Then

$$\sum_{k=1}^{p} \lambda_k \hat{y}_k \leq \sum_{k=1}^{p} \lambda_k y_k \text{ for all } y \in \mathcal{Y}.$$

Suppose that $\hat{y} \notin \mathcal{Y}_{wN}$. Then there is some $y' \in \mathcal{Y}$ with $y'_k < \hat{y}_k$, $k = 1, \ldots, p$. Thus,

$$\sum_{k=1}^{p} \lambda_k y'_k < \sum_{k=1}^{p} \lambda_k \hat{y}_k,$$

because at least one of the weights λ_k must be positive. This contradiction implies the result. □

For $\mathbb{R}^p_{\geqq}$-convex sets we can prove the converse inclusion.

Theorem 3.5. *If* $\mathcal{Y}$ *is* $\mathbb{R}^p_{\geqq}$*-convex, then* $\mathcal{Y}_{wN} = \mathcal{S}_(\mathcal{Y})$.

Proof. Due to Theorem 3.4 we only have to show $\mathcal{Y}_{wN} \subset \mathcal{S}_(\mathcal{Y})$. We first observe that $\mathcal{Y}_{wN} \subset (\mathcal{Y} + \mathbb{R}^p_{>})_{wN}$ (The proof of this fact is the same as that of Proposition 2.3, replacing $\mathbb{R}^p_{\geqq}$ by $\mathbb{R}^p_{>}$).

Therefore, if $\hat{y} \in \mathcal{Y}_{wN}$, we have

$$(\mathcal{Y}_{wN} + \mathbb{R}^p_{>} - \hat{y}) \cap (-\mathbb{R}^p_{>}) = \emptyset.$$

This means that the intersection of the relative interiors of the two convex sets $\mathcal{Y} + \mathbb{R}^p_{>} - \hat{y}$ and $-\mathbb{R}^p_{>}$ is empty. By Theorem 3.2 there is some $\lambda \in \mathbb{R}^p \setminus \{0\}$ such that

$$\langle \lambda, y + d - \hat{y} \rangle \geq 0 \geq \langle \lambda, -d' \rangle \tag{3.10}$$

for all $y \in \mathcal{Y}, d \in \mathbb{R}^p_{>}, d' \in \mathbb{R}^p_{>}$.

Since $\langle \lambda, -d' \rangle \leq 0$ for all $d' \in \mathbb{R}^p_{>}$ we can choose $d' = e_k + \varepsilon e$ – where e_k is the k-th unit vector, $e = (1, \ldots, 1) \in \mathbb{R}^p$ is a vector of all ones, and $\varepsilon > 0$ is arbitrarily small – to see that $\lambda_k \geq 0, k = 1, \ldots, p$. On the other hand, choosing $d = \varepsilon e$ in $\langle \lambda, y + d - \hat{y} \rangle \geq 0$ implies

$$\langle \lambda, y \rangle + \varepsilon \langle \lambda, e \rangle \geq \langle \lambda, \hat{y} \rangle \tag{3.11}$$

for all $y \in \mathcal{Y}$ and thus

$$\langle \lambda, y \rangle > \langle \lambda, \hat{y} \rangle. \tag{3.12}$$

Therefore $\lambda \in \mathbb{R}^p_{\geq}$ and $\hat{y} \in \mathcal{S}(\lambda, \mathcal{Y}) \subset \mathcal{S}_(\mathcal{Y})$. □

With Theorems 3.4 and 3.5 we have the first extension of inclusion (3.7), namely

$$\mathcal{S}(\mathcal{Y}) \subset \mathcal{S}_{(}\mathcal{Y}) \subset \mathcal{Y}_{wN} \tag{3.13}$$

in general and

$$\mathcal{S}(\mathcal{Y}) \subset \mathcal{S}_{(}\mathcal{Y}) = \mathcal{Y}_{wN} \tag{3.14}$$

for $\mathbb{R}_{\geqq}$-convex sets.

Next we relate $\mathcal{S}(\mathcal{Y})$ and $\mathcal{S}_{(}\mathcal{Y})$ to $\mathcal{Y}_N$.

Theorem 3.6. *Let $\mathcal{Y} \subset \mathbb{R}^p$. Then $\mathcal{S}(\mathcal{Y}) \subset \mathcal{Y}_N$.*

Proof. Let $\hat{y} \in \mathcal{S}(\mathcal{Y})$. Then there is some $\lambda \in \mathbb{R}^p_>$ satisfying $\sum_{k=1}^p \lambda_k \hat{y}_k \leq \sum_{k=1}^p \lambda_k y_k$ for all $y \in \mathcal{Y}$.

Suppose $\hat{y} \notin \mathcal{Y}_N$. Hence there must be $y' \in \mathcal{Y}$ with $y' \leq y$. Multiplying componentwise by the weights gives $\lambda_k y'_k \leq \lambda_k \hat{y}_k$ for all $k = 1, \ldots, p$ and strict inequality for one k. This strict inequality together with the fact that all λ_k are positive implies $\sum_{k=1}^p \lambda_k y'_k < \sum_{k=1}^p \lambda_k \hat{y}_k$, contradicting $\hat{y} \in \mathcal{S}(\mathcal{Y})$. □

Corollary 3.7. *$\mathcal{Y}_N \subset \mathcal{S}_{(}\mathcal{Y})$ if $\mathcal{Y}$ is an $\mathbb{R}^p_{\geqq}$-convex set.*

Proof. This result is an immediate consequence of Theorem 3.5 since $\mathcal{Y}_N \subset \mathcal{Y}_{wN} = \mathcal{S}_{(}\mathcal{Y})$. □

Theorem 3.6 and Corollary 3.7 yield

$$\mathcal{S}(\mathcal{Y}) \subset \mathcal{Y}_N; \quad \mathcal{S}_{(}\mathcal{Y}) \subset \mathcal{Y}_{wN} \tag{3.15}$$

in general and

$$\mathcal{S}(\mathcal{Y}) \subset \mathcal{Y}_N \subset \mathcal{S}_{(}\mathcal{Y}) = \mathcal{Y}_{wN} \tag{3.16}$$

for $\mathbb{R}^p_{\geqq}$-convex sets.

Theorem 3.6 can be extended by the following proposition.

Proposition 3.8. *If $\hat{y}$ is the unique element of $\mathcal{S}(\lambda, \mathcal{Y})$ for some $\lambda \in \mathbb{R}^p_{\geq}$ then $\hat{y} \in \mathcal{Y}_N$.*

Proof. The easy proof is left to the reader, see Exercise 3.2.

In Exercise 3.3 the reader is asked for examples where the inclusions in (3.15) and (3.16) are strict, demonstrating that these are the strongest relationship between weighted sum optimal points and (weakly) nondominated points that can be proved for general and $\mathbb{R}^p_{\geqq}$-convex sets, without additional assumptions, like the uniqueness of Proposition 3.8

Let us now summarize the analogies of the results of this section in terms of the decision space, i.e. (weakly) efficient solutions of multicriteria optimization problems.

Proposition 3.9. *Suppose that $\hat{x}$ is an optimal solution of the weighted sum optimization problem*

$$\min_{x \in \mathcal{X}} \sum_{k=1}^{p} \lambda_k f_k(x). \tag{3.17}$$

with $\lambda \in \mathbb{R}^p_{\geqq}$. Then the following statements hold.

1. *If $\lambda \in \mathbb{R}^p_{\geq}$ then $\hat{x} \in \mathcal{X}_{wE}$.*
2. *If $\lambda \in \mathbb{R}^p_{>}$ then $\hat{x} \in \mathcal{X}_{E}$.*
3. *If $\lambda \in \mathbb{R}^p_{\geq}$ and $\hat{x}$ is a unique optimal solution of (3.17) then $\hat{x} \in \mathcal{X}_{sE}$.*

Proof. The assertions follow directly from Theorem 3.4, Theorem 3.6, and Proposition 3.8 with the uniqueness of $\hat{x}$, respectively. □

Proposition 3.10. *Let $\mathcal{X}$ be a convex set, and let f_k be convex functions, $k = 1, \ldots, p$. If $\hat{x} \in \mathcal{X}_{wE}$ there is some $\lambda \in \mathbb{R}^p_{\geqq}$ such that $\hat{x}$ is an optimal solution of* (3.17).

The proof follows from Theorem 3.5. Note that there is no distinction between $\mathcal{X}_{wE}$ and $\mathcal{X}_E$ here, an observation that we shall regrettably have to make for almost all methods to find efficient solutions. This problem is caused by the (possibly) strict inclusions in (3.16). Therefore the examples of Exercise 3.3 and the usual trick of identifying decision and objective space should convince the reader that this problem cannot be avoided.

At the end of this section, we point out that Exercises 3.4 and 3.7 show how to generalize the weighted sum scalarization for nondominated points with respect to a convex and pointed cone $\mathcal{C}$.

Remembering that $\mathcal{Y}_{pN} \subset \mathcal{Y}_N$, we continue our investigations by looking at relationships between $\mathcal{Y}_{pN}$ and $\mathcal{S}(\mathcal{Y})$.

3.2 Weighted Sum Scalarization and Proper Efficiency

Here we will establish the relationships between properly nondominated points (in Benson's or Geoffrion's sense) and optimal points of weighted sum scalarizations with positive weights. The main result shows that these points coincide for convex sets. A deeper result shows that in this situation the difference between nondominated and properly nondominated points is small: The set of properly nondominated points is dense in the nondominated set.

From now on we denote the set of properly efficient points in Geoffrion's sense by $\mathcal{Y}_{pE}$.. Note that due to Theorem 2.48 Geoffrion's and Benson's definitions are equivalent for efficiency defined by $\mathbb{R}^p_{\geqq}$.

Unless otherwise stated, $\mathcal{X}_{pN}$ will denote the set of properly efficient solutions of a multicriteria optimization problem in Geoffrion's sense. Our first result shows that an optimal solution of (3.2) is a properly efficient solution of (3.1) if $\lambda > 0$.

Theorem 3.11 (Geoffrion (1968)). *Let $\lambda_k > 0$, $k = 1, \ldots, p$ with $\sum_{k=1}^{p} \lambda_k = 1$ be positive weights. If $\hat{x}$ is an optimal solution of* (3.2) *then $\hat{x}$ is a properly efficient solution of* (3.1)

Proof. Let $\hat{x}$ be an optimal solution of (3.2). To show that $\hat{x}$ is efficient suppose there exists some $x' \in \mathcal{X}$ with $f(x') \leq f(\hat{x})$. Positivity of the weights λ_k and $f_i(x') < f_i(\hat{x})$ for some $i \in \{1, \ldots, p\}$ imply the contradiction

$$\sum_{k=1}^{p} \lambda_k f_k(x') < \sum_{k=1}^{p} \lambda_k f_k(\hat{x}). \tag{3.18}$$

To show that $\hat{x}$ is properly efficient, we choose an appropriately large number M such that assuming there is a trade-off bigger than M yields a contradiction to optimality of $\hat{x}$ for the weighted sum problem. Let

$$M := (p-1) \max_{i,j} \frac{\lambda_j}{\lambda_i}. \tag{3.19}$$

Suppose that $\hat{x}$ is not properly efficient. Then there exist $i \in \{1, \ldots, p\}$ and $x \in \mathcal{X}$ such that $f_i(x) < f_i(\hat{x})$ and $f_i(\hat{x}) - f_i(x) > M(f_j(x) - f_j(\hat{x}))$ for all $j \in \{1, \ldots, p\}$ such that $f_j(\hat{x}) < f_j(x)$. Therefore

$$f_i(\hat{x}) - f_i(x) > \frac{p-1}{\lambda_i} \lambda_j (f_j(x) - f_j(\hat{x})) \tag{3.20}$$

for all $j \neq i$ by the choice of M (note that the inequality is trivially true if $f_j(\hat{x}) > f_j(x)$). Multiplying each of these inequalities by $\lambda_i/(p-1)$ and summing them over $j \neq i$ yields

$$\begin{aligned}
\lambda_i(f_i(\hat{x}) - f_i(x)) &> \sum_{j \neq i} \lambda_j (f_j(x) - f_j(\hat{x})) \\
\Rightarrow \lambda_i f_i(\hat{x}) - \lambda_i f_i(x) &> \sum_{j \neq i} \lambda_j f_j(x) - \sum_{j \neq i} \lambda_j f_j(\hat{x}) \\
\Rightarrow \lambda_i f_i(\hat{x}) + \sum_{j \neq i} \lambda_j f_j(\hat{x}) &> \lambda_i f_i(x) + \sum_{j \neq i} \lambda_j f_j(x) \\
\Rightarrow \sum_{i=1}^{p} \lambda_i f_i(\hat{x}) &> \sum_{i=1}^{p} \lambda_i f_i(x),
\end{aligned}$$

contradicting optimality of $\hat{x}$ for (3.2). Thus $\hat{x}$ is properly efficient. □

Theorem 3.11 immediately yields $\mathcal{S}(\mathcal{Y}) \subset \mathcal{Y}_{pE}$, strengthening the left part of (3.15).

Corollary 3.12. *Let* $\mathcal{Y} \subset \mathbb{R}^p$. *Then* $\mathcal{S}(\mathcal{Y}) \subset \mathcal{Y}_{pN}$.

Now that we have a sufficient condition for proper nondominance and proper efficiency, the natural question is, whether this condition is also necessary. In general it is not. We shall illustrate this graphically.

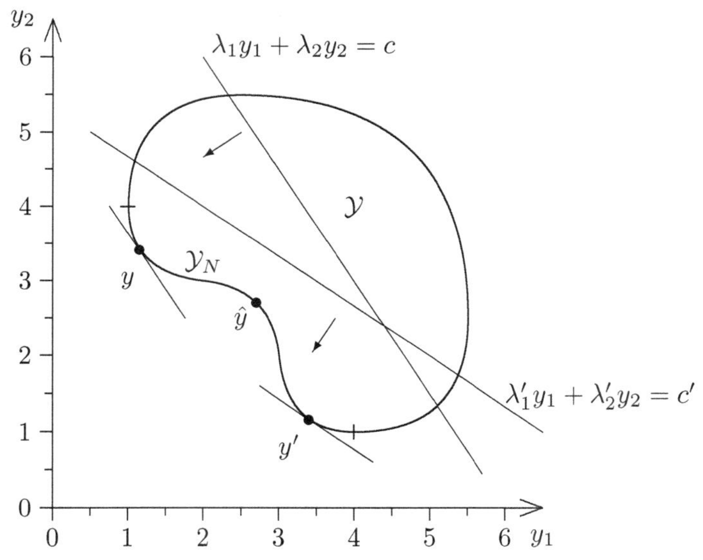

Fig. 3.3. Properly nondominated $\hat{y} \in \mathcal{Y}_N$.

In Figure 3.3, the feasible set in objective space for a nonconvex problem is shown ($\mathcal{Y}$ is not $\mathbb{R}_{\geqq}$-convex). Since all objective vectors $y = (f_1(x), \ldots, f_p(x))$, which attain the same value $c = \sum_{k=1}^{p} \lambda_k f_k(x)$ of the weighted sum objective, are located on a straight line, the minimization problem (3.4) amounts to pushing this line towards the origin, until the intersects $\mathcal{Y}$ only on the boundary of $\mathcal{Y}$. In Figure 3.3 this is illustrated for two weighting vectors (λ_1, λ_2) and (λ_1', λ_2'), that lead to the nondominated points y and y'. It is obvious that the third point $\hat{y} \in \mathcal{Y}_N$ is properly nondominated, but none of its preimages x under f can be an optimal solution of (3.2) for any choice of $(\lambda_1, \ldots, \lambda_p) \in \mathbb{R}^p_>$.

The converse of Theorem 3.11 and Corollary 3.12 can be shown for $\mathbb{R}^p_{\geqq}$-convex sets. We shall give a prove in objective space using Benson's definition and a proof in decision space that uses Geoffrion's definition.

Theorem 3.13. *If $\mathcal{Y}$ is $\mathbb{R}^p_{\geqq}$-convex then $\mathcal{Y}_{pE} \subset \mathcal{S}(\mathcal{Y})$.*

Proof. Let $\hat{y} \in \mathcal{Y}_{pE}$, i.e.

$$\mathrm{cl}(\mathrm{cone}(\mathcal{Y} + \mathbb{R}^p_{\geqq} - \hat{y})) \cap (-\mathbb{R}^p_{\geqq}) = \{0\}. \tag{3.21}$$

By definition, $\mathrm{cl}(\mathrm{cone}(\mathcal{Y} + \mathbb{R}^p_{\geqq} - \hat{y}))$ is a closed convex cone.

The idea of the proof is that if there exists a $\lambda \in \mathbb{R}^p_{>}$ such that

$$\langle \lambda, d \rangle \geq 0 \text{ for all } d \in \mathrm{cl}(\mathrm{cone}(\mathcal{Y} + \mathbb{R}_{\geqq} - y^*)) =: \mathcal{K} \tag{3.22}$$

we get, in particular,

$$\langle \lambda, y - y^* \rangle \geq 0 \text{ for all } y \in \mathcal{Y}, \tag{3.23}$$

i.e. $\langle \lambda, y \rangle \geq \langle \lambda, \hat{y} \rangle$ for all $y \in \mathcal{Y}$ and thus $\hat{y} \in \mathcal{S}(\mathcal{Y})$. This is true, because $\mathcal{Y} - \hat{y} \subset \mathrm{cl}(\mathrm{cone}(\mathcal{Y} + \mathbb{R}^p_{\geqq} - \hat{y}))$. We now prove the existence of $\lambda \in \mathbb{R}^p_{\geqq}$ with property (3.22).

Assume no such λ exists. Both $\mathbb{R}^p_{>}$ and

$$\mathcal{K}^\circ := \{\mu \in \mathbb{R}^p : \langle \mu, d \rangle \geq 0 \text{ for all } d \in \mathcal{K}\} \tag{3.24}$$

are convex sets and because of our assumption have nonintersecting relative interiors. Therefore we can apply Theorem 3.2 to get some nonzero $y^* \in \mathbb{R}^p_{\geq}$ and $\beta \in \mathbb{R}$ such that

$$\langle y^*, \mu \rangle \leq \beta \text{ for all } \mu \in \mathbb{R}^p_{>} \tag{3.25}$$

$$\langle y^*, \mu \rangle \geq \beta \text{ for all } \mu \in K^\circ. \tag{3.26}$$

Using $\mu' = \alpha\mu$ for some arbitrary but fixed $\mu \in K^\circ$ and letting $\alpha \to \infty$ in (3.26) we get $\beta = 0$. Therefore

$$\langle y^*, \mu \rangle \leq 0 \text{ for all } \mu \in \mathbb{R}^p_{>}. \tag{3.27}$$

Selecting $\mu = \varepsilon e + e_k = (\varepsilon, \ldots, \varepsilon, 1, \varepsilon, \ldots, \varepsilon)$ and letting $\varepsilon \to 0$ in (3.27) we obtain $y^*_k \leq 0$ for all $k = 1, \ldots, p$, i.e.

$$y^* \in -\mathbb{R}^p_{\geq}. \tag{3.28}$$

Let

$$\mathcal{K}^{\circ\circ} := \{y \in \mathbb{R}^p : \langle y, \mu \rangle \geq 0 \text{ for all } \mu \in \mathcal{K}^\circ\}. \tag{3.29}$$

According to (3.27), $y^* \in K^{\circ\circ}$. Once we have shown that $\mathcal{K}^{\circ\circ} \subset \mathrm{cl}\,\mathcal{K} = \mathcal{K}$ we know that

$$y^* \in \mathcal{K}. \tag{3.30}$$

Finally (3.28) and (3.30) imply that $y^* \in \mathcal{K} \cap (-\mathbb{R}^p_{\geqq})$ with $y^* \neq 0$ contradicting the proper nondominance conditions (3.21). Therefore the desired λ satisfying (3.22) exists.

To complete the proof, we have to show that $\mathcal{K}^{\circ\circ} \subset \operatorname{cl} \mathcal{K} = \mathcal{K}$. Let $y \in \mathbb{R}^p$, $y \notin \mathcal{K}$. Using Theorem 3.3 to separate $\{y\}$ and $\mathcal{K}$ we get $y^* \in \mathbb{R}^p$, $y^* \neq 0$ and $\alpha \in \mathbb{R}$ with $\langle d, y^* \rangle > \alpha$ for all $d \in \mathcal{K}$ and $\langle y, y^* \rangle < \alpha$. Then $0 \in \mathcal{K}$ implies $\alpha < 0$ and therefore $\langle y, y^* \rangle < 0$. Taking $d = \alpha d'$ for some arbitrary but fixed d' and letting $\alpha \to \infty$ we get $\langle d, y^* \rangle \geq 0$ for all $d \in \mathcal{K}$, i.e $y^* \in \mathcal{K}^\circ$. So $\langle y, y^* \rangle < 0$ implies $y \notin \mathcal{K}^{\circ\circ}$. Hence $\mathcal{K}^{\circ\circ} \subset \mathcal{K}$. □

The properties of and relationships among $\mathcal{K}, \mathcal{K}^\circ$, and $\mathcal{K}^{\circ\circ}$ we have used here are true for cones $\mathcal{K}$ in general, not just for the one used above. See Exercise 3.6 for more details. Let us now illustrate Theorem 3.13.

Example 3.14. Consider the set $\mathcal{Y} = \{(y_1, y_2) : y_1^2 + y_2^2 \leq 1\}$. Here

$$\mathcal{Y}_N = \{(y_1, y_2) : y_1^2 + y_2^2 = 1, \ y_1 \leq 0, \ y_2 \leq 0\}, \tag{3.31}$$

$$\mathcal{Y}_{pN} = \mathcal{Y}_N \setminus \{(-1, 0), (0, -1)\}. \tag{3.32}$$

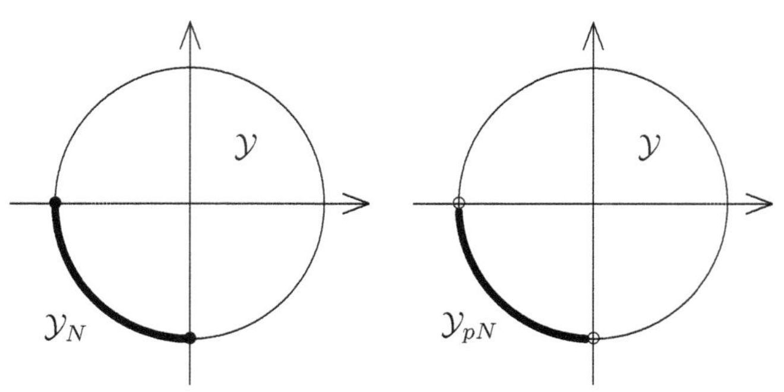

Fig. 3.4. Sets $\mathcal{Y}_N$ and $\mathcal{Y}_{pN}$.

All properly nondominated points $\hat{y}$ are optimal points of weighted sum scalarizations (indeed, the weights correspond to the normals of the tangents to the circle at $\hat{y}$). The two boundary points of $\mathcal{Y}_N$ are not properly nondominated. But $(-1, 0)$ and $(0, -1)$ are unique optimal solutions of

$$\min_{y \in \mathcal{Y}} \lambda_1 y_1 + \lambda_2 y_2 \tag{3.33}$$

for $\lambda = (1, 0)$ and $\lambda = (0, 1)$, respectively, and therefore belong to the nondominated set $\mathcal{Y}_N$, see Proposition 3.8. □

Theorem 3.15 (Geoffrion (1968)). *Let $\mathcal{X} \subset \mathbb{R}^n$ be convex and assume $f_k : \mathcal{X} \to \mathbb{R}$ are convex for $k = 1, \ldots, p$. Then $\hat{x} \in \mathcal{X}$ is properly efficient if and only if $\hat{x}$ is an optimal solution of* (3.2), *with strictly positive weights $\lambda_k, k = 1, \ldots, p$.*

Proof. Due to Theorem 3.11 we only have to prove necessity of the condition. Let $\hat{x} \in \mathcal{X}$ be properly efficient. Then, by definition, there exists a number $M > 0$ such that for all $i = 1, \ldots, p$ the system

$$\begin{aligned} f_i(x) &< f_i(\hat{x}) \\ f_i(x) + M f_j(x) &< f_i(\hat{x}) + M f_j(\hat{x}) \text{ for all } j \neq i \end{aligned} \tag{3.34}$$

has no solution. To see that, simply rearrange the inequalities in (2.41).

A property of convex functions, which we state as Theorem 3.16 below implies that for the ith such system there exist $\lambda_k^i \geq 0$, $k = 1, \ldots, p$ with $\sum_{k=1}^p \lambda_k^i = 1$ such that for all $x \in \mathcal{X}$ the following inequalities holds.

$$\begin{aligned} &\lambda_i^i f_i(x) + \sum_{k \neq i} \lambda_k^i (f_i(x) + M f_k(x)) \geq \lambda_i^i f_i(\hat{x}) + \sum_{k \neq i} \lambda_k^i (f_i(\hat{x}) + M f_k(\hat{x})) \\ \Leftrightarrow\ &\lambda_i^i f_i(x) + \sum_{k \neq i} \lambda_k^i f_i(x) + M \sum_{j \neq i} \lambda_j^i f_k(x) \geq \lambda_i^i f_i(\hat{x}) + \sum_{k \neq i} \lambda_k^i f_i(\hat{x}) + M \sum_{k \neq i} \lambda_j^i f_k(\hat{x}) \\ \Rightarrow\ &\sum_{k=1}^p \lambda_k^i f_i(x) + M \sum_{k \neq i} \lambda_k^i f_k(x) \geq \sum_{k=1}^p \lambda_k^i f_i(\hat{x}) + M \sum_{k \neq i} \lambda_k^i f_j(\hat{x}) \\ \Leftrightarrow\ &f_i(x) + M \sum_{k \neq i} \lambda_k^i f_k(x) \geq f_i(\hat{x}) + M \sum_{k \neq i} \lambda_k^i f_k(\hat{x}) \end{aligned}$$

We have such an inequality for each $i = 1, \ldots, p$ and now simply sum over i to obtain

$$\begin{aligned} &\sum_{i=1}^p f_i(x) + M \sum_{i=1}^p \sum_{k \neq i} \lambda_k^i f_k(x) \geq \sum_{i=1}^p f_i(\hat{x}) + M \sum_{k=1}^p \sum_{k \neq i} \lambda_j^i f_k(\hat{x}) \\ \Rightarrow &\sum_{k=1}^p \left(1 + M \sum_{i \neq k} \lambda_k^i\right) f_k(x) \geq \sum_{k=1}^p \left(1 + M \sum_{i \neq k} \lambda_k^i\right) f_k(x^*) \end{aligned}$$

for all $x \in \mathcal{X}$.

We can now normalize the values $(1 + M \sum_{k \neq i} \lambda_k^i)$, so that they sum up to one to obtain positive $\lambda_i, i = 1, \ldots, p$ for which $\hat{x}$ is optimal in (3.2). □

The theorem, which we have used, is the following. For a proof we refer to Mangasarian (1969, p. 65).

Theorem 3.16. *Let $\mathcal{X} \subset \mathbb{R}^n$ be a convex set, let $h_k : \mathbb{R}^n \to \mathbb{R}$ be convex functions, $k = 1, \ldots, p$. Then, if the system $h_k(x) < 0$, $k = 1, \ldots, p$ has no solution $x \in \mathcal{X}$, there exist $\lambda_k \geq 0$, $\sum_{k=1}^{p} \lambda_k = 1$ such that all $x \in \mathcal{X}$ satisfy*

$$\sum_{k=1}^{p} \lambda_k h_k(x) \geq 0. \tag{3.35}$$

With these results on proper nondominance and proper efficiency we can extend (3.15) and (3.16) as follows:

$$\mathcal{S}(\mathcal{Y}) \subset \mathcal{Y}_{pE} \subset \mathcal{Y}_E \text{ and } \mathcal{S}_(\mathcal{Y}) \subset \mathcal{Y}_{wE} \tag{3.36}$$

holds for general sets, whereas for $\mathbb{R}^p_{\geqq}$-convex sets

$$\mathcal{S}(\mathcal{Y}) = \mathcal{Y}_{pE} \subset \mathcal{Y}_E \subset \mathcal{Y}_{wE} = \mathcal{S}_(\mathcal{Y}). \tag{3.37}$$

A closer inspection of the inclusions reveals that the gap between $\mathcal{Y}_{wE}$ and $\mathcal{Y}_E$ might be quite large, even in the convex cases (see Example 2.27 for an illustration). This is not possible for the gap between $\mathcal{Y}_{pE}$ and $\mathcal{Y}_E$.

Theorem 3.17 (Hartley (1978)). *If $\mathcal{Y} \neq \emptyset$ is $\mathbb{R}^p_{\geqq}$-closed and $\mathbb{R}^p_{\geqq}$-convex, the following inclusions hold:*

$$\mathcal{S}(\mathcal{Y}) \subset \mathcal{Y}_N \subset \operatorname{cl} \mathcal{S}(\mathcal{Y}) = \operatorname{cl} \mathcal{Y}_{pN}. \tag{3.38}$$

Proof. The only inclusion we have to show is $\mathcal{Y}_N \subset \operatorname{cl} \mathcal{S}(\mathcal{Y})$. Since $\mathcal{Y}_N = (\mathcal{Y} + \mathbb{R}^p_{\geqq})_N$ and $\mathcal{S}(\mathcal{Y}) = \mathcal{S}(\mathcal{Y} + \mathbb{R}^p_{\geqq})$, we only prove it for a closed convex set $\mathcal{Y}$. Without loss of generality we shall also assume that $\hat{y} = 0 \in \mathcal{Y}_N$.

The proof proceeds in two steps: First we show the result for compact $\mathcal{Y}$, applying a minimax theorem to the scalar product on two compact convex sets. We shall then prove the general case by reduction to the compact case.

Case 1: $\mathcal{Y}$ is compact and convex.

Choose $d \in \mathbb{R}^p_{>}$ and $\mathcal{C}(\varepsilon) := \varepsilon d + \mathbb{R}^p_{\geqq}$ for $0 < \varepsilon \in \mathbb{R}$. If ε is sufficiently small, $\mathcal{C}(\varepsilon) \cap \mathcal{B}(0,1)$ is nonempty. Thus, both $\mathcal{Y}$ and $\mathcal{C}(\varepsilon) \cap \mathcal{B}(0,1)$ are nonempty, convex, and compact.

Applying the Sion-Kakutani minimax theorem (Theorem 3.18 below) to $\Phi = \langle \cdot, \cdot \rangle$ with $\mathcal{C} = \mathcal{C}(\varepsilon) \cap \mathcal{B}(0,1)$ and $\mathcal{D} = \mathcal{Y}$ we get the existence of $y(\varepsilon) \in \mathcal{Y}$ and $\lambda(\varepsilon) \in \mathcal{C}(\varepsilon) \cap \mathcal{B}(0,1)$ such that

$$\langle \lambda, y(\varepsilon) \rangle \leq \langle \lambda(\varepsilon), y(\varepsilon) \rangle \leq \langle \lambda(\varepsilon), y \rangle \text{ for all } y \in \mathcal{Y}, \text{ for all } \lambda \in \mathcal{C}(\varepsilon) \cap \mathcal{B}(0,1) \tag{3.39}$$

From (3.39) using $0 \in \mathcal{Y}$ we obtain $\langle \lambda, y(\varepsilon) \rangle \leq 0$ for all $\lambda \in \mathcal{C}(\varepsilon) \cap \mathcal{B}(0,1)$. Because $\mathcal{Y}$ is compact there exists a sequence $\varepsilon^k \to 0$ such that $y^k := y(\varepsilon^k) \to y' \in \mathcal{Y}$ for $k \to \infty$.

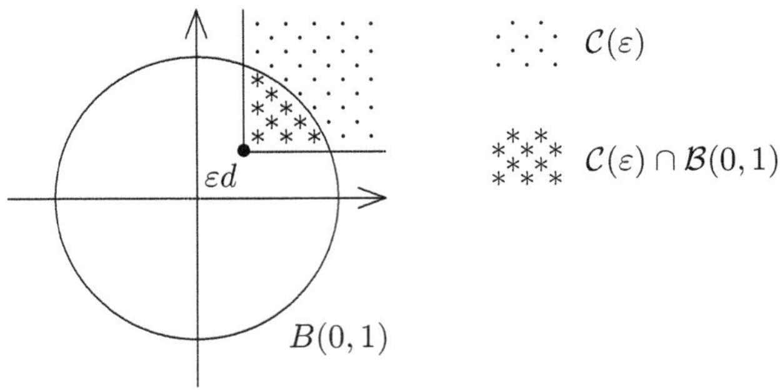

Fig. 3.5. Illustration of the first case in the proof of Theorem 3.17.

Furthermore, for each $\lambda \in \mathbb{R}^p_> \cap \mathcal{B}(0,1)$ there is some $\varepsilon' > 0$ such that $\lambda \in \mathcal{C}(\varepsilon) \cap \mathcal{B}(0,1)$ for all $\varepsilon \leq \varepsilon'$ and therefore $\langle \lambda, y^k \rangle \leq 0$ when k is large enough. The convergence $y^k \to y'$ then implies $\langle \lambda, \hat{y} \rangle \leq 0$ for all $\lambda \in \mathbb{R}^p_>$. This implies $y' \in -\mathbb{R}^p_{\geqq}$. Thus, $y' \leqq 0$ but since $\hat{y} = 0 \in \mathcal{Y}_N$ we must have $y' = 0$.

Next we show that $y' = \hat{y} = 0 \in \operatorname{cl} \mathcal{S}(\mathcal{Y})$. To this end let $\lambda^k := \lambda(\varepsilon^k)/\|\lambda(\varepsilon^k)\| \in \mathbb{R}^p_> \cap \operatorname{bd} \mathcal{B}(0,1)$, where $\lambda(\varepsilon^k)$ is the λ associated with ε^k and $y(\varepsilon^k)$ to satisfy (3.39). Therefore we have

$$\langle \lambda^k, y(\varepsilon^k) \rangle \leq \langle \lambda^k, y \rangle \text{ for all } y \in \mathcal{Y}, \tag{3.40}$$

i.e. $y^k = y(\varepsilon^k) \subset \mathcal{S}(\lambda^k, \mathcal{Y}) \subset \mathcal{S}(\mathcal{Y})$. Since $y' = \lim y^k$ this implies $\hat{y} = y' \in \operatorname{cl} \mathcal{S}(\mathcal{Y})$.

Case 2: $\mathcal{Y}$ is closed and convex (but not necessarily compact).

Again let $\hat{y} = 0 \in \mathcal{Y}_N$. $\mathcal{Y} \cap \mathcal{B}(0,1)$ is nonempty, convex, and compact and $0 \in (\mathcal{Y} \cap \mathcal{B}(0,1))_N$. Case 1 yields the existence of $\lambda^k \in \mathbb{R}^p_>$, $\|\lambda^k\| = 1$, and $y^k \in \mathcal{S}(\lambda^k, \mathcal{Y} \cap B(0,1))$ with $y^k \to 0$. We show that $y^k \in \mathcal{S}(\lambda^k, \mathcal{Y})$, which completes the proof.

Note that for k large enough $y^k \in \operatorname{int} \mathcal{B}$ (since $y^k \to 0$) and suppose $y' \in \mathcal{Y}$ exists with $\langle \lambda^k, y' \rangle < \langle \lambda^k, y^k \rangle$. Then $\alpha y' + (1-\alpha) y^k \in \mathcal{Y} \cap B(0,1)$ for sufficiently small α (see Figure 3.6).

This implies

$$\langle \lambda^k, \alpha y' + (1-\alpha) y^k \rangle = \alpha \langle \lambda^k, y' \rangle + (1-\alpha) \langle \lambda^k, y^k \rangle < \langle \lambda^k, y^k \rangle, \tag{3.41}$$

contradicting $y^k \in \mathcal{S}(\lambda^k, \mathcal{Y})$. □

The Sion-Kakutani minimax theorem that we used is stated for completeness. For a proof we refer to Stoer and Witzgall (1970, p. 232).

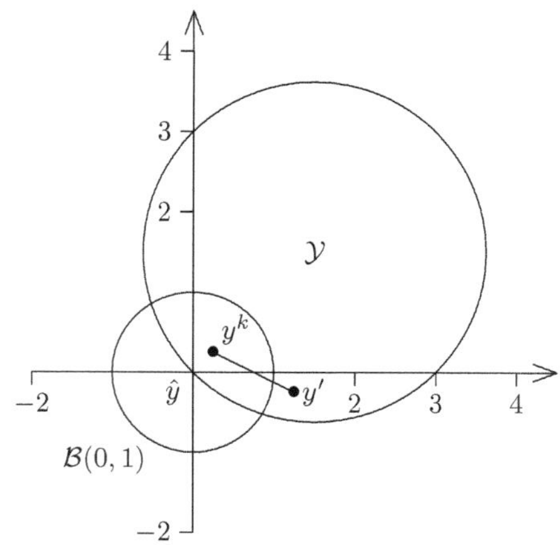

Fig. 3.6. Illustration of the second case in the proof of Theorem 3.17.

Theorem 3.18 (Sion-Kakutani minimax theorem). *Let $\mathcal{C}, \mathcal{D} \subset \mathbb{R}^p$ be nonempty, compact, convex sets and $\Phi : \mathcal{C} \times \mathcal{D} \to \mathbb{R}$ be a continuous mapping such that $\Phi(\cdot, d)$ is convex for all $d \in \mathcal{D}$ and $\Phi(c, \cdot)$ is concave for all $c \in \mathcal{C}$. Then*

$$\max_{d\in\mathcal{D}} \min_{c\in\mathcal{C}} \Phi(c,d) = \min_{c\in\mathcal{C}} \max_{d\in\mathcal{D}} \Phi(c,d). \tag{3.42}$$

Although Theorem 3.17 shows that $\mathcal{Y}_N \subset \operatorname{cl} \mathcal{Y}_{pN}$, the inclusion $\operatorname{cl} \mathcal{Y}_{pN} \subset \mathcal{Y}_N$ is not always satisfied.

Example 3.19 (Arrow et al. *(1953)).*

Consider the set $\mathcal{Y}' = \{(y_1, y_2, y_3) : (y_1 - 1)^2 + (y_2 - 1)^2 = 1,\ y_1 \leq 1,\ y_2 \leq 1,\ y_3 = 1\}$ and define

$$Y := \operatorname{conv}\left(\mathcal{Y}' \cup \{(1,0,0)\}\right), \tag{3.43}$$

shown from different angles in Figure 3.7.

$\mathcal{Y}$ is closed and convex. Note that $\hat{y} = (1,0,1) \notin \mathcal{Y}_N$ because $(1,0,0) \leq \hat{y}$. From Theorem 3.13 we know $\mathcal{Y}_{pN} = \mathcal{S}(\mathcal{Y})$. We show that all $y' \in Y'$ with $y'_1 < 1, y'_2 < 1$ are properly efficient.

Let $y' = (1 - \cos\theta, 1 - \sin\theta, 1)$ for $0 < \theta < \pi/2$ and $\lambda = (1 - \alpha)(\cos\theta, \sin\theta, 0) + \alpha(0,0,1)$ with $0 < \alpha < 1$ so that $\lambda \in \mathbb{R}^p_>$.

We compute $\langle \lambda, y - y' \rangle$ for $y = (1 - \cos\theta', 1 - \sin\theta', 1)$, $0 \leq \theta' \leq \pi/2$:

$$\begin{aligned} \langle \lambda, y - y' \rangle &= (1-\alpha)\left[\cos\theta(\cos\theta - \cos\theta') + \sin\theta(\sin\theta - \sin\theta')\right] \\ &= (1-\alpha)(1 - (\cos\theta\cos\theta' + \sin\theta\sin\theta')) \\ &= (1-\alpha)(1 - \cos(\theta - \theta')) \geq 0. \end{aligned} \tag{3.44}$$

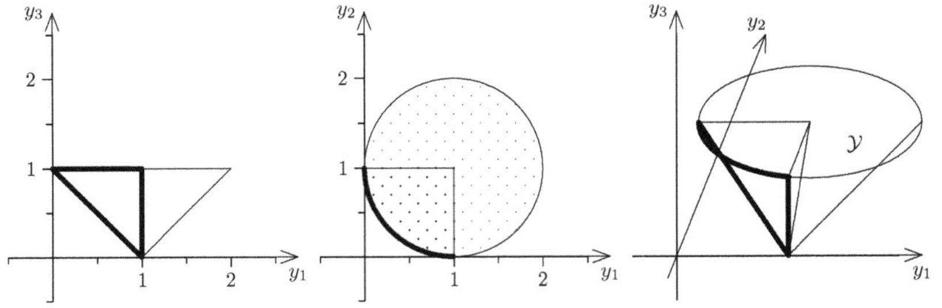

Fig. 3.7. The set $\mathcal{Y}$ in Example 3.19.

Furthermore, for $y = (1, 0, 0)$ we get

$$\begin{aligned} \langle \lambda, (1,0,0) - y' \rangle &= (1-\alpha)\left[\cos^2\theta - \sin\theta(1-\sin\theta)\right] - \alpha \\ &= (1-\alpha)(1-\sin\theta) - \alpha \quad > 0 \end{aligned} \tag{3.45}$$

for small α. So by taking convex combinations of (3.44) and (3.45) we get $\langle \lambda, y - \overline{y} \rangle \geq 0$ for all $y \in \mathcal{Y}$ and thus $y' \in \mathcal{S}(\mathcal{Y})$. In addition, for $\theta \to 0$ we get $y' \to \hat{y}$ which is therefore in $\operatorname{cl} \mathcal{S}(\mathcal{Y})$. □

3.3 Optimality Conditions

In this section we prove necessary and sufficient conditions for weak and proper efficiency of solutions of a multicriteria optimization problem. These results follow along the lines of Karush-Kuhn-Tucker optimality conditions known from single objective nonlinear programming. We use the results to prove the yet missing link in Figure 2.17 and we give an example that shows that Kuhn and Tucker's and Geoffrion's definitions of properly efficient solutions do not always coincide.

We recall the Karush-Kuhn-Tucker necessary and sufficient optimality conditions in single objective nonlinear programming, see e.g. Bazaraa *et al.* (1993).

Theorem 3.20. *Let $f, g_j : \mathbb{R}^n \to \mathbb{R}$ be continuously differentiable functions and consider the single objective optimization problem*

$$\min\{f(x) : g_j(x) \leqq 0, j = 1, \ldots, m\}. \tag{3.46}$$

Denote $\mathcal{X} := \{x \in \mathbb{R}^n : g_j(x) \leq 0, j = \{1, \ldots, m\}\}$.

- *If $\hat{x} \in \mathcal{X}$ is a (locally) optimal solution of* (3.46) *there is some $\hat{\mu} \in \mathbb{R}^m_{\geqq}$ such that*

$$\nabla f(\hat{x}) + \sum_{j=1}^{m} \hat{\mu}_j \nabla g_j(\hat{x}) = 0, \tag{3.47}$$

$$\sum_{j=1}^{m} \hat{\mu}_j g_j(\hat{x}) = 0. \tag{3.48}$$

- *If f, g_j are convex and there are $\hat{x} \in \mathcal{X}$ and $\hat{\mu} \in \mathbb{R}^m_{\geqq}$ such that* (3.47) *and* (3.48) *hold then $\hat{x}$ is a locally, thus globally, optimal solution of* (3.46).

We start with conditions for weak efficiency.

Theorem 3.21. *Suppose that the KT constraint qualification (see Definition 2.50) is satisfied at $\hat{x} \in \mathcal{X}$. If $\hat{x}$ is weakly efficient there exist $\hat{\lambda} \in \mathbb{R}^p_{\geq}$ and $\hat{\mu} \in \mathbb{R}^m_{\geqq}$ such that*

$$\sum_{k=1}^{p} \hat{\lambda}_k \nabla f_k(\hat{x}) + \sum_{j=1}^{m} \hat{\mu}_j \nabla g_j(\hat{x}) = 0 \tag{3.49}$$

$$\sum_{j=1}^{m} \hat{\mu}_j g_j(\hat{x}) = 0 \tag{3.50}$$

$$\hat{\lambda} \geq 0 \tag{3.51}$$

$$\hat{\lambda} \geqq 0 \tag{3.52}$$

Proof. Let $\hat{x} \in \mathcal{X}_{wE}$. We first show that there can be no $d \in \mathbb{R}^n$ such that

$$\langle \nabla f_k(\hat{x}), d \rangle < 0 \text{ for all } k = 1, \ldots, p \tag{3.53}$$

$$\langle \nabla g_j(\hat{x}), d \rangle < 0 \text{ for all } j \in \mathcal{J}(\hat{x}) := \{j : g_j(\hat{x}) = 0\}. \tag{3.54}$$

We then apply Motzkin's theorem of the alternative (Theorem 3.22) to obtain the multipliers $\hat{\lambda}_k$ and $\hat{\mu}_j$.

Suppose that such a $d \in \mathbb{R}^n$ exists. From the KT constraint qualification there is a continuously differentiable function $\theta : [0, \bar{t}] \to \mathbb{R}^n$ such that $\theta(0) = \hat{x}$, $g(\theta(t)) \leq 0$ for all $t \in [0, \bar{t}]$, and $\theta'(0) = \alpha d$ with $\alpha > 0$. Thus,

$$f_k(\theta(t)) = f_k(\hat{x}) + t\langle \nabla f_k(\hat{x}), \alpha d \rangle + o(t) \tag{3.55}$$

and using $\langle \nabla f_k(\hat{x}), d \rangle < 0$ it follows that $f_k(\theta(t)) < f_k(\hat{x})$, $k = 1, \ldots, p$ for t sufficiently small, which contradicts $\hat{x} \in \mathcal{X}_{wE}$.

It remains to show that (3.53) and (3.54) imply the conditions of (3.49) – (3.52). This is achieved by using matrices $B = (\nabla f_k(\hat{x}))_{k=1,\ldots,p}$, $C =$

$(\nabla g_j(\hat{x}))_{j\in\mathcal{J}(\hat{x})}$, $D = 0$ with $l = |\mathcal{J}(\hat{x})|$ in Theorem 3.22. Then, since (3.53) and (3.54) have no solution $d \in \mathbb{R}^n$, according to Theorem 3.22 there must be $y^1 =: \hat{\lambda}, y^2 =: \hat{\mu}$, and y^3 such that $B^T y^1 + C^T y^2 = 0$ with $y^1 \geq 0$ and $y^2 \geqq 0$, i.e.

$$\sum_{k=1}^{p} \hat{\lambda}_k \nabla f_k(\hat{x}) + \sum_{j=\mathcal{J}(\hat{x})}^{m} \hat{\mu}_j \nabla g_j(\hat{x}) = 0.$$

We complete the proof by setting $\hat{\mu}_j = 0$ for $j \in \{1, \ldots, m\} \setminus \mathcal{J}(\hat{x})$. □

Theorem 3.22 (Motzkin's theorem of the alternative). *Let B, C, D be $p \times n$, $l \times n$ and $o \times n$ matrices, respectively. Then either*

$$Bx < 0, \ Cx \leqq 0, \ Dx = 0$$

has a solution $x \in \mathbb{R}^n$ or

$$B^T y^1 + C^T y^2 + D^T y^3 = 0, \ y^1 \geq 0, \ y^2 \geqq 0 \tag{3.56}$$

has a solution $y^1 \in \mathbb{R}^p$, $y^2 \in \mathbb{R}^l$, $y^3 \in \mathbb{R}^o$, but never both.

A proof of Theorem 3.22 can be found in (Mangasarian, 1969, p.28).

For convex functions, we also have a sufficient condition for weakly efficient solutions.

Corollary 3.23. *Under the assumptions of Theorem 3.21 and the additional assumption that all functions f_k and g_j are convex (3.49) – (3.52) with $\hat{\lambda} \geq 0$ and $\hat{\mu} \geqq 0$ in Theorem 3.21 are sufficient for $\hat{x}$ to be weakly efficient.*

Proof. By the second part of Theorem 3.20 and Theorem 3.21, (3.49) – (3.52) imply that $\hat{x}$ is an optimal solution of the single objective optimization problem

$$\min_{x\in\mathcal{X}} \sum_{k=1}^{p} \hat{\lambda}_k f_k(x).$$

Since $\hat{\lambda} \in \mathbb{R}_{\geq}$ this implies that $\hat{x} \in \mathcal{X}_{wE}$ according to the first statement of Proposition 3.9. □

Next, we prove similar conditions for properly efficient solutions in Kuhn and Tucker's sense and in Geoffrion's sense.

Kuhn and Tucker's definition of proper efficiency (Definition 2.49) is based on the system of inequalities (3.57) – (3.59)

$$\langle \nabla f_k(\hat{x}), d\rangle \leq 0 \quad \forall\, k = 1, \ldots, p \tag{3.57}$$

$$\langle \nabla f_i(\hat{x}), d\rangle < 0 \quad \text{for some } i \in \{1, \ldots, p\} \tag{3.58}$$

$$\langle \nabla g_j(\hat{x}), d\rangle \leq 0 \quad \forall\, j \in \mathcal{J}(\hat{x}) = \{j = 1, \ldots, m : \ g_j(\hat{x}) = 0\} \tag{3.59}$$

having no solution. We apply Tucker's theorem of the alternative, given below, to show that a dual system of inequalities then has a solution. This system yields a necessary condition for proper efficiency in Kuhn and Tucker's sense.

Theorem 3.24 (Tucker's theorem of the alternative). *Let B, C and D be $p \times n$, $l \times n$ and $o \times n$ matrices. Then either*

$$Bx \leq 0, \ Cx \leqq 0, \ Dx = 0 \tag{3.60}$$

has a solution $x \in \mathbb{R}^n$ or

$$B^T y^1 + C^T y^2 + D^T y^3 = 0, \ y^1 > 0, \ y^2 \geqq 0 \tag{3.61}$$

has a solution $y^1 \in \mathbb{R}^p$, $y^2 \in \mathbb{R}^l$, $y^3 \in \mathbb{R}^o$, but never both.

A proof of Theorem 3.24 can be found in Mangasarian (1969, p.29).

Theorem 3.25. *If $\hat{x}$ is properly efficient in Kuhn and Tucker's sense there exist $\hat{\lambda} \in \mathbb{R}^p$, $\hat{\mu} \in \mathbb{R}^m$ such that*

$$\sum_{k=1}^{p} \hat{\lambda}_k \nabla f_k(\hat{x}) + \sum_{j=1}^{m} \hat{\mu}_j \nabla g_j(\hat{x}) = 0 \tag{3.62}$$

$$\sum_{j=1}^{m} \hat{\mu}_j g_j(\hat{x}) = 0 \tag{3.63}$$

$$\hat{\lambda} > 0 \tag{3.64}$$

$$\hat{\mu} \geqq 0. \tag{3.65}$$

Proof. Because $\hat{x}$ is properly efficient in Kuhn and Tucker's sense there is no $d \in \mathbb{R}^n$ satisfying (3.57) – (3.59).

We apply Theorem 3.24 to the matrices

$$\begin{aligned} B &= (\nabla f_k(\hat{x}))_{k=1,\ldots,p} \\ C &= (\nabla g_j(\hat{x}))_{j \in \mathcal{J}(\hat{x})} \\ D &= 0 \end{aligned}$$

with $l = |J(\hat{x})|$. Since (3.57) – (3.59) do not have a solution $d \in \mathbb{R}^n$ we obtain $y^1 =: \hat{\lambda}$, $y^2 =: \hat{\mu}$ and y^3 with $\hat{\lambda}_k > 0$ for $k = 1, \ldots, p$, $\hat{\mu}_j \geq 0$ for $j \in \mathcal{J}(\hat{x})$ satisfying

$$\sum_{k=1}^{p} \hat{\lambda}_k \nabla f_k(\hat{x}) + \sum_{j \in \mathcal{J}(\hat{x})} \hat{\mu}_j \nabla g_j(\hat{x}) = 0. \tag{3.66}$$

Letting $\hat{\mu}_j = 0$ for all $j \in \{1, \ldots, m\} \setminus \mathcal{J}(\hat{x})$, the proof is complete. □

With Theorem 3.25 providing necessary conditions for Kuhn-Tucker proper efficiency and Theorem 2.51, which shows that Geoffrion's proper efficiency implies Kuhn and Tucker's under the constraint qualification we obtain Corollary 3.26 as an immediate consequence.

Corollary 3.26. *If $\hat{x}$ is properly efficient in Geoffrion's sense and the KT constraint qualification is satisfied at $\hat{x}$ then* (3.62) – (3.65) *are satisfied.*

For the missing link in the relationships of proper efficiency definitions we use the single objective Karush-Kuhn-Tucker sufficient optimality conditions of Theorem 3.20 and apply them to the weighted sum problem. We obtain the following theorem.

Theorem 3.27. *Assume that $f_k, g_j : \mathbb{R}^n \to \mathbb{R}$ are convex, continuously differentiable functions. Suppose that there are $\hat{x} \in \mathcal{X}$, $\hat{\lambda} \in \mathbb{R}^p$ and $\hat{\mu} \in \mathbb{R}^m$ satisfying* (3.62) – (3.65). *Then $\hat{x}$ is properly efficient in the sense of Geoffrion.*

Proof. Let $f(x) := \sum_{k=1}^p \hat{\lambda}_k \nabla f_k(x)$, which is a convex function. By the second part of Theorem 3.20 $\hat{x}$ is an optimal solution of $\min_{x \in \mathcal{X}} \sum_{k=1}^p \hat{\lambda}_k f_k(x)$. Since $\hat{\lambda}_k > 0$ for $k = 1, \ldots, p$ Theorem 3.15 yields that $\hat{x}$ is properly efficient in the sense of Geoffrion. □

We can derive two corollaries, the first one shows that for convex problems proper efficiency in Kuhn and Tucker's sense implies proper efficiency in Geoffrion's sense.

Corollary 3.28. *(See Theorem 2.52) Let $f_k, g_j : \mathbb{R}^n \to \mathbb{R}$ be convex, continuously differentiable functions and suppose $\hat{x}$ is properly efficient in Kuhn and Tucker's sense. Then $\hat{x}$ is properly efficient in Geoffrion's sense.*

Proof. The result follows from Theorem 3.25 and Theorem 3.27. □

The second corollary provides sufficient conditions for proper efficiency in Kuhn and Tucker's sense. It follows immediately from Theorems 3.27 and 2.51.

Corollary 3.29. *If, in addition to the assumptions of Theorem 3.27 the KT constraint qualification is satisfied at $\hat{x}$,* (3.62) – (3.65) *are sufficient for $\hat{x}$ to be properly efficient in Kuhn and Tucker's sense.*

We close the section by examples showing that Geoffrion's and Kuhn-Tucker's definitions are different in general.

Example 3.30 (Geoffrion (1968)). In the following problem, $\hat{x} = 0$ is properly efficient according to Kuhn and Tucker's definition, but not according to Geoffrion's definition. Consider

$$\min f(x) = (f_1(x), f_2(x)) = (-x^2, x^3)$$
$$\text{subject to } x \in \mathcal{X} = \{x \in \mathbb{R} : x \geq 0\}.$$

Figure 3.8 shows graphs of the objective functions and the feasible set in objective space, $\mathcal{Y} = f(\mathcal{X})$ as graph of $y_2(y_1) = (-y_1)^{\frac{3}{2}}$. The only constraint is given by $g(x) = -x \leq 0$.

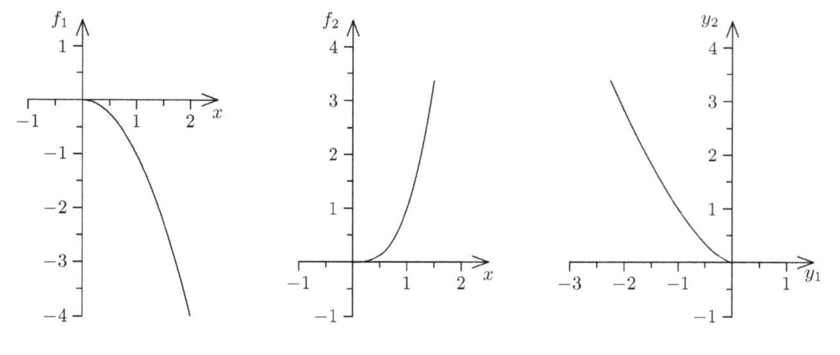

Fig. 3.8. Objective functions in Example 3.30.

To see that Definition 2.49 is satisfied compute

$$\nabla f(x) = (-2x, 3x^2) \qquad \nabla f_1(\hat{x}) = (0, 0)$$
$$\nabla g(x) = -1 \qquad \nabla g(\hat{x}) = -1$$

and choose $\hat{\lambda} = (1, 1)$, $\hat{\mu} = 0$ which satisfies (3.62) – (3.65).

To see that Definition 2.39 is not satisfied, let $\varepsilon > 0$ and compute the trade-off

$$\frac{f_1(\hat{x}) - f_1(\varepsilon)}{f_2(\varepsilon) - f_2(\hat{x})} = \frac{0 + \varepsilon^2}{\varepsilon^3 - 0} = \frac{1}{\varepsilon} \xrightarrow{\varepsilon \to 0} \infty.$$

□

The reader is asked to come up with an example, where a feasible solution $\hat{x}$ is properly efficient in Geoffrion's sense, but not in Kuhn and Tucker's sense, see Exercise 3.5.

We have shown necessary and sufficient conditions for weakly and strictly efficient solutions. Why are there none for efficient solutions? The answer is that the ones that can be proved are included in the above results. Observe that, because $\mathcal{X}_E \subset \mathcal{X}_{wE}$, the necessary condition of Theorem 3.21 holds for efficient solutions, too. On the other hand, because $\mathcal{S}(\mathcal{Y}) = \mathcal{X}_{pE} \subset \mathcal{X}_E$ for convex problems, the sufficient condition of Theorem 3.27 are sufficient for $\hat{x}$ to be efficient, too. Note that the essential difference between the conditions for weak and proper efficiency is $\hat{\lambda} \geq 0$ versus $\hat{\lambda} > 0$. We can therefore not expect any further results of this type for efficient solutions. This is pretty much the same situation we have encountered in Sections 3.1 and 3.2, where for convex problems we have been able to characterize weakly nondominated and properly nondominated points through weighted sum scalarization with $\lambda \geq 0$ and $\lambda > 0$, respectively.

3.4 Connectedness of Efficient and Nondominated Sets

We have discussed existence of nondominated points and efficient solutions and we have seen how the different concepts of efficiency relate to weighted sum scalarization. In this section, we use scalarizations to prove a topological property of the efficient and nondominated sets, connectedness. Connectedness is an important property, when it comes to determining these sets. If $\mathcal{Y}_N$ or $\mathcal{X}_E$ is connected, the whole nondominated or efficient set can possibly be explored starting from a single nondominated/efficient point using local search ideas. Connectedness will also make the task of selecting a final compromise solution from among the set of efficient solutions $\mathcal{X}_E$ easier, as there are no "gaps" in the efficient set.

In Figure 3.9 two sets $\mathcal{Y}$ are shown, one of which has a connected nondominated set and one of which has not.

Apparently, connectedness cannot be expected, when $\mathcal{Y}$ is not $\mathbb{R}^p_{\geqq}$-convex.

Definition 3.31. *A set* $\mathcal{S} \subset \mathbb{R}^p$ *is called* not connected *if it can be written as* $\mathcal{S} = \mathcal{S}_1 \cup \mathcal{S}_2$, *with* $\mathcal{S}_1, \mathcal{S}_2 \neq \emptyset$, $\operatorname{cl} \mathcal{S}_1 \cap \mathcal{S}_2 = \mathcal{S}_1 \cap \operatorname{cl} \mathcal{S}_2 = \emptyset$. *Equivalently,* $\mathcal{S}$ *is* not connected *if there exist open sets* $\mathcal{O}_1, \mathcal{O}_2$ *such that* $\mathcal{S} \subset \mathcal{O}_1 \cup \mathcal{O}_2$, $\mathcal{S} \cap \mathcal{O}_1 \neq \emptyset$, $\mathcal{S} \cap \mathcal{O}_2 \neq \emptyset$, $\mathcal{S} \cap \mathcal{O}_1 \cap \mathcal{O}_2 = \emptyset$. *Otherwise,* $\mathcal{S}$ *is called* connected.

In the proofs of the following theorems, we use some facts about connected sets which we state without proof here.

Lemma 3.32. *1. If* $\mathcal{S}$ *is connected and* $\mathcal{S} \subset \mathcal{U} \subset \operatorname{cl} \mathcal{S}$ *then* $\mathcal{U}$ *is connected.*
2. If $\{\mathcal{S}_i : i \in \mathcal{I}\}$ *is a family of connected sets with* $\cap_{i \in \mathcal{I}} \mathcal{S}_i \neq \emptyset$ *then* $\cup_{i \in \mathcal{I}} \mathcal{S}_i$ *is connected.*

We derive a preliminary result, considering $\mathcal{S}(\lambda, \mathcal{Y})$ and $\mathcal{S}(\mathcal{Y})$. From Theorem 3.17 we know $\mathcal{S}(\mathcal{Y}) \subset \mathcal{Y}_N \subset \operatorname{cl} \mathcal{S}(\mathcal{Y})$ for $\mathbb{R}^p_{\geqq}$-convex sets $\mathcal{Y}$. We prove

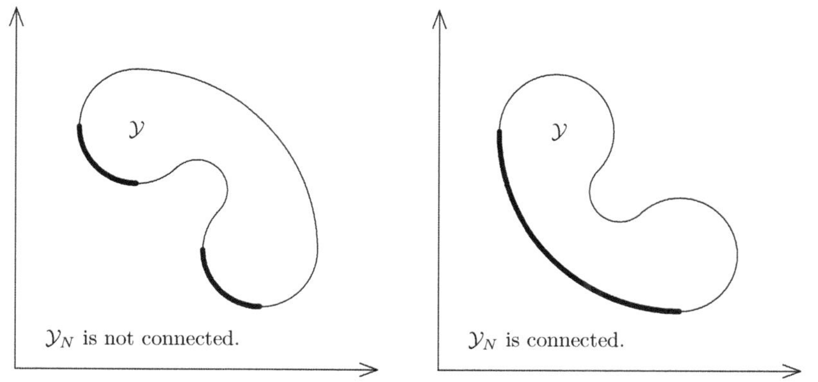

Fig. 3.9. Connectedness of $\mathcal{Y}_N$.

connectedness of $\mathcal{S}(\mathcal{Y})$ in the case that $\mathcal{Y}$ is compact, which implies the connectedness of $\mathcal{Y}_N$ with Lemma 3.32.

Proposition 3.33. *If $\mathcal{Y}$ is compact and convex then $\mathcal{S}(\mathcal{Y})$ is connected.*

Proof. Suppose $\mathcal{S}(\mathcal{Y})$ is not connected. Then we have open sets $\mathcal{Y}_1, \mathcal{Y}_2$ such that $\mathcal{Y}_i \cap \mathcal{S}(\mathcal{Y}) \neq \emptyset$ for $i = 1, 2$, $\mathcal{Y}_1 \cap \mathcal{Y}_2 \cap \mathcal{S}(\mathcal{Y}) = \emptyset$, and $\mathcal{S}(\mathcal{Y}) \subset \mathcal{Y}_1 \cup \mathcal{Y}_2$. Let

$$\mathcal{L}_i := \{\lambda \in \mathbb{R}^p_> : \mathcal{S}(\lambda, \mathcal{Y}) \cap \mathcal{Y}_i \neq \emptyset\}, \ i = 1, 2. \tag{3.67}$$

Because $\mathcal{S}(\lambda, \mathcal{Y})$ is convex and every convex set is connected, we know that $\mathcal{S}(\lambda, \mathcal{Y})$ is connected. Therefore

$$\mathcal{L}_i = \{\lambda \in \mathbb{R}^p_> : \mathcal{S}(\lambda, \mathcal{Y}) \subset \mathcal{Y}_i\}, \ i = 1, 2 \tag{3.68}$$

and $\mathcal{L}_1 \cap \mathcal{L}_2 = \emptyset$. But since $\mathcal{Y}_i \cap \mathcal{S}(\mathcal{Y}) \neq \emptyset$ we also have $\mathcal{L}_i \cap \mathbb{R}^p_> \neq \emptyset$ for $i = 1, 2$. From $\mathcal{S}(\mathcal{Y}) \subset \mathcal{Y}_1 \cup \mathcal{Y}_2$ it follows that $\mathbb{R}^p_> \subset \mathcal{L}_1 \cup \mathcal{L}_2$ (in fact, these sets are equal). By Lemma 3.34 below the sets $\mathcal{L}_i$ are open, which implies the absurd statement that $\mathbb{R}^p_{\geqq}$ is not connected. □

Lemma 3.34. *The sets $\mathcal{L}_i = \{\lambda \in \mathbb{R}^p_> : \ \mathcal{S}(\mathcal{Y}) \subset \mathcal{Y}_i\}$ in the proof of Proposition 3.33 are open.*

Proof. We will show the Lemma for $\mathcal{L}_1$, which by symmetry is enough. If $\mathcal{L}_1$ is not open there must be $\hat{\lambda} \in \mathcal{L}_1$ and $\{\lambda^k, k \geq 1\} \subset \mathbb{R}^p_> \setminus \mathcal{L}_1 = \mathcal{L}_2$ such that $\lambda^k \to \hat{\lambda}$.

Let $y^k \in \mathcal{S}(\lambda^k, \mathcal{Y})$, $k \geq 1$. Since $\mathcal{Y}$ is compact, we can assume (taking a subsequence if necessary) that $y^k \to \hat{y} \in \mathcal{Y}$ and $\hat{y} \in \mathcal{S}(\hat{\lambda}, \mathcal{Y})$. Note that otherwise there would be $y' \in \mathcal{Y}$ such that $\langle \hat{\lambda}, y' \rangle < \langle \hat{\lambda}, \hat{y} \rangle$ and by continuity of the scalar product, we would have $\langle \lambda^k, y' \rangle < \langle \lambda^k, y^k \rangle$ for sufficiently large k, contradicting $y^k \in \mathcal{S}(\lambda^k, \mathcal{Y})$.)

Now we have $y^k \in \mathcal{S}(\lambda^k, \mathcal{Y}) \subset (\mathcal{Y}_2 \cap \mathcal{S}(\mathcal{Y}))$ and $\mathcal{Y}_1 \cap \mathcal{Y}_2 \cap \mathcal{S}(\mathcal{Y}) = \emptyset$ so $y^k \in \mathcal{Y}_1^c$ for each $k \geq 1$. Since $\mathcal{Y}_1^c$ is closed, $\hat{y} = \lim y^k \in \mathcal{Y}_1^c$, i.e. $\hat{y} \notin \mathcal{Y}_1$ contradicting $\hat{\lambda} \in \mathcal{L}_1$. □

Theorem 3.35 (Naccache (1978)). *If $\mathcal{Y}$ is closed, convex, and $\mathbb{R}^p_{\geqq}$-compact then $\mathcal{Y}_N$ is connected.*

Proof. We will first construct compact and convex sets $\mathcal{Y}(\alpha), \alpha \in \mathbb{R}$, for which Proposition 3.33 is applicable. We apply Theorem 3.17 to get that $\mathcal{Y}(\alpha)_N \subset \operatorname{cl} \mathcal{S}(\mathcal{Y}(\alpha))$ and apply Lemma 3.32 to see that sets $\mathcal{Y}(\alpha)_N$ are connected. It is then easy to derive the claim of the theorem by showing $\mathcal{Y}_N = \cup_{\alpha \geq \hat{\alpha}} \mathcal{Y}(\alpha)_N$ for some $\hat{\alpha}$ with$\cap_{\alpha \geq \hat{\alpha}} \mathcal{Y}(\alpha)_N \neq \emptyset$ and applying Lemma 3.32 again.

To construct $\mathcal{Y}(\alpha)$ choose $d \in \mathbb{R}^p_{>}$ and define $y(\alpha) = \alpha d$, $\alpha \in \mathbb{R}$. We claim that for all $y \in \mathbb{R}^p$ there is a real number $\alpha > 0$ such that $y \in y(\alpha) - \mathbb{R}^p$ (see Figure 3.10).

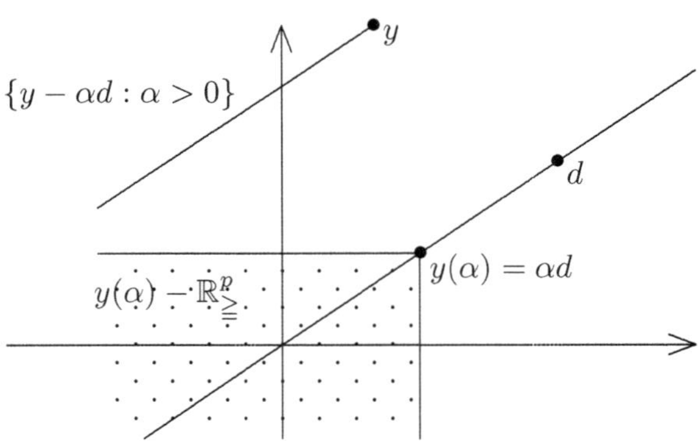

Fig. 3.10. The claim in the proof of Theorem 3.35.

To see this, observe that if it were not true there would be no $d' \in \mathbb{R}^p_{\geqq}$ such that $y = \alpha d - d'$, or $y - \alpha d = -d'$. Thus, we would have two nonempty convex sets $\{y - \alpha d : \alpha > 0\}$ and $-\mathbb{R}^p_{\geqq}$ which can be separated according to Theorem 3.2. Doing so provides some $y^* \in \mathbb{R}^p \setminus \{0\}$ with

$$\langle y^*, y - \alpha d \rangle \geq 0 \text{ for all } \alpha > 0, \tag{3.69}$$

$$\langle y^*, -d' \rangle \leq 0 \text{ for all } d' \in \mathbb{R}^p_{\geqq}. \tag{3.70}$$

Hence $\langle y^*, d' \rangle \geq 0$ for all $d' \in \mathbb{R}^p_{\geq}$, in particular $\langle y^*, d \rangle > 0$ because $d \in \mathbb{R}^p_{>}$. But then $\langle \lambda, y - \alpha d \rangle < 0$ for α sufficiently large, a contradiction to (3.69).

With the claim proved, we can choose $y \in \mathcal{Y}_N$ and appropriate $\hat{\alpha} > 0$ such that $y \in y(\hat{\alpha}) - \mathbb{R}^p_{\geqq}$, which means that $(y(\hat{\alpha}) - \mathbb{R}^p_{\geqq}) \cap \mathcal{Y}_N \neq \emptyset$. We define

$$\mathcal{Y}(\alpha) := \left[\left(y(\alpha) - \mathbb{R}^p_{\geqq}\right) \cap \mathcal{Y}\right]. \tag{3.71}$$

With this notation, the claim above implies in particular that

$$\mathcal{Y}_N = \bigcup_{\alpha \geq \hat{\alpha}} Y(\alpha)_N. \tag{3.72}$$

Because $\mathcal{Y}(\alpha)$ is convex and compact ($\mathcal{Y}$ is $\mathbb{R}^p_{\geqq}$-compact) we can apply Theorem 3.17 to get

$$\mathcal{S}(\mathcal{Y}(\alpha)) \subset \mathcal{Y}(\alpha)_N \subset \mathcal{Y}\alpha_{pN}.$$

Thus, Proposition 3.33 and the first part of Lemma 3.32 imply that $\mathcal{Y}(\alpha)_N$ is connected.

Observing that $\mathcal{Y}(\alpha)_N \supset \mathcal{Y}(\hat{\alpha})_N$ for $\alpha > \hat{\alpha}$, i.e. $\cap_{\alpha \geq \hat{\alpha}} \mathcal{Y}(\alpha)_N = \mathcal{Y}(\hat{\alpha})_N \neq \emptyset$ we have expressed $\mathcal{Y}_N$ as a union of a family of connected sets with nonempty intersection (see Figure 3.11. The second part of Lemma 3.32 proves that $\mathcal{Y}_N$ is connected. □

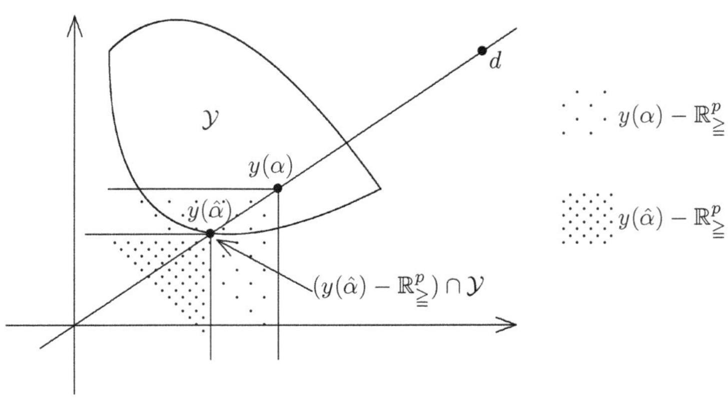

Fig. 3.11. Finding the sets $E(\alpha)$ in the proof of Theorem 3.35.

With Theorem 3.35 we have a criterion for connectedness in the objective space. What about the decision space? If we assume convexity of f, it is possible to show that $\mathcal{X}_{wE}$ is connected. Let $\mathcal{X} \subset \mathbb{R}^n$ be convex and compact and $f_k : \mathbb{R}^n \to \mathbb{R}$ be convex. We will use Theorem 3.5 ($\mathcal{Y}_{wN} = \mathcal{S}(\mathcal{Y})$) and the following fact:

Lemma 3.36. *Let $f : \mathbb{R}^n \to \mathbb{R}$ be convex on the closed convex $\mathcal{X}$. Then the set $\{\hat{x} \in \mathcal{X} : f(\hat{x}) = \inf_{x \in \mathcal{X}} f(x)\}$ is closed and convex.*

We also need a theorem providing a result on connectedness of preimages of sets, taken from Warburton (1983), where a proof can be found.

Theorem 3.37. *Let $\mathcal{V} \subset \mathbb{R}^n, \mathcal{W} \subset \mathbb{R}^p$, and assume that $\mathcal{V}$ is compact and $\mathcal{W}$ is connected. Furthermore, let $g : \mathcal{V} \times \mathcal{W} \to \mathbb{R}$ be continuous. Denote by $\mathcal{X}(w) = \operatorname{argmin}\{g(v,w) : v \in \mathcal{V}\}$. If $\mathcal{X}(w)$ is connected for all $w \in \mathcal{W}$ then $\cup_{w \in \mathcal{W}} \mathcal{X}(w)$ is connected.*

Theorem 3.38. *Let $\mathcal{X}$ be a compact convex set and assume that $f_k : \mathbb{R}^n \to \mathbb{R}, k = 1, \ldots, p$ are convex. Then $\mathcal{X}_{wE}$ is connected.*

Proof. Since the objective functions f_k are continuous and $\mathcal{X}$ is compact, $\mathcal{Y} = f(\mathcal{X})$ is compact. Using Theorem 3.5 we have $\mathcal{Y}_{wN} = \mathcal{S}(\mathcal{Y})$. In terms of f and $\mathcal{X}$ this means

$$\begin{aligned} \mathcal{X}_{wP} &= \bigcup_{\lambda \in \mathbb{R}_{\geq}} \{\hat{x} : \textstyle\sum_{k=1}^p \lambda_k f_k(\hat{x}) \leq \sum_{k=1}^p \lambda_k f_k(x) \text{ for all } x \in \mathcal{X}\} \\ &=: \bigcup_{\lambda \in \mathbb{R}_{\geq}} \mathcal{X}(\lambda). \end{aligned} \tag{3.73}$$

Noting that $\langle f(\cdot), \cdot \rangle : \mathcal{X} \times \mathbb{R}_{\geq} \to \mathbb{R}$ is continuous, that $\mathbb{R}^p_{\geq}$ is connected, that $\mathcal{X}$ is compact, and that by Lemma 3.36 $\mathcal{X}(\lambda)$ is nonempty and convex (hence connected) we can apply Theorem 3.37 to get that $\mathcal{X}_{wE}$ is connected. □

We remark that the proof works in the same way to see that $\mathcal{X}_{pE}$ is connected under the same assumptions. This is true, because as in (3.73), we can write

$$\mathcal{X}_{pE} = \bigcup_{\lambda \in \mathbb{R}^p_{>}} \mathcal{X}(\lambda). \tag{3.74}$$

and as we observed, $\mathcal{X}(\lambda)$ is connected (convex), and of course $\mathbb{R}^p_{>}$ is connected.

To derive a connectedness result for $\mathcal{X}_E$ we need an additional Lemma.

Lemma 3.39. *Let $f : \mathcal{X} \subset \mathbb{R}^n \longrightarrow \mathbb{R}^p$ be a continuous function and let $\tilde{\mathcal{Y}} \subset \mathbb{R}^p$ be such that $f^{-1}(\operatorname{cl} \tilde{\mathcal{Y}}) \subset \mathcal{X}$. Then*

$$f^{-1}(\operatorname{cl} \tilde{\mathcal{Y}}) = \operatorname{cl}(f^{-1}(\tilde{\mathcal{Y}})). \tag{3.75}$$

Theorem 3.40. *Let $\mathcal{X} \subset \mathbb{R}^n$ be a convex and compact set. Assume that all objective functions f_k are convex. Then $\mathcal{X}_E$ is connected.*

Proof. Because $\mathcal{X}$ is compact and convex and f_k are convex and continuous, $\mathcal{Y} = f(\mathcal{X})$ is also compact and convex. Thus, from Theorem 3.17

$$\mathcal{S}(\mathcal{Y}) \subset \mathcal{Y}_N \subset \operatorname{cl} \mathcal{S}(\mathcal{Y}). \tag{3.76}$$

Therefore, taking preimages and applying Theorem 3.13 and Corollary 3.12 ($\mathcal{Y}_{pN} = \mathcal{S}(\mathcal{Y})$) we get

$$\mathcal{X}_{pE} \subset \mathcal{X}_E \subset f^{-1}\left(\operatorname{cl} \mathcal{S}(\mathcal{Y})\right). \tag{3.77}$$

We apply Lemma 3.39 to $\tilde{\mathcal{Y}} = \mathcal{S}(\mathcal{Y})$ to get $f^{-1}(\operatorname{cl} \mathcal{S}(\mathcal{Y})) = \operatorname{cl}(f^{-1}(\mathcal{S}(\mathcal{Y}))) = \operatorname{cl} \mathcal{X}_{pE}$ and obtain

$$\mathcal{X}_{pE} \subset \mathcal{X}_E \subset \operatorname{cl} \mathcal{X}_{pE}. \tag{3.78}$$

The result now follows from Lemma 3.32. □

For once deriving results on $\mathcal{Y}$ from results on $\mathcal{X}$, we note the consequences of Theorem 3.38 and Theorem 3.40 for $\mathcal{Y}_{wN}, \mathcal{Y}_N$, and $\mathcal{Y}_{pE}$.

Corollary 3.41. *If $\mathcal{X}$ is a convex, compact set and $f_k : \mathbb{R}^n \to \mathbb{R}, k = 1, \ldots, p$ are convex functions then $\mathcal{Y}_{wN}$, $\mathcal{Y}_N$, and $\mathcal{Y}_{pN}$ are connected*

Proof. The image of a connected set under a continuous mapping is connected. □

That a relaxation of convexity, namely quasi-convexity, is not sufficient to prove connectedness of $\mathcal{X}_E$ can be seen from Exercise 3.11

3.5 Notes

Equations (3.37) and (3.38) imply

$$\mathcal{Y}_{pE} \subset \mathcal{Y}_N \subset \operatorname{cl} \mathcal{Y}_{pN}$$

for $\mathbb{R}^p_{\geqq}$-convex and $\mathbb{R}^p_{\geqq}$-closed sets. Results of this type are called Arrow-Barankin-Blackwell theorems, after the first theorem of this type for closed convex sets, proved by Arrow *et al.* (1953). This has been generalized to orders defined by closed cones $\mathcal{C}$. Hirschberger (2002) shows that the convexity is not essential and the result remains true if *calY* is closed and $\mathcal{Y}_{pN} \neq \emptyset$.

The necessary and sufficient conditions for proper efficiency in Kuhn and Tucker's sense go back to Kuhn and Tucker (1951). Fritz-John type necessary conditions for efficiency have been proved in Da Cuhna and Polak (1967). All of the conditions we have mentioned here are first order conditions. There is of course also literature on second order necessary and sufficient conditions for efficiency. For this type of conditions it is usually assumed that the objective functions $f_k, k = 1, \ldots, p$ and the constraint functions $g_j, j = 1, \ldots, m$ of the MOP are twice continuously differentiable.

Several necessary and sufficient second-order conditions for the MOP are developed by Wang (1991). Cambini *et al.* (1997) establish second order conditions for MOPs with general convex cones while Cambini (1998) develops second order conditions for MOPs with the componentwise order. Aghezzaf (1999) and Aghezzaf and Hachimi (1999) develop second-order necessary conditions. Recent works include Bolintinéanu and El Maghri (1998), Bigi and Castellani (2000), Jimenez and Novo (2002).

There is some literature on the connectedness of nondominated sets. Bitran and Magnanti (1979) show $\mathcal{Y}_N$ and $\mathcal{Y}_{pN}$ are connected if $\mathcal{Y}$ is compact and convex. Luc (1989) proves connectedness results for $calY_{wN}$ if $\mathcal{Y}$ is $\mathcal{C}$-compact and convex. Danilidis *et al.* (1997) consider problems with three objectives, and Hirschberger (2002) shows that the convexity is not essential: if $calC$ and $\mathcal{Y}$ are closed, $\mathcal{Y}_N$ and $\mathcal{Y}_{pN}$ are connected. $\mathcal{Y}_{wN}$ is connected if in addition $\mathcal{Y}_N$ is nonempty.

Exercises

3.1. Prove that if $\mathcal{Y}$ is closed then $\operatorname{cl}\mathcal{S}(\mathcal{Y}) \subset \mathcal{S}_0(\mathcal{Y})$. Hint: Choose sequences λ_k, y^k such that $y^k \in \operatorname{Opt}(\lambda_k, \mathcal{Y})$ and show that $\lambda_k \to \hat{\lambda}$ and $y^k \to \hat{y}$ with $\hat{y} \in \operatorname{Opt}(\hat{\lambda}, \mathcal{Y}), \hat{\lambda} \geq 0$.

3.2. Prove Proposition 3.8, i.e. show that if $\hat{y}$ is the unique element of $\operatorname{Opt}(\lambda, \mathcal{Y})$ for some $\lambda \in \mathbb{R}^p_{\geq}$ then $\hat{y} \in \mathcal{Y}_N$.

3.3. Give one example of a set $\mathcal{Y} \in \mathbb{R}^2$ for each of the following situations:

1. $\mathcal{S}_0(\mathcal{Y}) \subset \mathcal{Y}_{wN}$ with strict inclusion.
2. $\mathcal{S}(\mathcal{Y}) \subset \mathcal{Y}_N \subset \mathcal{S}_0(\mathcal{Y})$ with both inclusions strict,
3. $\mathcal{S}(\mathcal{Y}) \cup \mathcal{S}_0'(\mathcal{Y}) = \mathcal{Y} = \mathcal{S}_0(\mathcal{Y})$, where

$$\mathcal{S}_0'(\mathcal{Y}) = \left\{ y' \in \mathcal{Y} : y' \text{ is the unique element of } \operatorname{Opt}(\lambda, \mathcal{Y}), \lambda \in \mathbb{R}^p_{\geq} \right\}.$$

3.4. Let $\mathcal{Y} = \left\{(y_1, y_2) : y_1^2 + y_2^2 \leq 1\right\}$ and $\mathcal{C} = \left\{(y_1, y_2) = y_2 \leq \frac{1}{2} y_1\right\}$.

1. Show that $\hat{y} = (-1, 0)$ is properly nondominated in Benson's sense, i.e.

$$(\operatorname{cl}(\operatorname{cone}(\mathcal{Y} + \mathcal{C} - \hat{y}))) \cap (-\mathcal{C}) = \{0\}.$$

2. Show that $\hat{y} \in \operatorname{Opt}(\lambda, \mathcal{Y})$ for some $\lambda \notin \mathbb{R}^p_{>}$ and verify that this $\lambda \in \mathcal{C}^{so}$, where

$$\mathcal{C}^{so} = \{\mu \in \mathbb{R}^p : \langle \mu, d \rangle > 0 \text{ for all } d \in \mathcal{C}\}.$$

This result shows that proper nondominance is related to weighted sum scalarization with weighting vectors in $\mathcal{C}^{so}$.

3.5 (Tamura and Arai (1982)). Let

$$\begin{aligned} \mathcal{X} &= \left\{(x_1, x_2) \in \mathbb{R}^2 : -x_1 \leq 0,\ -x_2 \leq 0,\ (x_1 - 1)^3 + x_2 \leq 0\right\} \\ f_1(x) &= -3x_1 - 2x_2 + 3 \\ f_2(x) &= -x_1 - 3x_2 + 1. \end{aligned}$$

Graph $\mathcal{X}$ and $\mathcal{Y} = f(\mathcal{X})$. Show that $\hat{x} = (1, 0)$ is properly efficient in Geoffrion's sense, but not in Kuhn-Tucker's sense. (You may equivalently use Benson's instead of Geoffrion's definition.)

3.6. Let $\mathcal{C} \subset \mathbb{R}^p$ be a cone. The polar cone $\mathcal{C}^\circ$ of $\mathcal{C}$ is defined as follows:

$$\mathcal{C}^\circ := \{y \in \mathbb{R}^p : \langle y, d \rangle \geq 0 \text{for all } d \in \mathcal{C} \setminus \{0\}\}.$$

Prove the following:

1. $\mathcal{C}^\circ$ is a closed convex cone containing 0.
2. $\mathcal{C} \subset (\mathcal{C}^\circ)^\circ =: \mathcal{C}^{\circ\circ}$.
3. $\mathcal{C}_1 \subset \mathcal{C}_2 \Rightarrow \mathcal{C}_2^\circ \subset \mathcal{C}_1^\circ$.
4. $\mathcal{C}^\circ = (\mathcal{C}^{\circ\circ})^\circ$.

3.7. This exercise is about comparing weighted sum scalarizations with weighting vectors from polar cones and $\mathcal{C}$-nondominance. Let $\mathcal{C}$ be a convex pointed cone and $\lambda \in \mathcal{C}^\circ$ and define

$$\mathrm{Opt}_{\mathcal{C}}(\lambda, \mathcal{Y}) := \left\{ \hat{y} \in \mathcal{Y} : \langle \lambda, \hat{y} \rangle = \min_{y \in \mathcal{Y}} \langle \lambda, y \rangle \right\}.$$

1. Show that
$$\mathcal{S}_{\mathcal{C}^\circ}(\mathcal{Y}) := \bigcup_{\lambda \in \mathcal{C}^\circ \setminus \{0\}} \mathrm{Opt}(\lambda, \mathcal{Y}) \subset \mathcal{Y}_{\mathcal{C}wN},$$
where $\hat{y} \in \mathcal{Y}_{\mathcal{C}wN}$ if $(\mathcal{Y} + \operatorname{int} \mathcal{C} - \hat{y}) \cap (-\operatorname{int} \mathcal{C}) = \emptyset$
2. Let $\mathcal{C}^{s\circ}$ be as in Exercise 3.6. Show
$$\mathcal{S}_{\mathcal{C}^{s\circ}}(\mathcal{Y}) := \bigcup_{\lambda \in \mathcal{C}^{s\circ}} \mathrm{Opt}(\lambda, \mathcal{Y}) \subset \mathcal{Y}_{\mathcal{C}N}.$$

Hint: Look at the proofs of Theorems 3.4 and 3.7, respectively.

3.8 (Wiecek (1995)). Consider the problem

$$\begin{aligned} \min\ & \left[(x_1 - 2)^2 + (x_2 - 1)^2,\ x_1^2 + (x_2 - 3)^2\right] \\ \text{s.t. } & g_1(x) = x_1^2 - x_2 \leq 0 \\ & g_2(x) = x_1 + x_2 - 2 \leq 0 \\ & g_3(x) = -x_1 \leq 0 \end{aligned}$$

Use the conditions of Theorem 3.25 to find at least one candidate for a properly efficient solution $\hat{x}$ (in the sense of Kuhn and Tucker). Try to determine all candidates.

3.9. Prove that $\hat{x} \in \mathcal{X}$ is efficient if and only if the optimal value of the optimization problem

$$\begin{aligned} \min \sum_{k=1}^{p} & f_k(x) \\ \text{subject to } & f_k(x) \leq f_k(\hat{x}) \\ & x \in \mathcal{X} \end{aligned}$$

is $\sum_{k=1}^{p} f_k(x^0)$.

3.10. Use Karush-Kuhn-Tucker conditions for single objective optimization (see Theorem 3.20) and Exercise 3.9 to derive optimality conditions for efficient solutions.

3.11. A function $f : \mathbb{R}^n \to \mathbb{R}$ is called quasi-convex if $f(\alpha x^1 + (1-\alpha)x^2) \leq \max\{f(x^1), f(x^2)\}$ for all $\alpha \in (0,1)$. It is well known that f is quasi-convex if and only if $L_{\leq}(f(x))$ is convex for all x (this is a nice exercise on level sets).

Give an Example of a multicriteria optimization problem with $\mathcal{X} \subset \mathbb{R}$ convex, $f_k : \mathbb{R} \to \mathbb{R}$ quasi-convex such that $\mathcal{X}_E$ is not connected. Hint: Monotone increasing or decreasing functions are quasi-convex, in particular those with horizontal parts in the graph.

4

Scalarization Techniques

The traditional approach to solving multicriteria optimization problems of the Pareto class is by scalarization, which involves formulating a single objective optimization problem that is related to the MOP

$$\min_{x \in \mathcal{X}} (f_1(x), \ldots, f_p(x)) \tag{4.1}$$

by means of a real-valued scalarizing function typically being a function of the objective functions of the MOP (4.1), auxiliary scalar or vector variables, and/or scalar or vector parameters. Sometimes the feasible set of the MOP is additionally restricted by new constraint functions related to the objective functions of the MOP and/or the new variables introduced.

In Chapter 3 we introduced the "simplest" method to solve multicriteria problems, the weighted sum method, where we solve

$$\min_{x \in \mathcal{X}} \sum_{k=1}^{p} \lambda_k f_k(x). \tag{4.2}$$

The weighted sum problem (4.2) uses the vector of weights $\lambda \in \mathbb{R}^p_{\geqq}$ as a parameter. We have seen that the method enables computation of the properly efficient and weakly efficient solutions for convex problems by varying λ. The following Theorem summarizes the results.

Theorem 4.1. *1. Let $\hat{x} \in \mathcal{X}$ be an optimal solution of (4.2). The following statements hold.*
- *If $\lambda > 0$ then $\hat{x} \in \mathcal{X}_{pE}$.*
- *If $\lambda \geq 0$ then $\hat{x} \in \mathcal{X}_{wE}$.*
- *If $\lambda \geq 0$ and $\hat{x}$ is a unique optimal solution of (4.2) then $\hat{x} \in \mathcal{X}_{sE}$.*

2. Let $\mathcal{X}$ be a convex set and $f_k, k = 1, \ldots, p$ be convex functions. Then the following statements hold.

- *If* $\hat{x} \in \mathcal{X}_{pE}$ *then there is some* $\lambda > 0$ *such that* $\hat{x}$ *is an optimal solution of* (4.2).
- *If* $\hat{x} \in \mathcal{X}_{wE}$ *then there is some* $\lambda \geq 0$ *such that* $\hat{x}$ *is an optimal solution of* (4.2).

For nonconvex problems, however, it may work poorly. Consider the following example.

Example 4.2. Let $\mathcal{X} = \{x \in \mathbb{R}^2_{\geqq} : x_1^2 + x_2^2 \geq 1\}$ and $f(x) = x$. In this case $\mathcal{X}_E = \{x \in \mathcal{X} : x_1^2 + x_2^2 = 1\}$, yet $\hat{x}^1 = (1, 0)$ and $\hat{x}^2 = (1, 0)$ are the only feasible solutions that are optimal solutions of (4.2) for any $\lambda \geq 0$.

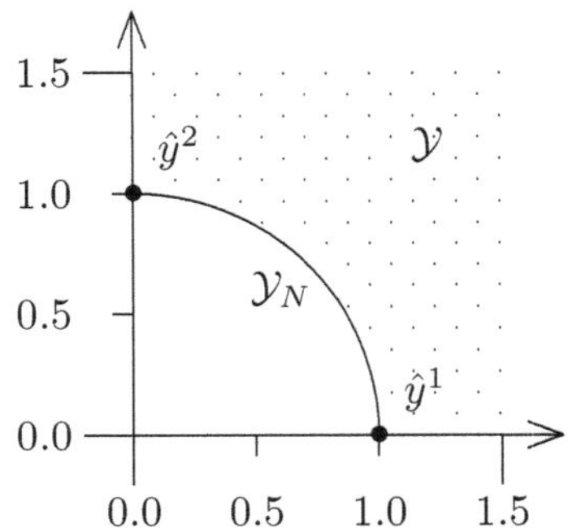

Fig. 4.1. The weighted sum method fails for nonconvex problems.

□

In this chapter we introduce some other scalarization methods, which are also applicable when $\mathcal{Y}$ is not $\mathbb{R}^p_{\geqq}$-convex.

4.1 The ε-Constraint Method

Besides the weighted sum approach, the ε-constraint method is probably the best known technique to solve multicriteria optimization problems. There is no aggregation of criteria, instead only one of the original objectives is minimized, while the others are transformed to constraints. It was introduced byHaimes *et al.* (1971), and an extensive discussion can be found in Chankong and Haimes (1983).

We substitute the multicriteria optimization problem (4.1) by the ε-constraint problem

$$\begin{aligned} &\min_{x \in \mathcal{X}} f_j(x) \\ &\text{subject to } f_k(x) \le \varepsilon_k \quad k = 1, \ldots, p \quad k \ne j, \end{aligned} \tag{4.3}$$

where $\varepsilon \in \mathbb{R}^p$. The component ε_j is irrelevant for (4.3), but the convention to include it will be convenient later.

Figure 4.2 illustrates a bicriterion problem, where an upper bound constraint is put on $f_1(x)$. The optimal values of (4.3) problem with $j = 2$ for two values of ε_1 are indicated. These show that the constraints $f_k(x) \le \varepsilon_k$ might or might not be active at an optimal solution of (4.3).

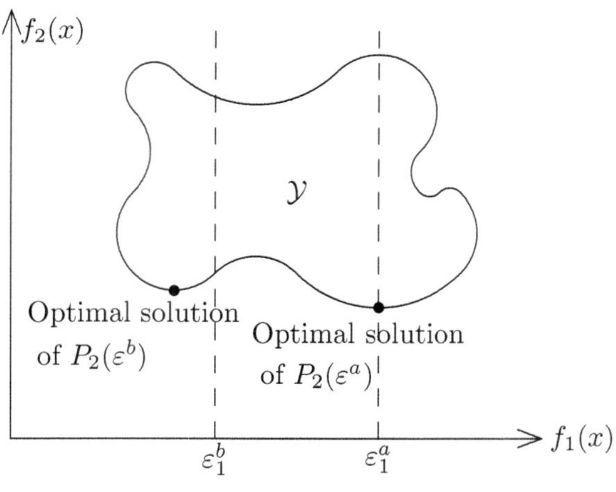

Fig. 4.2. Optimal solutions of ε-constraint problems (4.3).

To justify the approach we show that optimal solutions of (4.3) problems are at least weakly efficient. A necessary and sufficient condition for efficiency shows that this method works for general problems, no convexity assumption is needed. We will also prove a result relating (4.3) to the weighted sum problem (4.2).

Proposition 4.3. *Let $\hat{x}$ be an optimal solution of* (4.3) *for some j. Then $\hat{x}$ is weakly efficient.*

Proof. Assume $\hat{x} \notin \mathcal{X}_{wE}$. Then there is an $x \in \mathcal{X}$ such that $f_k(x) < f_k(\hat{x})$ for all $k = 1, \ldots, p$. In particular, $f_j(x) < f_j(\hat{x})$. Since $f_k(x) < f_k(\hat{x}) \le \varepsilon_k$ for $k \ne j$, the solution x is feasible for (4.3). This is a contradiction to $\hat{x}$ being an optimal solution of (4.3). □

In order to strengthen Proposition 4.3 to obtain efficiency we require the optimal solution of (4.3) to be unique. Note the similarity to Theorem 3.4 and Proposition 3.8 for the weighted sum scalarization.

Proposition 4.4. *Let $\hat{x}$ be a unique optimal solution of* (4.3) *for some j. Then $\hat{x} \in \mathcal{X}_{sE}$ (and therefore $\hat{x} \in \mathcal{X}_E$).*

Proof. Assume there is some $x \in \mathcal{X}$ with $f_k(x) \leq f_k(\hat{x}) \leq \varepsilon_k$ for all $k \neq j$. If in addition $f_j(x) \leq f_j(\hat{x})$ we must have $f_j(x) = f_j(\hat{x})$ because $\hat{x}$ is an optimal solution of (4.3). So x is an optimal solution of (4.3). Thus, uniqueness of the optimal solution implies $x = \hat{x}$ and $\hat{x} \in \mathcal{X}_{sE}$. □

In general, efficiency of $\hat{x}$ is related to $\hat{x}$ being an optimal solution of (4.3) for all $j = 1, \ldots, p$ with the same ε used in all of these problems.

Theorem 4.5. *The feasible solution $\hat{x} \in \mathcal{X}$ is efficient if and only if there exists an $\hat{\varepsilon} \in \mathbb{R}^p$ such that $\hat{x}$ is an optimal solution of* (4.3) *for all $j = 1, \ldots, p$.*

Proof. "$\Longrightarrow$" Let $\hat{\varepsilon} = f(\hat{x})$. Assume $\hat{x}$ is not an optimal solution of (4.3) for some j. Then there must be some $x \in \mathcal{X}$ with $f_j(x) < f_j(\hat{x})$ and $f_k(x) \leq \hat{\varepsilon}_k = f_k(\hat{x})$ for all $k \neq j$, that is, $\hat{x} \notin \mathcal{X}_E$.
"$\Longleftarrow$" Suppose $\hat{x} \notin \mathcal{X}_E$. Then there is an index $j \in \{1, \ldots, p\}$ and a feasible solution $x \in \mathcal{X}$ such that $f_j(x) < f_j(\hat{x})$ and $f_k(x) \leq f_k(\hat{x})$ for $k \neq j$. Therefore $\hat{x}$ cannot be an optimal solution of (4.3) for any ε for which it is feasible. Note that any such ε must have $f_k(\hat{x}) \leq \varepsilon_k$ for $k \neq j$. □

Theorem 4.5 shows that with appropriate choices of ε all efficient solutions can be found. However, as the proof shows, these ε_j values are equal to the actual objective values of the efficient solution one would like to find. A confirmation or check of efficiency is obtained rather than the discovery of efficient solutions.

We denote by

$$\mathcal{E}_j := \{\varepsilon \in \mathbb{R}^p : \{x \in \mathcal{X} : f_k(x) \leq \varepsilon_k,\ k \neq j\} \neq \emptyset\}$$

the set of right hand sides for which (4.3) is feasible and by

$$\mathcal{X}_j(\varepsilon) := \{x \in \mathcal{X} : x \text{ is an optimal solution of (4.3)}\}$$

for $\varepsilon \in \mathcal{E}_j$ the set of optimal solutions of (4.3). From Theorem 4.5 and Proposition 4.3 we have that for each $\varepsilon \in \cap_{j=1}^p \mathcal{E}_j$

$$\bigcap_{j=1}^{p} \mathcal{X}_j(\varepsilon) \subset \mathcal{X}_E \subset \mathcal{X}_j(\varepsilon) \subset \mathcal{X}_{wE} \tag{4.4}$$

for all $j = 1, \ldots, p$ (cf. (3.37) for weighted sum scalarization).

Our last result in this section provides a link between the weighted sum method and the ε-constraint method.

Theorem 4.6 (Chankong and Haimes (1983)).

1. *Suppose $\hat{x}$ is an optimal solution of* $\min_{x \in \mathcal{X}} \sum_{k=1}^{p} \lambda_k f_k(x)$. *If $\lambda_j > 0$ there exists $\hat{\varepsilon}$ such that $\hat{x}$ is an optimal solution of* (4.3), *too.*
2. *Suppose $\mathcal{X}$ is a convex set and $f_k : \mathbb{R}^n \to \mathbb{R}$ are convex functions. If $\hat{x}$ is an optimal solution of* (4.3) *for some j, there exists $\hat{\lambda} \in \mathbb{R}^p_{\geq}$ such that $\hat{x}$ is optimal for* $\min_{x \in \mathcal{X}} \sum_{k=1}^{p} \hat{\lambda}_k f_k(x)$.

Proof. 1. As in the previous proof we show that we can set $\hat{\varepsilon} = f(\hat{x})$. From optimality of $\hat{x}$ for a weighted sum problem we have

$$\sum_{k=1}^{p} \lambda_k (f_k(x) - f_k(\hat{x})) \geq 0$$

for all $x \in \mathcal{X}$. Suppose $\hat{x}$ is not optimal for (4.3) with right hand sides $\hat{\varepsilon}$. The contradiction follows from the fact that for any $x' \in \mathcal{X}$ with $f_j(x') < f_j(\hat{x})$ and $f_k(x') \leq f_k(\hat{x})$ for $k \neq j$

$$\lambda_j (f_j(x') - f_j(\hat{x})) + \sum_{k \neq j} \lambda_k (f_k(x^0) - f_k(\hat{x})) < 0 \tag{4.5}$$

because $\lambda_j > 0$.

2. Suppose $\hat{x}$ solves (4.3) optimally. Then there is no $x \in \mathcal{X}$ satisfying $f_j(x) < f_j(\hat{x})$ and $f_k(x) \leq f_k(\hat{x}) \leq \varepsilon_k$ for $k \neq j$. Using convexity of f_k we apply Theorem 3.16 to conclude that there must be some $\hat{\lambda} \in \mathbb{R}^p_{\geq}$ such that $\sum_{k=1}^{p} \hat{\lambda}_k (f_k(x) - f_k(\hat{x})) \geq 0$ for all $x \in \mathcal{X}$. Since $\hat{\lambda} \in \mathbb{R}^p_{\geq}$ we get

$$\sum_{k=1}^{p} \hat{\lambda}_k f_k(x) \geq \sum_{k=1}^{p} \hat{\lambda}_k f_k(\hat{x}) \tag{4.6}$$

for all $x \in \mathcal{X}$. Therefore $\hat{\lambda}$ is the desired weighting vector. □

A further result in this regard, showing when an optimal solution of the weighted sum problem is also an optimal solution of the (4.3) problem for all $j = 1, \ldots, p$ is given as Exercise 4.1.

4.2 The Hybrid Method

It is possible to combine the weighted sum method with the ε-constraint method. In that case, the scalarized problem to be solved has a weighted sum objective and constraints on *all* objectives. Let x^0 be an arbitrary feasible point for an MOP. Consider the following problem:

$$\begin{aligned} \min \sum_{k=1}^{p} \lambda_k f_k(x) & \\ \text{subject to } f_k(x) \leq f_k(x^0) & \quad k = 1, \ldots, p \\ x \in \mathcal{X} & \end{aligned} \tag{4.7}$$

where $\lambda \in \mathbb{R}^p_>$.

Theorem 4.7. *Guddat* et al. *(1985) Let $\lambda \in \mathbb{R}^p_>$. A feasible solution $x^0 \in \mathcal{X}$ is an optimal solution of problem (4.7) if and only if $x^0 \in \mathcal{X}_E$.*

Proof. Let $x^0 \in \mathcal{X}$ be efficient. Then there is no $x \in \mathcal{X}$ such that $f(x) \leq f(x^0)$. Thus any feasible solution of (4.7) satisfies $f(x) = f(x^0)$ and is an optimal solution.

Let x^0 be an optimal solution of (4.7). If there were an $x \in \mathcal{X}$ such that $f(x) \leq f(x^0)$ the positive weights would imply

$$\sum_{k=1}^{p} \lambda_k f_k(x) < \sum_{k=1}^{p} \lambda_k f_k(x^0).$$

Thus x^0 is efficient. □

4.3 The Elastic Constraint Method

For the ε-constraint method we have no results on properly efficient solutions. In addition, the scalarized problem (4.3) may be hard to solve in practice due to the added constraints $f_k(x) \leq \varepsilon_k$. In order to address this problem we can "relax" these constraints by allowing them to be violated and penalizing any violation in the objective function. Ehrgott and Ryan (2002) used this idea to develop the e;elastic constraint scalarization

$$\begin{aligned} \min f_j(x) + \sum_{k \neq j} \mu_k s_k & \\ \text{subject to } f_k(x) - s_k \leq \varepsilon_k & \quad k \neq j \\ s_k \geq 0 & \quad k \neq j \\ x \in \mathcal{X}, & \end{aligned} \tag{4.8}$$

where $\mu_k \geq 0, k \neq j$. The feasible set of (4.8) in x variables is $\mathcal{X}$, i.e. the feasible set of the original multicriteria optimization problem (4.1).

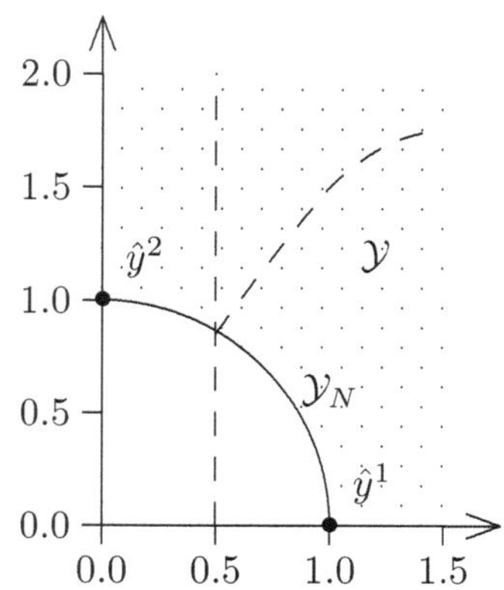

Fig. 4.3. Feasible set and objective function of Problem (4.8).

Note that if $(\hat{x}, \hat{s})$ is an optimal solution of (4.8), then we may without loss of generality assume that $\hat{s}_k = \max\{0, \varepsilon_k - f_k(\hat{x})\}$.

In Figure 4.3 (4.8) for $j = 2$ is illustrated for the bicriterion problem of Example 4.2. The vertical dotted line marks the value $\varepsilon_1 = 0.5$. The dotted curve shows the objective function of (4.8) as a function of component y_1 of nondominated points $\mathcal{Y}_N$. The idea of the method is that, by penalizing violations of the constraint $f_1(x) \leq \varepsilon_1$, a minimum is attained with the constraint active. As can be seen here, the minimum of (4.8) will be attained at $x = (0.5, 0.5)$.

We obtain the following results:

Proposition 4.8. *Let $(\hat{x}, \hat{s})$ be an optimal solution of* (4.8) *with $\mu \geqq 0$. Then $\hat{x} \in \mathcal{X}_{wE}$.*

Proof. Suppose $\hat{x}$ is not weakly efficient. Then there is some $x \in \mathcal{X}$ such that $f_k(x) < f_k(\hat{x}), k = 1, \ldots, p$. Then $(x, \hat{s})$ is feasible for (4.8) with an objective value that is smaller than that of $(\hat{x}, \hat{s})$. □

Under additional assumptions we get stronger results.

Proposition 4.9. *If $\hat{x}$ is unique in an optimal solution of* (4.8)*, then $\hat{x} \in \mathcal{X}_{sE}$ is a strictly efficient solution of the MOP.*

Proof. Assume that $x \in \mathcal{X}$ is such that $f_k(x) \leq f_k(\hat{x}), k = 1, \ldots, p$. Then $(x, \hat{s})$ is a feasible solution of (4.8). Since the objective function value of $(x, \hat{s})$ is not worse than that of $(\hat{x}, \hat{s})$, uniqueness of $\hat{x}$ implies that $x = \hat{x}$. □

The following example shows that even if $\mu > 0$ an optimal solution of (4.8) may be just weakly efficient.

Example 4.10. Consider

$$\mathcal{X} = \left\{(x_1, x_2) \in \mathbb{R}^2_{\geqq} : (x_1 - 1)^2 + (x_2 - 1)^2 \leq 1\right\} + \mathbb{R}^2_{\geqq}$$

and $f(x) = x$. Let $\varepsilon_1 > 1$. Then $(\hat{x}_1, \hat{x}_2, \hat{s}_1) = (\hat{x}_1, 0, 0)$ is an optimal solution of (4.8) with $j = 2$ for all $1 \leq \hat{x}_1 \leq \varepsilon_1$. If $\hat{x}_1 > 1$ this solution is weakly efficient, but not efficient. This result is independent of the choice of μ. □

The problem here is the possible existence of weakly efficient solutions that satisfy the constraints $f_k(x) \leq \varepsilon_k$ for all $k \neq j$. If, however, all ε_k are chosen in such a way that no merely weakly efficient solution satisfies the ε-constraints, an optimal solution of (4.8) with $\mu > 0$ will yield an efficient solution.

We now turn to the problem of showing that (properly) efficient solutions are optimal solutions of (4.8) for appropriate choices of k, ε, and μ. The following corollary follows immediately from Theorem 4.5 by choosing $\varepsilon = f(\hat{x})$, $\hat{s} = 0$ and $\mu_k = \infty$ for all $k = 1, \ldots, p$.

Corollary 4.11. *Let $\hat{x} \in \mathcal{X}_E$. Then there exist $\varepsilon, \mu \geqq 0$ and $\hat{s}$ such that $(\hat{x}, \hat{s})$ is an optimal solution of (4.8) for all $j \in \{1, \ldots, p\}$.*

A more careful analysis shows that for properly efficient solutions, we can do without the infinite penalties.

Theorem 4.12. *Let $\mathcal{Y}_N$ be externally stable. Let $\hat{x} \in \mathcal{X}_{pE}$ be properly efficient. Then, for every $j \in \{1, \ldots, p\}$ there are $\varepsilon, \hat{s}, \mu^j$ with $\mu^j_k < \infty$ for all $k \neq j$ such that $(\hat{x}, \hat{s})$ is an optimal solution of (4.3) for all $\mu \in \mathbb{R}^{p-1}$, $\mu \geqq \mu^j$.*

Proof. We choose $\varepsilon_k := f_k(\hat{x}), k = 1, \ldots, p$. Thus, we can choose $\hat{s} = 0$. Let $j \in \{1, \ldots, p\}$. Because $\hat{x}$ is properly efficient there is $M > 0$ such that for all $x \in \mathcal{X}$ with $f_j(x) < f_j(\hat{x})$ there is $k \neq j$ such that $f_k(\hat{x}) < f_k(x)$ and $\frac{f_j(\hat{x}) - f_j(x)}{f_k(x) - f_k(\hat{x})} < M$.

We define μ^j by $\mu^j_k := \max(M, 0)$ for all $k \neq j$.

Let $x \in \mathcal{X}$ and $s \in \mathbb{R}$ be such that $s_k = \max\{0, f_k(x) - \varepsilon_k\} = \max\{0, f_k(x) - f_k(\hat{x})\}$ for all $k \neq j$, i.e. the smallest possible value it can take. We need to show that

$$f_j(x) + \sum_{k \neq j} \mu_k s_k \geq f_j(\hat{x}) + \sum_{k \neq j} \mu_k \hat{s}_k = f_k(\hat{x}). \tag{4.9}$$

First, we prove that we can assume $x \in \mathcal{X}_E$ in (4.9). Otherwise there is $x' \in \mathcal{X}_E$ with $f(x') \leq f(x)$ (because $\mathcal{Y}_N$ is externally stable, see Definition

2.20) and s' with $s'_k = \max\{0, f_k(x') - \varepsilon_k\}$. Since $s' \leqq s$ we get that $f_j(x') + \sum_{k \neq j} \mu_k s'_k \leq f_k(x) + \sum_{k \neq j} \mu_k s_k$ for any $\mu \geqq 0$.

Now let $x \in \mathcal{X}_E$. We consider the case $f_j(x) \geq f_j(\hat{x})$. Then

$$f_j(x) + \sum_{k \neq j} \mu_k s_k > f_k(\hat{x}) + 0 = f_j(\hat{x}) + \sum_{k \neq j} \mu_k^j \hat{s}_k$$

for any $\mu \geq 0$.

Now consider the case $f_j(x) < f_j(\hat{x})$ and let $\mathcal{I}(x) := \{k \neq j : f_k(x) > f_k(\hat{x})\}$. As both $\hat{x}$ and x are efficient, $\mathcal{I}(x) \neq \emptyset$. Furthermore, we can assume $s_k = 0$ for all $k \notin \mathcal{I}(x), k \neq j$. Let $k' \in \mathcal{I}(x)$. Then

$$\begin{aligned}
f_j(x) + \sum_{k \neq j} \mu_k s_k &\geq f_j(x) + \sum_{k \neq j} \mu_k^j s_k \\
&\geq f_j(x) + \sum_{k \in \mathcal{I}(x)} \frac{f_j(\hat{x}) - f_j(x)}{f_k(x) - f_k(\hat{x})} s_k \\
&\geq f_j(x) + \frac{f_j(\hat{x}) - f_j(x)}{f_{k'}(x) - f_{k'}(\hat{x})} s_{k'} \\
&= f_j(x) + \frac{f_j(\hat{x}) - f_j(x)}{f_{k'}(x) - f_{k'}(\hat{x})} (f_{k'}(x) - f_{k'}(\hat{x})) \\
&= f_j(\hat{x}) = f_j(\hat{x}) + \sum_{k \neq j} \mu_k \hat{s}_k.
\end{aligned}$$

This follows from $\mu_k \geq \mu_k^j$, the definition of μ_k^j, nonnegativity of all terms, $s_k = f_k(x) - f_k(\hat{x})$ for $k \in \mathcal{I}(x)$ and $\hat{s} = 0$. □

We can also see, that for $x \in \mathcal{X}_E \setminus \mathcal{X}_{pE}$ finite values of μ are not sufficient.

Example 4.13. Let $p = 2$ and $\mathcal{X} = \{x \in \mathbb{R}^2 : (x_1 - 1)^2 + (x_2 - 1)^2 \leq 1\}$ with $f(x) = x$. Then $(1, 0)$ and $(0, 1)$ are efficient, but not properly efficient. The scalarization

$$\begin{aligned}
\min\ & x_2 + \mu s \\
\text{subject to } & x_1 - s \leq 0 \\
& x \in \mathcal{X}
\end{aligned}$$

is equivalent to (has the same optimal solution x as)

$$\min\{x_2 + \mu x_1 : (x_1 - 1)^2 + (x_2 - 1)^2 = 1\}.$$

It is easy to see that the unique optimal solution is given by $x_1 = 1 - \sqrt{1 - \frac{1}{\mu + 1}}$ and it is necessary that $\mu \to \infty$ to get $x_1 \to 0$.

Note, however, that in order to obtain $(0, 1)$, we can also consider

$$\begin{aligned} &\min x_1 + \mu s \\ &\text{subject to } x_2 - s \leq 1 \\ &\qquad\qquad\quad x \in \mathcal{X}. \end{aligned}$$

It is clear that $x_1 = 0, x_2 = 1, s = 0$ is an optimal solution of this problem for any $\mu \geq 0$. □

It is worth noting that in the elastic constraint method ε-constraints of (4.3) are relaxed in a manner similar to penalty function methods in nonlinear programming. This may help solving the scalarized problem in practice.

4.4 Benson's Method

The method and results described in this section are from Benson (1978). The idea is to choose some initial feasible solution $x^0 \in \mathcal{X}$ and, if it is not itself efficient, produce a dominating solution that is. To do so nonnegative deviation variables $l_k = f_k(x^0) - f_k(x)$ are introduced, and their sum maximized. This results in an x dominating x^0, if one exists, and the objective ensures that it is efficient, pushing x as far from x^0 as possible.

The substitute problem (4.10) for given x^0 is

$$\begin{aligned} &\max \sum_{k=1}^{p} l_k \\ &\text{subject to } f_k(x^0) - l_k - f_k(x) = 0 \quad k = 1, \ldots, p \\ &\qquad l \geqq 0 \\ &\qquad x \in \mathcal{X}. \end{aligned} \tag{4.10}$$

An illustration in objective space (Figure 4.4) demonstrates the idea. The initial feasible, but dominated, point $f(x^0)$ has values greater than the efficient point $f(\hat{x})$. Maximizing the total deviation $\hat{l}_1 + \hat{l}_2$, the intention is to find a dominating solution, which is efficient.

First of all, solving (4.10) is a check for efficiency of the initial solution x^0 itself. We will see this result again later, when we deal with linear problems in Chapter 6.

Theorem 4.14. *The feasible solution $x^0 \in \mathcal{X}$ is efficient if and only if the optimal objective value of* (4.10) *is* 0.

Proof. Let (x, l) be a feasible solution of (4.10). Because of the nonnegativity constraint $l_k \geq 0$ for $k = 1, \ldots, p$ and the definition of l_k as $f_k(x^0) - f_k(x)$ we have

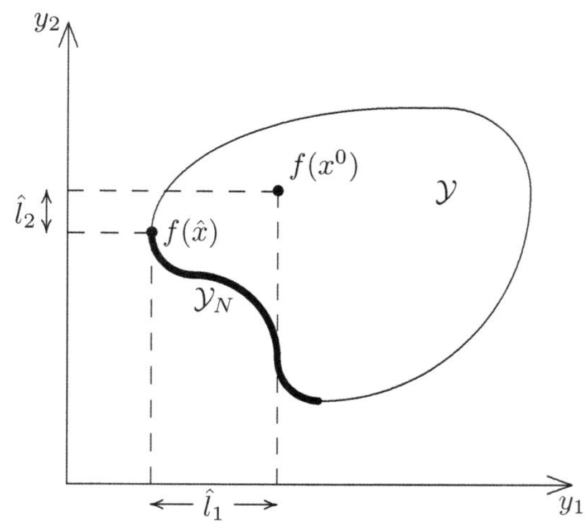

Fig. 4.4. Illustration of Benson's problem (4.10).

$$\sum_{k=1}^{p} l_k = 0 \Longleftrightarrow l_k = 0 \quad k = 1, \ldots, p$$

$$\Longleftrightarrow f_k(x^0) = f_k(x) \quad k = 1, \ldots, p$$

Thus, if the optimal value is 0, and $x \in \mathcal{X}$ is such that $f(x) \leqq f(x^0)$ it must hold that $f(x) = f(x^0)$, i.e. x^0 is efficient. If, on the other hand, x^0 is efficient, the feasible set of (4.10) consists of those (x, l) for which $x \in \mathcal{X}$ and $f(x) = f(x^0)$ and thus $l = 0$. □

That the initial solution x^0 is efficient cannot be expected in general. The strength of the method lies in the fact that whenever problem (4.10) has a finite optimal solution value, the optimal solution is efficient. Under convexity assumptions, we can even show that when the objective function of (4.8) is unbounded, no properly efficient solutions exist. From an application point of view, this constitutes a pathological situation: all efficient solutions will have unbounded trade-offs. However, this can only happen in situations where existence of efficient solutions is not guaranteed in general.

Proposition 4.15. *If problem* (4.10) *has an optimal solution* $(\hat{x}, \hat{l})$ *(and the optimal objective value is finite) then* $\hat{x} \in \mathcal{X}_E$.

Proof. Suppose $\hat{x} \notin \mathcal{X}_E$. Then there is some $x' \in \mathcal{X}$ such that $f_k(x') \leq f_k(\hat{x})$ for all $k = 1, \ldots, p$ and $f_j(x') < f_j(\hat{x})$ for at least one j. We define $l' := f(x^0) - f(x')$. Then (x', l') is feasible for (4.10) because

$$l'_k = f_k(x^0) - f_k(x') \geq f_k(x^0) - f_k(\hat{x}) = \hat{l}_k \geq 0. \tag{4.11}$$

Furthermore, $\sum_{k=1}^p l'_k > \sum_{k=1}^p \hat{l}_k$ as $l'_j > \hat{l}_j$. This is impossible because $(\hat{x}, \hat{l})$ is an optimal solution of (4.10). □

The question of what happens if (4.10) is unbounded can be answered under convexity assumptions.

Theorem 4.16 (Benson (1978)). *Assume that the functions $f_k, k = 1, \ldots, p$ are convex and that $\mathcal{X} \subset \mathbb{R}^n$ is a convex set. If (4.10) has no finite optimal objective value then $\mathcal{X}_{pE} = \emptyset$.*

Proof. Since (4.10) is unbounded, for every real number $M \geq 0$ we can find $x^M \in \mathcal{X}$ such that $l = f(x^0) - f(x^M) \geq 0$ and

$$\sum_{k=1}^p l_k = \sum_{k=1}^p (f_k(x^0) - f_k(x^M)) > M. \tag{4.12}$$

Assume that $\hat{x}$ is properly efficient in Geoffrion's sense. From Theorem 3.15 we know that there are weights $\lambda_k > 0$ for $k = 1, \ldots, p$ such that $\hat{x}$ is an optimal solution of $\min_{x \in \mathcal{X}} \sum_{k=1}^p \lambda_k f_k(x)$. Therefore $\sum_{k=1}^p \lambda_k (f_k(x) - f_k(\hat{x})) \geq 0$ for all $x \in \mathcal{X}$, and in particular

$$\sum_{k=1}^p \lambda_k (f_k(x^0) - f_k(\hat{x})) \geq 0. \tag{4.13}$$

We define $\hat{\lambda} := \min\{\lambda_1, \ldots, \lambda_p\} > 0$ and for some arbitrary, but fixed $M' \geq 0$ let $M := M'/\hat{\lambda}$. From (4.12) we know that for this M there is some $x^M \in \mathcal{X}$ satisfying $f_k(x^0) - f_k(x^M) \geq 0$ for all $k = 1, \ldots, p$ and

$$\hat{\lambda} \sum_{k=1}^p (f_k(x^0) - f_k(x^M)) > \hat{\lambda} M = \frac{M'}{\hat{\lambda}} \cdot \hat{\lambda} = M'. \tag{4.14}$$

This implies that

$$M' < \sum_{k=1}^p \hat{\lambda}(f_k(x^0) - f_k(x^M)) \leq \sum_{k=1}^p \lambda_k (f_k(x^0) - f_k(x^M)) \tag{4.15}$$

is true for all $M' \geq 0$ because of the definition of $\hat{\lambda}$ and because M' was chosen arbitrarily. We can therefore use $M' = \sum_{k=1}^p \lambda_k (f_k(x^0) - f_k(\hat{x}))$ to get

$$\sum_{k=1}^p \lambda_k (f_k(x^0) - f_k(\hat{x})) < \sum_{k=1}^p \lambda_k (f_k(x^0) - f_k(x^M)), \tag{4.16}$$

i.e. $\sum_{k=1}^p \lambda_k f_k(x^M) < \sum_{k=1}^p \lambda_k f_k(\hat{x})$, contradicting optimality of $\hat{x}$ for the weighted sum problem. □

Recalling that $\mathcal{Y}_N \subset \operatorname{cl} \mathcal{Y}_{pN}$ if in addition to convexity $\mathcal{Y} = f(\mathcal{X})$ is $\mathbb{R}^p_{\geqq}$-closed (Theorem 3.17) we can strengthen Theorem 4.16 to emptiness of $\mathcal{X}_E$.

Corollary 4.17. *Assume $\mathcal{X} \subset \mathbb{R}^n$ is convex, $f_k : \mathbb{R}^n \to \mathbb{R}$ are convex for $k = 1, \ldots, p$ and $f(\mathcal{X})$ is $\mathbb{R}^p_{\geqq}$-closed. If (4.10) is unbounded then $\mathcal{X}_E = \emptyset$.*

Proof. From Theorem 3.17 we know that $\mathcal{Y}_N \subset \operatorname{cl} S(\mathcal{Y}) = \operatorname{cl} \mathcal{Y}_{pN}$. From Theorem 4.16 $\mathcal{Y}_{pN} = \emptyset$ whence $\operatorname{cl} \mathcal{Y}_{pE} = \emptyset$ and $\mathcal{Y}_N = \emptyset$. Thus $\mathcal{X}_E = \emptyset$. □

Example 4.18 (Wiecek (1995)). Consider the multicriteria optimization problem with a single variable

$$\begin{aligned} \min\ &\left(x^2 - 4, (x-1)^4\right) \\ \text{subject to } &-x - 100 \leq 0. \end{aligned}$$

Benson's problem (4.10) in this case is

$$\begin{aligned} \max\ l_1 + l_2 & \\ \text{subject to } -x - 100 &\leq 0 \\ (x^0)^2 - 4 - l_1 - x^2 + 4 &= 0 \\ (x^0 - 1)^4 - l_2 - (x-1)^4 &= 0 \\ l &\geqq 0. \end{aligned}$$

We solve the problem for two choices of x^0. First, consider $x^0 = 0$. We obtain

$$\max l_1 + l_2 \tag{4.17}$$
$$\text{subject to } -x - 100 \leq 0 \tag{4.18}$$
$$x^2 + l_1 = 0 \tag{4.19}$$
$$1 - l_2 - (x-1)^4 = 0 \tag{4.20}$$
$$l_1, l_2 \geqq 0 \tag{4.21}$$

From (4.19) and (4.21) $l_1 = 0$ and $x = 0$. Then (4.20) and (4.21) imply $l_2 = 0$. Therefore $x = 0, l = (0, 0)$ is the only feasible solution of (4.10) with $x^0 = 0$ and Theorem 4.14 implies that $x^0 = 0 \in \mathcal{X}_E$.

The (strictly, weakly) efficient sets for this problem here are all equal to [0, 1] (use the result in Exercise 2.8 to verify this). Therefore let us try (4.10) with an initial solution $x^0 = 2$, to see if $x^0 \notin \mathcal{X}_E$ can be confirmed, and to find a dominating efficient solution.

The problem becomes

$$\begin{aligned} \max l_1 + l_2 \\ \text{subject to } -x - 100 \leq 0 \\ -x^2 + 4 - l_1 = 0 \\ 1 - (x-1)^4 - l_2 = 0 \\ l_1, l_2 \geqq 0. \end{aligned}$$

From the constraints we deduce $0 \leq l_1 \leq 4$ and $0 \leq l_2 \leq 1$. Therefore the optimal objective value is bounded, and according to Proposition 4.15 an optimal solution of (4.10) with $x^0 = 2$ is efficient. Because $x = 0, l_1 = 4, l_2 = 0$ is feasible for (4.10), the optimal objective value is nonzero. Theorem 4.14 implies that $x^0 = 2$ is not efficient. The (unique) optimal solution of the problem is $\hat{x} \approx 0.410$.

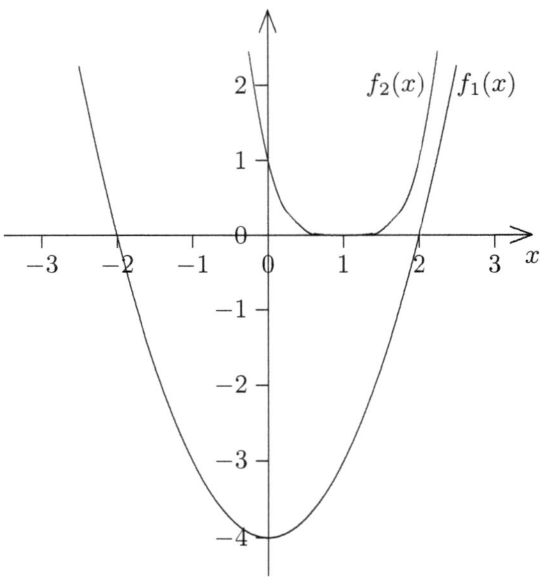

Fig. 4.5. Objective functions in Example 4.18.

□

4.5 Compromise Solutions – Approximation of the Ideal Point

The best possible outcome of a multicriteria problem would be the ideal point y^I (see Definition 2.22). Yet when the objectives are conflicting the ideal values are impossible to obtain. However, the ideal point can serve as a reference point, with the goal to seek for solutions as close as possible to the ideal point. This is the basic idea of compromise programming.

Given a distance measure

$$d : \mathbb{R}^p \times \mathbb{R}^p \to \mathbb{R}_{\geqq}, \tag{4.22}$$

the compromise programming problem is given by

$$\min_{x \in \mathcal{X}} d(f(x), y^I). \tag{4.23}$$

In this text, we will only consider metrics derived from norms as distance measures, i.e. $d(y^1, y^2) = \|y^1 - y^2\|$. In particular for $y^1, y^2, y^3 \in \mathcal{Y}$: d is symmetric $d(y^1, y^2) = d(y^2, y^1)$, satisfies the triangle inequality $d(y^1, y^2) \leq d(y^1, y^3) + d(y^3, y^2)$, and $d(y^1, y^2) = 0$ if and only if $y^1 = y^2$.

The compromise programming problem (4.23) has a nice interpretation in terms of the level sets $\{y \in \mathbb{R}^p : \|y - y^I\| \leq c\}$. These sets contain all points of distance c or less to the ideal point y^I. Therefore the goal of the compromise programming problem is to find the smallest value c such that the intersection of the corresponding level set with $\mathcal{Y} = f(\mathcal{X})$ is nonempty. Figure 4.6 illustrates this perspective for the l_1 distance $\|y^1 - y^2\|_1 := \sum_{k=1}^p |y_k^1 - y_k^2|$, the l_∞ distance $\|y^1 - y^2\|_\infty := \max_{k=1}^p |y_k^1 - y_k^2|$, and a distance measure d derived from a norm γ with asymmetric level sets.

Whether an optimal solution of problem (4.23) is efficient depends on properties of the distance measure d, and therefore on properties of norm $\|\cdot\|$, from which d is derived.

Definition 4.19. *1. A norm $\|\cdot\| : \mathbb{R}^p \to \mathbb{R}_{\geqq}$ is called* monotone, *if $\|y^1\| \leq \|y^2\|$ holds for all $y^1, y^2 \in \mathbb{R}^p$ with $|y_k^1| \leq |y_k^2|$, $k = 1, \ldots, p$ and moreover $\|y^1\| < \|y^2\|$ if $|y_k^1| < |y_k^2|$, $k = 1, \ldots, p$.*
2. A norm $\|\cdot\|$ is called strictly monotone, *if $\|y^1\| < \|y^2\|$ holds whenever $|y_k^1| \leq |y_k^2|$, $k = 1, \ldots, p$ and $|y_j^1| \neq |y_j^2|$ for some j.*

With definition 4.19 we can prove the following basic results.

Theorem 4.20. *1. If $\|\cdot\|$ is monotone and $\hat{x}$ is an optimal solution of* (4.23) *then $\hat{x}$ is weakly efficient. If $\hat{x}$ is a unique optimal solution of* (4.23) *then $\hat{x}$ is efficient.*

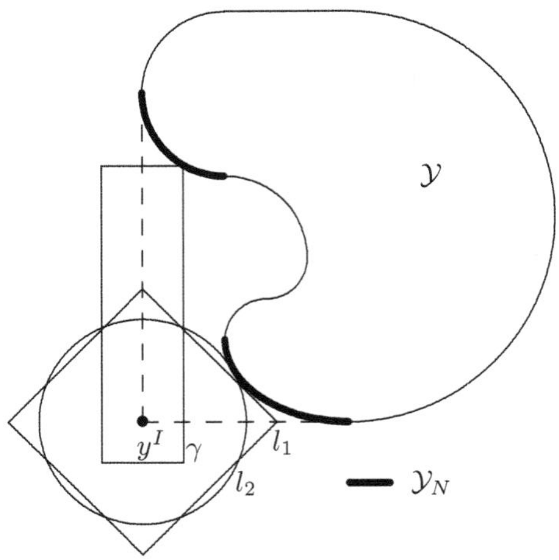

Fig. 4.6. Level Sets $\{y : \|y - y^I\| \leq c\}$ for different norms.

2. *If $\|\cdot\|$ is strictly monotone and $\hat{x}$ is an optimal solution of* (4.23) *then $\hat{x}$ is efficient.*

Proof. 1. Suppose $\hat{x}$ is an optimal solution of (4.23) and $\hat{x} \notin \mathcal{X}_{wE}$. Then there is some $x' \in \mathcal{X}$ such that $f(x') < f(\hat{x})$. Therefore $0 \leq f_k(x') - y^I_k < f_k(\hat{x}) - y^0_k$ for $k = 1, \ldots, p$ and

$$\|f(x') - y^I\| < \|f(\hat{x}) - y^0\|, \tag{4.24}$$

a contradiction.
Now assume that $\hat{x}$ is a unique optimal solution of (4.23) and that $\hat{x} \notin \mathcal{X}_E$. Then there is some $x' \in \mathcal{X}$ such that $f(x') \leq f(\hat{x})$. Therefore $0 \leq f_k(x) - y^I_k \leq f_k(\hat{x}) - y^I_k$ for $k = 1, \ldots, p$ with one strict inequality, and

$$\|f(x) - y^I\| \leq \|f(\hat{x}) - y^I\|. \tag{4.25}$$

From optimality of $\hat{x}$ equality must hold, which contradicts the uniqueness of $\hat{x}$.

2. Suppose $\hat{x}$ is an optimal solution of (4.23) and $\hat{x} \notin \mathcal{X}_E$. Then there are $x' \in \mathcal{X}$ and $j \in \{1, \ldots, p\}$ such that $f_k(x') \leq f_k(\hat{x})$ for $k = 1, \ldots, p$ and $f_j(x') < f_j(\hat{x})$. Therefore $0 \leq f_k(x) - y^I_k \leq f_k(\hat{x}) - y^I_k$ for all $k = 1, \ldots, p$ and $0 \leq f_j(x) - y^I_j < f_j(\hat{x}) - y^I_j$. Again the contradiction

$$\|f(x) - y^0\| < \|f(\hat{x}) - y^0\| \tag{4.26}$$

follows. □

The most important class of norms is the class of l_p-norms $\|\cdot\| = \|\cdot\|_\mathsf{p}$, i.e.

$$\|y\|_\mathsf{p} = \left(\sum_{k=1}^{p} |y_k|^\mathsf{p}\right)^{\frac{1}{\mathsf{p}}} \tag{4.27}$$

for $1 \leq \mathsf{p} \leq \infty$. The l_p norm $\| \ \|_\mathsf{p}$ is strictly monotone for $1 \leq \mathsf{p} < \infty$ and monotone for $\mathsf{p} = \infty$. The special cases $\mathsf{p} = 1$ with $\|y\| = \sum_{k=1}^{p} |y_k|$ and $\mathsf{p} = \infty$ with $\|y\| = \max_{k=1}^{p} |y_k|$ are of major importance.

As long as we just minimize the distance between a feasible point in objective space and the ideal point, we will find one (weakly) efficient solution for each choice of a norm. The results can be strengthened if we allow weights in the norms. From now on we only consider l_p-norms. The weighted compromise programming problems are

$$\min_{x \in \mathcal{X}} \left(\sum_{k=1}^{p} \lambda_k (f_k(x) - y_k^I)^\mathsf{p}\right)^{\frac{1}{\mathsf{p}}} \tag{4.28}$$

for general p, and

$$\min_{x \in \mathcal{X}} \max_{k=1,\ldots,p} \lambda_k (f_k(x) - y_k^I), \tag{4.29}$$

for $\mathsf{p} = \infty$.

Here we assume, as usual, that the vector of weights $\lambda \in \mathbb{R}^p_{\geq}$ is nonnegative and nonzero. Note that the functions $\|\cdot\|_\mathsf{p}^\lambda : \mathbb{R}^p \to \mathbb{R}_{\geqq}$ are not necessarily norms if some of the weights λ_k are zero. It is also of interest to observe that for $\mathsf{p} = 1$ (4.28) can be written as

$$\min_{x \in \mathcal{X}} \sum_{k=1}^{p} (\lambda_k f_k(x) - y_k^I) = \min_{x \in \mathcal{X}} \left(\sum_{k=1}^{p} \lambda_k f_k(x)\right) - \sum_{k=1}^{p} \lambda_k y_k^I.$$

Hence weighted sum scalarization can be seen as a special case of weighted compromise programming. We can therefore exclude this case from now on. The emphasis on the distinction between $1 < \mathsf{p} < \infty$ and $\mathsf{p} = \infty$ is justified for two reasons: The latter is the most interesting case, and the most widely used, and the results are often different from those for $\mathsf{p} < \infty$.

For (4.28) and (4.29) we can prove some basic statements analogous to Theorem 4.20.

Theorem 4.21. *An optimal solution $\hat{x}$ of* (4.28) *with $\mathsf{p} < \infty$ is efficient if one of the following conditions holds.*

1. $\hat{x}$ is a unique optimal solution of (4.28).

2. $\lambda_k > 0$ *for all* $k = 1, \ldots, p$.

Proof. Assume $\hat{x}$ is a minimizer of (4.28) but $\hat{x} \notin \mathcal{X}_E$. Then there is some $x' \in \mathcal{X}$ dominating $\hat{x}$.

1. In this case, x' must also be an optimal solution of (4.28), which due to $x \neq \hat{x}$ is impossible.
2. From $\lambda > 0$ we have $0 < \lambda_k(f_k(x) - y_k^I) \leq \lambda_k(f_k(\hat{x}) - y_k^I)$ for all $k = 1, \ldots, p$ with strict inequality for some k. Taking power p and summing up preserves strict inequality, which contradicts $\hat{x}$ being an optimal solution of (4.28). □

Proposition 4.22. *Let* $\lambda > 0$ *be a strictly positive weight vector. Then the following statements hold.*

1. If $\hat{x}$ *is an optimal solution of* (4.29) *then* $\hat{x} \in \mathcal{X}_{wE}$.
2. If $\mathcal{Y}_N$ *is externally stable (see Definition 2.20) and* (4.29) *has an optimal solution then at least one of its optimal solutions is efficient.*
3. If (4.29) *has a unique optimal solution* $\hat{x}$*, then* $\hat{x} \in \mathcal{X}_E$.

Proof. 1. The proof is standard and left out. See the proofs of Theorems 4.20 and 4.21.
2. Assume that (4.29) has optimal solutions, but none of them is is efficient. Let $\hat{x}$ be an optimal solution of (4.29). Because $\mathcal{Y}_N$ is externally stable there must be an $x \in \mathcal{X}_E$ with $f(x) \leq f(\hat{x})$. Then $\lambda_k(f_k(x) - y_k^I) \leq \lambda_k(f_k(\hat{x}) - y_k^I)$ for $k = 1, \ldots, p$, which means x is optimal for (4.29), too.
3. This part can be shown as usual. If $\mathcal{Y}_N$ is externally stable it follows directly from the second statement. □

Actually, all the results we proved so far remain valid, if the ideal point y^I is replaced by any other reference point y^R, as long as this reference point is chosen to satisfy $y^R \leqq y^I$.

Definition 4.23. *A point* $y^U := y_i^I - \varepsilon$*, where* $\varepsilon \in \mathbb{R}^p_>$ *has small positive components is called a* utopia point.

Note that not even minimizing the single objectives independently of one another will yield the utopia values: $f_k(x) > y_k^U$ for all feasible solutions $x \in \mathcal{X}$ and all $k = 1, \ldots, p$. The advantage of using utopia points instead of ideal points will become clear from the following theorems. The first complements Proposition 4.22 by a necessary and sufficient condition for weak efficiency.

Theorem 4.24 (Choo and Atkins (1983)). *A feasible solution* $\hat{x} \in \mathcal{X}$ *is weakly efficient if and only if there is a weight vector* $\lambda > 0$ *such that* $\hat{x}$ *is an optimal solution of the problem*

$$\min_{x\in\mathcal{X}} \max_{k=1,\ldots,p} \lambda_k(f_k(x) - y_k^U). \tag{4.30}$$

Proof. "$\Longleftarrow$" The proof of sufficiency is the same standard proof as that of the first part of Proposition 4.22.

"$\Longrightarrow$" We define appropriate weights and show that they do the job. Let $\lambda_k := 1/(f_k(\hat{x}) - y_k^U)$. These weights are positive and finite. Suppose $\hat{x}$ is not optimal for (4.30). Then there is a feasible $x \in \mathcal{X}$ such that

$$\max_{k=1,\ldots,p} \lambda_k(f_k(x) - y_k^U) < \max_{k=1,\ldots,p} \frac{1}{f_k(\hat{x}) - y_k^U}(f_k(\hat{x}) - y_k^U) = 1$$

and therefore

$$\lambda_k(f_k(x) - y_k^U) < 1 \text{ for all } k = 1, \ldots, p.$$

Dividing by λ_k we get $f_k(x) - y_k^U < f_k(\hat{x}) - y_k^U$ for all $k = 1, \ldots, p$ and thus $f(x) < f(\hat{x})$, contradicting $\hat{x} \in \mathcal{X}_{wE}$. □

With Theorem 4.24 we have a complete characterization of weakly efficient solutions for general, nonconvex problems. However, as for the ε-constraint method, we have to accept the drawback that in practice the result will only be useful as a check for weak efficiency, because $f(\hat{x})$ is needed to define the weights to prove optimality of $\hat{x}$. It should also be noted that if y^U is replaced by y^I in Theorem 4.24 then not even

$$\mathcal{Y}_{pN} \subset \bigcup_{\lambda\in\mathbb{R}^p_>} \left\{\hat{y} : \max_{k=1,\ldots,p} \lambda_k|\hat{y}_k - y_k^I| \leq \max_{k=1,\ldots,p} \lambda_k|y_k - y_k^I| \text{ for all } y \in \mathcal{Y}\right\}$$

is true, see Exercise 4.8.

We are now able to prove the main result of this section. It is the formal extension of the main result on the weighted sum scalarization in Chapter 3. We have noted earlier that (4.28) contains the weighted sum problem as a special case (setting $\mathsf{p} = 1$). For this special case we have seen in Theorem 3.17 that for $\mathbb{R}^p_{\geqq}$-convex and $\mathbb{R}^p_{\geqq}$-bounded $\mathcal{Y}$

$$\mathcal{S}(\mathcal{Y}) \subset \mathcal{Y}_{pN} \subset \mathcal{Y}_N \subset \text{cl}(\mathcal{S}(\mathcal{Y})).$$

For the general problem (4.28) we can therefore expect more general results, when convexity is relaxed. Theorem 4.25 is this generalization. Before we can prove the theorem, we have to introduce some notation to enhance readability of the proof.

Let

$$\Lambda := \left\{ \lambda \in \mathbb{R}^p_{\geqq} : \sum_{k=1}^{p} \lambda_k = 1 \right\}$$

$$\Lambda^0 := \operatorname{ri} \Lambda = \left\{ \lambda \in \mathbb{R}^p_{>} : \sum_{k=1}^{p} \lambda_k = 1 \right\}.$$

For $\lambda \in \Lambda$ and $y \in \mathcal{Y}$ we shall write

$$\lambda \odot y = (\lambda_1 y_1, \dots, \lambda_p y_p).$$

Furthermore, in analogy to $\operatorname{Opt}(\lambda, \mathcal{Y})$ and $S(\mathcal{Y})$, the set of best approximations of y^I for a certain weight λ and norm $\| \ \|_{\mathsf{p}}$ is denoted by

$$\mathcal{A}(\lambda, \mathsf{p}, \mathcal{Y}) := \left\{ \hat{y} \in \mathcal{Y} : \left\| \lambda \odot (\hat{y} - y^U) \right\|_{\mathsf{p}} = \min_{y \in \mathcal{Y}} \left\| w \odot (y - y^U) \right\|_{\mathsf{p}} \right\} \tag{4.31}$$

$$\mathcal{A}(\mathcal{Y}) := \bigcup_{\lambda \in \Lambda^0} \bigcup_{1 \leq \mathsf{p} < \infty} \mathcal{A}(\lambda, \mathsf{p}, \mathcal{Y}). \tag{4.32}$$

From Theorem 4.21 and Theorem 4.24 we already know that

$$\mathcal{A}(\mathcal{Y}) \subset \mathcal{Y}_N \subset \mathcal{Y}_{wN} = \bigcup_{\lambda \in \Lambda^0} \mathcal{A}(\lambda, \infty, \mathcal{Y}). \tag{4.33}$$

The main result will show that this can be strengthened to

$$\mathcal{A}(\mathcal{Y}) \subset \mathcal{Y}_{pE} \subset \mathcal{Y}_E \subset \operatorname{cl}(\mathcal{A}(\mathcal{Y})) \tag{4.34}$$

for $\mathbb{R}^p_{\geqq}$-closed sets $\mathcal{Y}$, a complete analogy to Theorem 3.17 for nonconvex sets.

In the proof of Theorem 4.25 some of the essential arguments are based on the following properties of l_{p}-norms:

(P1) $\|y\|_\infty \leq \|y\|_{\mathsf{p}}$ for all $1 \leq \mathsf{p} < \infty$ and all $y \in \mathbb{R}^p$,
(P2) $\|y\|_{\mathsf{p}} \to \|y\|_\infty$ as $\mathsf{p} \to \infty$ holds for any $y \in \mathbb{R}^p$,
(P3) $\| \cdot \|_{\mathsf{p}}$ is strictly monotone for all $1 \leq \mathsf{p} < \infty$.

Theorem 4.25 (Sawaragi *et al.* (1985)). *If $\mathcal{Y}$ is $\mathbb{R}^p_{\geqq}$-closed then*

$$\mathcal{A}(\mathcal{Y}) \subset \mathcal{Y}_{pN} \subset \mathcal{Y}_N \subset \operatorname{cl}(\mathcal{A}(\mathcal{Y})).$$

Proof. The proof is divided into two main parts, corresponding to the two inclusions $\mathcal{A}(\mathcal{Y}) \subset \mathcal{Y}_{pN}$ and $\mathcal{Y}_N \subset \operatorname{cl}(\mathcal{A}(\mathcal{Y}))$.

Part 1: $\mathcal{A}(\mathcal{Y}) \subset \mathcal{Y}_{pN}$. Let $\hat{y} \in \mathcal{A}(\mathcal{Y})$. By definition of $\mathcal{A}(\mathcal{Y})$ there is a positive weight vector $\lambda \in \Lambda^0$ and some $\mathsf{p} \in [1, \infty)$ such that

$$\left\| \lambda \odot (\hat{y} - y^U) \right\|_{\mathsf{p}} \leq \left\| \lambda \odot (y - y^U) \right\|_{\mathsf{p}} \tag{4.35}$$

for all $y \in \mathcal{Y}$. Let us assume that $\hat{y} \notin \mathcal{Y}_{pN}$, which according to Benson's definition 2.44 means that there are sequences $\{\beta_k\} \subset \mathbb{R}$, $\{y^k\} \subset \mathcal{Y}$, and $\{d^k\} \subset \mathbb{R}^p_{\geqq}$ with $\beta_k > 0$ and

$$\beta_k(y^k + d^k - \hat{y}) \to -d \text{ for some } d \in \mathbb{R}^p_{\geq}. \tag{4.36}$$

We distinguish the two cases $\{\beta_k\}$ bounded and $\{\beta_k\}$ unbounded and use (4.36) to construct a point $\tilde{y}$, respectively a sequence y^k, which do not satisfy (4.35).

$\{\beta_k\}$ bounded: In this case we can assume, without less of generality, that β_k converges to some number $\beta_0 \geq 0$ (taking a subsequence, if necessary). If $\beta_0 = 0$ the fact $y^k + d^k - \hat{y} \geq y^U - \hat{y}$ implies

$$\beta_k(y^k + d^k - \hat{y}) \geq \beta_k(y^U - \hat{y}). \tag{4.37}$$

Because the left hand side term in (4.37) converges to $-d$, and the right hand side term to 0, we get $-d \geq 0$, a contradiction.
If, on the other hand, $\beta_0 > 0$ we have that $y^k + d^k - \hat{y} \to (-d)/\beta_0$, which is nonzero, and $y^k + d^k \to \hat{y} - d/\beta_0$. Since $y^k + d^k \in \mathcal{Y} + \mathbb{R}^p_{\geqq}$ and this set is closed, it must be that the limit $\hat{y} - d/\beta_0 \in \mathcal{Y} + \mathbb{R}^p_{\geqq}$. From this observation we conclude that there is some $\tilde{y} \in \mathcal{Y}$ such that $\hat{y} \geq \tilde{y}$. Positive weights and strict monotonicity of the l_{p}-norm finally yield $\|\lambda \odot (\hat{y} - y^U)\|_{\mathsf{p}} > \|\lambda \odot (\tilde{y} - y^U)\|_{\mathsf{p}}$.

$\{\beta_k\}$ unbounded: Taking subsequences if necessary, we can here assume $\beta_k \to \infty$, which by the convergence in (4.36) gives $y^k + d^k - \hat{y} \to 0$. Because $\hat{y}_k > y^U_k$ for all $k = 1, \ldots, p$ we can find a sufficiently large $\beta' > 0$ so that

$$0 \leq \hat{y} - \frac{d}{\beta} - y^U < \hat{y} - y^U \tag{4.38}$$

for all $\beta > \beta'$. We use strict monotonicity of the norm and $\lambda > 0$ to obtain

$$\left\|\lambda \odot \left(\hat{y} - \frac{d}{\beta} - y^U\right)\right\|_{\mathsf{p}} < \left\|\lambda \odot (\hat{y} - y^U)\right\|_{\mathsf{p}} \tag{4.39}$$

for all $\beta > \beta'$. Since $\beta_k \to \infty$ we will have $\beta_k > \beta'$ for all $k \geq k_0$ with a sufficiently large k_0. Therefore

$$\begin{aligned} \left\|\lambda \odot (y^k + d^k - y^U)\right\|_{\mathsf{p}} &= \left\|\lambda \odot \left(y^k + d^k - \hat{y} + \tfrac{d}{\beta_k} + \hat{y} - \tfrac{d}{\beta_k} - y^U\right)\right\|_{\mathsf{p}} \\ &\leq \left\|\lambda \odot (y^k + d^k - \hat{y})\right\|_{\mathsf{p}} + \tfrac{\|\lambda \odot d\|_{\mathsf{p}}}{\beta_k} + \\ &\quad \left\|\lambda \odot \left(\hat{y} - \tfrac{d}{\beta_k} - y^U\right)\right\|_{\mathsf{p}}. \end{aligned} \tag{4.40}$$

We know that the first term on the right hand side of the inequality of (4.40) converges to 0. The sequence β_k being unbounded implies the second term converges to 0, too. Thus from (4.40) and (4.39)

$$\begin{aligned}\lim_{k\to\infty} \left\|\lambda \odot (y^k + d^k - y^U)\right\|_{\mathsf{p}} &\leq \lim_{k\to\infty} \left\|\lambda \odot \left(\hat{y} - \frac{d}{\beta_k} - y^U\right)\right\|_{\mathsf{p}} \\ &< \left\|\lambda \odot (\hat{y} - y^U)\right\|_{\mathsf{p}}.\end{aligned} \tag{4.41}$$

But since $y^k + d^k - y^U \geqq y^k - y^U \geq 0$, applying monotonicity of the norm once more, (4.41) implies $\lim_{k\to\infty} \|\lambda\odot(y^k - y^U)\|_{\mathsf{p}} < \|\lambda\odot(\hat{y} - y^U)\|_{\mathsf{p}}$.

Part 2: $\mathcal{Y}_N \subset \mathrm{cl}(\mathcal{A}(\mathcal{Y}))$. We prove this part by showing that for all $\hat{y} \in \mathcal{Y}_N$ and for all $\varepsilon > 0$ there is some $y^\varepsilon \in \mathcal{A}(\mathcal{Y})$ in an ε-neighbourhood of $\hat{y}$. Then, taking the closure of $\mathcal{A}(\mathcal{Y})$, the result follows. The ε-neighbourhood is defined according to the l_∞-norm.
I.e. let $\hat{y} \in \mathcal{Y}_N$ and let $\varepsilon > 0$. We show that there is some $y^\varepsilon \in \mathcal{A}(\mathcal{Y})$ with $\|y^\varepsilon - \hat{y}\|_\infty = \max_{k=1,\ldots,p} |y^\varepsilon_k - \hat{y}_k| < \varepsilon$.
First we proof an auxiliary claim: For each $\varepsilon > 0$ there is $y' > \hat{y}$ such that $\|y - \hat{y}\|_\infty < \varepsilon$ for all y in the section $(y' - \mathbb{R}^p_{\geqq}) \cap \mathcal{Y}$, see Figure 4.7. To see this, assume that for some $\varepsilon > 0$ there is no such y'. Then there must be a sequence $\{\hat{y}^k\} \subset \mathbb{R}^p$ with $\hat{y}^k \geq \hat{y}$, $\hat{y}^k \to \hat{y}$ such that for all k there is $y^k \in (\hat{y}^k - \mathbb{R}^p_{\geqq}) \cap \mathcal{Y}$ with $\|y^k - \hat{y}\| \geq \varepsilon$.

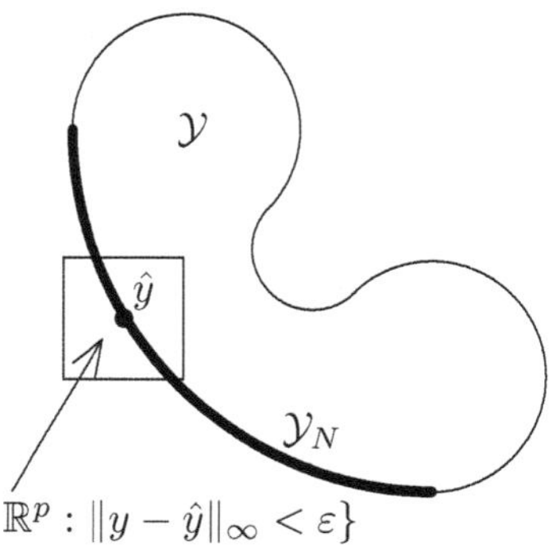

Fig. 4.7. ε-neighbourhoods of nondominated points in the l_∞-norm.

Because $\mathcal{Y}+\mathbb{R}^p_{\geqq}$ is closed and $\mathcal{Y} \subset y^U + \mathbb{R}_{\geqq}$, i.e. $\mathcal{Y}$ is bounded below we can assume without loss of generality that $y^k \to y'' + d''$, where $y'' \in \mathcal{Y}, d'' \geqq 0$ and $\|y''+d''-\hat{y}\|_\infty \geq \varepsilon$. On the other hand $y''+d'' \in (\hat{y}-\mathbb{R}^p_{\geqq})\cap(\mathcal{Y}+\mathbb{R}^p_{\geqq}) = \{\hat{y}\}$ (since $\hat{y} \in \mathcal{Y}_N$), a contradiction.

For y' from the claim we know $y^U \leq \hat{y} < y'$ and thus there is some $\lambda \in \Lambda^0$, and $\beta > 0$ such that $y' - y^U = \beta(1/\lambda_1, \ldots, 1/\lambda_p)$. Hence

$$\lambda_k(\hat{y}_k - y_k^U) < \lambda_k(y_k' - y_k^U) = \beta \tag{4.42}$$

for all $k = 1, \ldots, p$ and

$$\|\lambda \odot (\hat{y} - y^U)\|_\infty < \beta. \tag{4.43}$$

Choose $y(\mathsf{p}) \in \mathcal{A}(\lambda, \mathsf{p}, \mathcal{Y})$. Note that $\mathcal{A}(\lambda, \mathsf{p}, \mathcal{Y})$ is nonempty because $\mathcal{Y} + \mathbb{R}^p_{\geqq}$ is closed. We obtain

$$\begin{aligned} \|\lambda \odot (y(\mathsf{p}) - y^U)\|_\infty &\leq \|\lambda \odot (y(\mathsf{p}) - y^U)\|_\mathsf{p} \\ &\leq \|\lambda \odot (\hat{y} - y^U)\|_\mathsf{p} \\ &\rightarrow \|\lambda \odot (\hat{y} - y^U)\|_\infty \; < \; \beta, \end{aligned} \tag{4.44}$$

where we have used (P1), the definition of $\mathcal{A}(\lambda, \mathsf{p}, \mathcal{Y})$, and (P2), respectively.

This means we have $\|\lambda \odot (y(\mathsf{p}) - y^U)\|_\infty \leq \beta$, if p is sufficiently large. By the definition of the l_∞-norm

$$y_k(\mathsf{p}) - y_k^U \leq \frac{\beta}{\lambda_k} = y_k' - y_k^U \text{ for all } k = 1, \ldots, p, \tag{4.45}$$

i.e. $y(\mathsf{p}) \leq y'$ or $y(\mathsf{p}) \in (y' - \mathbb{R}^p_{\geqq}) \cap \mathcal{Y}$ and therefore, using the auxiliary claim, we can choose $y^\varepsilon := y(\mathsf{p})$ for sufficiently large p. □

We know that if $\mathcal{Y} + \mathbb{R}^p_{\geqq}$ is convex, $\mathsf{p} = 1$ will always work for $y(\mathsf{p}) \in \mathcal{A}(\lambda, \mathsf{p}, \mathcal{Y})$ and that $\mathsf{p} = \infty$ can be chosen for arbitrary sets. The proof of the second part of the theorem suggests that, if $\mathcal{Y}$ is not $\mathbb{R}_{\geq}$-convex, p has to be bigger than one. The value of p seems to be related to the degree of nonconvexity of $\mathcal{Y}$. An Example, where $1 < \mathsf{p} < \infty$ can be chosen to generate $\mathcal{Y}_N$ by solving (4.28) is given in Exercise 4.7.

At the end of this section we have two examples. The first one shows that the inclusion $\operatorname{cl} \mathcal{A}(\mathcal{Y}) \subset \mathcal{Y}_N$ may not be true. In the second we solve the problem from Example 4.18 by the compromise programming method.

Example 4.26. Let $\mathcal{Y} := \{y \in \mathbb{R}^2 : y_1^2 + (y_2 - 1)^2 \leq 1\} \cup \{y \in \mathbb{R}^2 : y_1 \geq 0,\ y_2 \geq -1\}$. Here the efficient set is $\mathcal{Y}_N = \{y \in \mathcal{Y} : y_1^2 + (y_2 - 1)^2 = 1, y_2 \leq 1; y_1 > -1\} \cup \{(0, -1)\}$, see Figure 4.8.

Therefore $0 \notin \mathcal{Y}_N$ but $0 \in \operatorname{cl} \mathcal{A}(\mathcal{Y})$. Note that the efficient points with $y_2 < 1$ and $y_1 < 0$ are all generated as optimal solutions of (4.28) with any choice of $y^U < (-1, -1)$ for appropriate λ and p. □

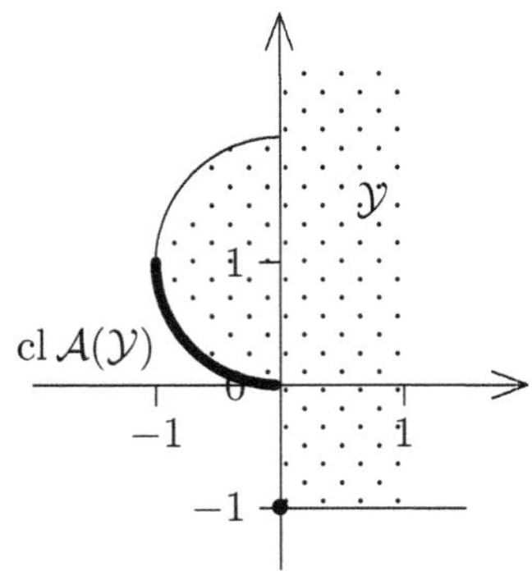

Fig. 4.8. cl $\mathcal{A}(\mathcal{Y})$ may contain dominated points.

Example 4.27. We apply the compromise programming method to the problem of Example 4.18:

$$\min(x^2 - 4, (x-1)^4)$$
$$\text{subject to } -x - 100 \leq 0.$$

Let $\lambda = (0.5, 0.5)$ and $\mathsf{p} = 2$. The ideal point is $y^I = (-4, 0)$ and we choose $y^U = (-5, -1)$. So (4.28) with $\mathsf{p} = 2$ and y^U as reference point is

$$\begin{aligned} \min \sqrt{\frac{1}{2}(x^2 - 4 + 5)^2 + \frac{1}{2}((x-1)^4 + 1)^2} \\ \text{subject to } -x - 100 \leq 0. \end{aligned} \tag{4.46}$$

Observing that the compromise programming objective is convex, that the problem is in fact unconstrained, and that the derivative of the objective function in (4.46) is zero if and only if the derivative of the term under the root is zero we set

$$\phi(x) = \frac{1}{2}(x^2+1)^2 + \frac{1}{2}((x-1)^4+1)^2$$

and compute

$$\begin{aligned} \phi'(x) &= (x^2+1)2x + ((x-1)^4+1) \cdot 4(x-1)^3 \\ &= 2x^3 + 2x + 4(x-1)^7 + 4(x-1)^3 \end{aligned}$$

From $\phi'(x) = 0$ we obtain $\hat{x} \approx 0.40563$ as unique minimizer. Theorem 4.21 confirms that $\hat{x} \in \mathcal{X}_E$. □

4.6 The Achievement Function Method

A certain class of real-valued functions $s_r : \mathbb{R}^p \to \mathbb{R}$, referred to as achievement functions, can be used to scalarize the MOP (4.1). The scalarized problem is given by

$$\begin{aligned} &\min s_R(f(x)) \\ &\text{subject to } x \in \mathcal{X}. \end{aligned} \tag{4.47}$$

Similar to distance functions discussed in Section 4.5 above, certain properties of achievement functions guarantee that problem (4.47) yields (weakly) efficient solutions.

Definition 4.28. *An achievement function $s_R : \mathbb{R}^p \to \mathbb{R}$ is said to be*

1. *increasing if for $y^1, y^2 \in \mathbb{R}^p$, $y^1 \leqq y^2$ then $s_R(y^1) \leq s_R(y^2)$,*
2. *strictly increasing if for $y^1, y^2 \in \mathbb{R}^p$, $y^1 < y^2$ then $s_R(y^1) < s_R(y^2)$,*
3. *strongly increasing if for $y^1, y^2 \in \mathbb{R}^p$, $y^1 \leq y^2$ then $s_R(y^1) < s_R(y^2)$.*

Theorem 4.29 (Wierzbicki (1986a,b)).

1. *Let an achievement function s_R be increasing. If $\hat{x} \in \mathcal{X}$ is a unique optimal solution of problem (4.47) then $\hat{x} \in \mathcal{X}_{sE}$.*
2. *Let an achievement function s_R be strictly increasing. If $\hat{x} \in \mathcal{X}$ is an optimal solution of problem (4.47) then $\hat{x} \in \mathcal{X}_{wE}$.*
3. *Let an achievement function s_R be strongly increasing. If $\hat{x} \in \mathcal{X}$ is an optimal solution of problem (4.47) then $\hat{x} \in \mathcal{X}_E$.*

We omit the proof, as it is very similar to the proofs of Theorems 4.20, 4.21 and Proposition 4.22, see Exercise 4.11.

Among many achievement functions satisfying the above properties we mention the strictly increasing function

$$s_R(y) = \max_{k=1,\ldots,p} \{\lambda_k(y_k - y_k^R)\}$$

and the strongly increasing functions

$$\begin{aligned} s_R(y) &= \max_{k=1,\ldots,p} \{\lambda_k(y_k - y_k^R)\} + \rho_1 \sum_{k=1}^{p} \lambda_k(y_k - y_k^R) \\ s_R(y) &= -\|y - y^R\|^2 + \rho_2 \|(y - y^R)_+\|^2, \end{aligned}$$

where $y^R \in \mathbb{R}^p$ is a reference point, $\lambda \in \mathbb{R}^p_>$ is a vector of positive weights, $\rho_1 > 0$ and sufficiently small, $\rho_2 > 1$ is a penalty parameter, and $(y - y^R)_+$ is a vector with components $\max\{0, y_k - r_k\}$ (Wierzbicki, 1986a,b).

4.7 Notes

In Guddat *et al.* (1985), Theorem 4.7 is also generalized for scalarizations in the form of problem (4.7) with an objective function being strictly increasing on $\mathbb{R}^p$ (cf. Definition 4.28). See also Exercise 4.10.

The formulation of the scalarized problem of Benson's method (4.10) has been used by Ecker and Hegner (1978) and Ecker and Kouada (1975) in multiobjective linear programming earlier. In fact, already Charnes and Cooper (1961) have formulated the problem and proved Theorem 4.14.

Some discussion of compromise programming that covers several aspects we neglected here can be found in Yu (1985). Two further remarks on the proof of Theorem 4.25 are in order. First, the statement remains true, if y^I is chosen as reference point. However, the proof needs modification (we have used $y > y^U$ in both parts). We refer to Sawaragi *et al.* (1985) for this extension. Second, we remark that the definition of the l_p-norms has never been used. Therefore the theorem is valid for any family of norms with properties (P1) – (P3). This fact has been used by several researchers to justify methods for generation of efficient solutions, e.g. Choo and Atkins (1983). Other norms used for compromise programming are include the augmented l_∞-norm in Steuer and Choo (1983); Steuer (1985) and the modified l_∞-norm by Kaliszewski (1987).

There are many more scalarization methods available in the literature than we can present here. They can roughly be classified as follows.

Weighting methods These include weighted sum method (Chapter 3), the weighted t-th power method White (1988), and the weighted quadratic method Tind and Wiecek (1999)

Constraint methods We have discussed the ε-constraint method (Section 4.1), the hybrid method (Section 4.2), the elastic constraint method (Section 4.3) and Benson's method (Section 4.4). See also Exercises 4.1, 4.2 and 4.10 for more.

Reference point methods The most important in this category are the compromise programming method of Section 4.5 and the (more general) achievement function method (Section 4.6). But goal programming (see e.g. Tamiz and Jones (1996)) and the weighted geometric mean approach of Lootsma *et al.* (1995) also fit in this category.

Direction based methods There is a wide variety of direction based methods, including the reference direction approach Korhonen and Wallenius (1988), the Pascoletti-Serafini method Pascoletti and Serafini (1984) and the gauge-based technique of Klamroth *et al.* (2002).

Of course, some methods can be associated with several of these categories. A survey with the most important results can be found in Ehrgott and Wiecek (2005).

Exercises

4.1. Suppose $\hat{x}$ is the unique optimal solution of

$$\min_{x \in \mathcal{X}} \sum_{k=1}^{p} \lambda_i f_i(x)$$

with $\lambda \in \mathbb{R}^p_{\geq}$. Then there exists some $\hat{\varepsilon} \in \mathbb{R}^p$ such that $\hat{x}$ is an optimal solution of (4.3) for all $j = 1, \ldots, p$.

4.2 (Corley (1980)). Show that $\hat{x} \in \mathcal{X}_E$ if and only if there are $\lambda \in \mathbb{R}^p_{>}$ and $\varepsilon \in \mathbb{R}^p$ such that $\hat{x}$ is an optimal solution of

$$\begin{aligned} &\min_{x \in \mathcal{X}} \sum_{k=1}^{p} \lambda_k f_k(x) \\ &\text{subject to } f(x) \leq \varepsilon. \end{aligned} \tag{4.48}$$

4.3. Show, by choosing the parameters μ and ε in (4.8) appropriately, that both the weighted sum problem (4.2) and the ε-constraint problem (4.3) are special cases of (4.8).

4.4. Consider the following bicriterion optimization problem.

$$\begin{aligned} \min \quad & -6x_1 - 4x_2 \\ \min \quad & \qquad -x_1 \\ \text{s.t.} \quad & \quad x_1 + x_2 \leq 100 \\ & \quad 2x_1 + x_2 \leq 150 \\ & \quad x_1, x_2 \geq 0. \end{aligned}$$

Use $\varepsilon = 0$ and solve the ε-constraint problem (4.3) with $j = 1$. Check if the optimal solution $\hat{x}$ of $P_1(0)$ is efficient using Benson's test (4.10).

4.5. Consider $\min_{x \in \mathcal{X}} (f_1(x), \ldots, f_p(x))$ and assume $0 < \min_{x \in \mathcal{X}} f_k(x)$ for all $k = 1, \ldots, p$. Prove that $x \in \mathcal{X}_{wE}$ if and only if x is an optimal solution of

$$\min_{x \in \mathcal{X}} \max_{k=1,\ldots,p} \lambda_k f_k(x)$$

for some $\lambda \in \mathbb{R}^p_{>}$.

4.6. Find an efficient solution of the problem of Exercise 4.4 using the compromise programming method. Use $\lambda = (1/2, 1/2)$ and find an optimal solution of (4.28) for $\mathsf{p} = 1, 2, \infty$.

4.7. Consider finding a compromise solution by maximizing the distance to the nadir point.

1. Let $\|\cdot\|$ be a norm. Show that an optimal solution of the problem
$$\begin{aligned}&\max_{x\in\mathcal{X}} \|f(x) - y^N\| \\ &\text{subject to } f_k(x) \leq y_k^n \quad k = 1, \ldots, p\end{aligned} \tag{4.49}$$
is weakly efficient. Give a condition under which an optimal solution of (4.49) is efficient.
2. Another possibility is to solve
$$\begin{aligned}&\max_{x\in\mathcal{X}} \min_{k=1,\ldots,p} |f_k(x) - y_k^N| \\ &\text{subject to} f_k(x) \leq y_k^n, \quad k = 1, \ldots, p.\end{aligned} \tag{4.50}$$
Prove that an optimal solution of (4.50) is weakly efficient.

4.8. Let $\mathcal{Y} = \{y \in \mathbb{R}^2 : y_1 + y_2 \geq 1,\ 0 \leq y_1 \leq 1\}$. Show that $\hat{y} = (0, 1) \in \mathcal{Y}_{pN}$ according to Benson's definition, but that there is no $\lambda \in \Lambda^0$ such that $\hat{y} \in \mathcal{A}(\lambda, \infty, \mathcal{Y})$, if y^I is used as reference point in (4.28).

4.9. Let $\mathcal{Y} = \left\{(y_1, y_2) \in \mathbb{R}^2_{\geqq} : y_1^2 + y_2^2 \geq 1\right\}$. Verify that there is $1 < p < \infty$ such that
$$\mathcal{Y}_N = \bigcup_{\lambda\in\Lambda^0} \mathcal{A}(\lambda, p, \mathcal{Y}).$$
Choose either y^I or y^U in the definition of $\mathcal{A}(\lambda, p, \mathcal{Y})$ and N_p^λ.

4.10 (Soland (1979)). A function $s : \mathbb{R}^p \to \mathbb{R}$ is called strongly increasing, if for $y^1, y^2 \in \mathbb{R}^p$ with $y^1 \leq y^2$ the inequality $s(y^1) < s(y^2)$ holds (see Definition 4.28).

Consider the following single objective optimization problem, where $\varepsilon \in \mathbb{R}^p$ and $f : \mathbb{R}^n \to \mathbb{R}^p$.
$$\begin{aligned}&\min s(f(x)) \\ &\text{subject to } x \in \mathcal{X} \\ &\qquad f(x) \leq \varepsilon.\end{aligned} \tag{4.51}$$

Let s be strongly increasing. Prove that $x \in \mathcal{X}_E$ if and only if there is $\varepsilon \in \mathbb{R}^p$ such that x is an optimal solution of (4.51) with finite objective value.

4.11. Prove Theorem 4.29.

4.12. An achievement function $s_R : \mathbb{R} \to \mathbb{R}$ is called order representing if s_R is strictly increasing (Definition 4.28) for any $y^R \in \mathbb{R}^p$ and in addition

$$\{y \in \mathbb{R}^p : s_R(y) < 0\} = y^R - \mathbb{R}^p_>$$

holds for all $y^R \in \mathbb{R}^p$. Which of the functions

$$s_R(y) = d(y, y^R) = \|y - y^R\|,$$
$$s_R(y) = \max_{k=1,\ldots,p} \{\lambda_k(y_k - y_k^R)\},$$
$$s_R(y) = \max_{k=1,\ldots,p} \{\lambda_k(y_k - y_k^R)\} + \rho_1 \sum_{k=1}^{p} \lambda_k(y_k - y_k^R)$$
$$s_R(y) = -\|y - y^R\|^2 + \rho_2\|(y - y^R)_+\|^2$$

is order representing?

4.13. Show that Benson's problem (4.10), the weighted sum scalarization (4.2) with $\lambda \in \mathbb{R}^p_>$, the compromise programming problem (4.28) with $1 \leq p < \infty$ and $\lambda \in \mathbb{R}^p_>$, and Corley's problem (4.48) (see Exercise 4.2) can all be seen as special cases of (4.51).

5

Other Definitions of Optimality – Nonscalarizing Methods

The concept of efficiency and its variants are by far the most important definitions of optimality in multicriteria optimization. Their extensive coverage in Chapters 2 to 4 reflects this fact. But as we have seen in Chapter 1 with the discussion of orders and the classification of multicriteria problems this is not the end of the story. Other choices of orders and model maps give rise to different classes of multicriteria optimization problems. In this chapter we shall discuss some of these. Specifically we address lexicographic optimality, max-ordering optimality, and finally a combination of the two, lexicographic max-ordering optimality. Lexicographic max-ordering defines a class of problems with many interesting features. Of particular interest will be the relationships between optimal solutions of these problems and efficient solutions. In this way they can be seen as nonscalarizing methods for finding efficient solutions. We do not study these problems out of curiosity about their theory, however.

Lexicographic optimization problems arise naturally when conflicting objectives exist in a decision problem but for reasons outside the control of the decision maker the objectives have to be considered in a hierarchical manner. Weber *et al.* (2002) describe the optimization of water resources planning for Lake Verbano (Lago Maggiore) in northern Italy. The goal is to determine an optimal policy for the management of the water supply over some planning horizon. The objectives are to maximize flood protection, minimize supply shortage for irrigation, and maximization of electricity generation. This order of objectives is prescribed by law, so that the problem indeed has a lexicographic nature. The actual formulation of the problem is via stochastic dynamic programming, which is beyond the scope of this book, and we omit it.

A common application of max-ordering problems is location planning. Ehrgott (2002) describes the problem of locating rescue helicopters in South Tyrol, Italy. The objective in this problem is to minimize the distance between potential accident sites and the closest helicopter location. In order to minimize worst case response times in an emergency, the problem can be formulated as follows. Let $x^h = (x_1^h, x_2^h), h \in \mathcal{H}$ denote variables that define helicopter locations and $(a_1^k, a_2^k), k \in 1, \dots, p$ the potential emergency sites. Optimal helicopter locations are found by solving

$$\min_{x \in \mathbb{R}^{2|\mathcal{H}|}} \max_{k \in 1,\dots,p} f_k(x)$$

where $f_k(x)$ is defined as

$$f_k(x) = \min_{h \in \mathcal{H}} w_k ||x^h - a^k||_2.$$

Georgiadis *et al.* (2002) describe the problem of picking routes and associated route bandwidth in a computer network so that bandwidth request is satisfied and the network is in a balanced state, i.e. the bandwidth allocation results in an even spreading of the load to various links of the network. They formulate this problem as a lexicographic max-ordering network flow problem. Variables x_{ij} denote the load on links ij. Let $C_{ij}(x_{ij})$be a function that describes the link cost and $b(i)$ be the bandwidth demand at a node of the network. Then the balanced bandwidth allocation problem is

$$\begin{aligned} &\min \operatorname{sort}(C_{ij}(x_{ij})) \\ \text{subject to } &\sum_j x_{ij} - \sum_j x_{ji} = b(i), \quad i \in N \\ &x_{ij} \geq 0. \end{aligned}$$

These examples should give an indication that the lexicographic, max-ordering, and lexicographic max-ordering classes are very relevant for practical applications.

Before we start our investigations, we state one general assumption. Throughout this chapter we shall assume that the single objective optimization problems $\min_{x \in \mathcal{X}} f_k(x)$ have optimal solutions for $k = 1, \dots, p$ and that $\mathcal{X}_E \neq \emptyset$, unless stated otherwise.

5.1 Lexicographic Optimality

In lexicographic optimization we consider the lexicographic order when comparing objective vectors in criterion space. As for efficiency, the model map is the identity map, so in terms of classification we deal with problems of the class $(\bullet/\text{id}/(\mathbb{R}^p, <_{\text{lex}}))$. An optimal solution $\hat{x}$ of such a problem is called lexicographically optimal and $f(\hat{x})$ is a lexicographically minimal vector in $\mathcal{Y} = f(\mathcal{X})$.

We can also write the lexicographic optimization problem with a "lexmin" operator as:

$$\operatorname*{lexmin}_{x \in \mathcal{X}}(f_1(x), \ldots, f_p(x)). \tag{5.1}$$

Definition 5.1. *A feasible solution $\hat{x} \in \mathcal{X}$ is* lexicographically optimal *or a* lexicographic solution *if there is no $x \in \mathcal{X}$ such that $f(x) <_{\text{lex}} f(\hat{x})$.*

Recall that $y^1 <_{\text{lex}} y^2$ if $y^1_q < y^2_q$ where $q = \min\{k : y^1_k \neq y^2_k\}$ and that the lexicographic order is total. Therefore, in addition to Definition 5.1, which is a "negative" definition of optimality, we can state that $\hat{x} \in \mathcal{X}$ is lexicographically optimal, if

$$f(\hat{x}) \leqq_{\text{lex}} f(x) \text{ for all } x \in \mathcal{X}.$$

First, we establish the relationship between lexicographically optimal solutions and efficient solutions.

Lemma 5.2. *Let $\hat{x} \in \mathcal{X}$ be such that $f(\hat{x}) \leqq_{\text{lex}} f(x)$ for all $x \in \mathcal{X}$. Then $\hat{x} \in \mathcal{X}_E$.*

Proof. Suppose that $\hat{x}$ is not efficient. Then there is an $x \in \mathcal{X}$ such that $f(x) \leq f(\hat{x})$. So for some $k \in \{1, \ldots, p\}$ we have $f_k(x) < f_k(\hat{x})$. Defining $q := \min\{k : f_k(x) < f_k(\hat{x})\}$ we get that $f_k(x) = f_k(\hat{x})$ for $k = 1, \ldots, q-1$ and $f_q(x) < f_q(\hat{x})$. Therefore $f(x) <_{\text{lex}} f(\hat{x})$ contradicting lexicographic optimality of $\hat{x}$. □

While the essential feature of efficiency is the existence of tradeoff between objectives, lexicographic optimality implies a ranking of the objectives in the sense that optimization of f_k is only considered once optimality for objectives $\{1, \ldots, k-1\}$ has been established. That means objective 1 has the highest priority, and only in the case of multiple optimal solutions objectives f_2 and further objectives are considered. This priority ranking implies the absence of tradeoffs between criteria. An improvement in an objective f_k can never compensate the deterioration of any $f_i, i < k$.

The hierarchy among criteria allows us to solve lexicographic optimization problems sequentially, minimizing one objective f_k at a time and using optimal objective values of $f_i, i < k$ as constraints, as shown in Algorithm 5.1.

Algorithm 5.1 (Lexicographic Optimization)
Input: Feasible set $\mathcal{X}$ and objective functions f.
Initialization: Define $\mathcal{X}_1 := \mathcal{X}$ and $k := 1$.

Solve the single objective optimization problem

$$\min_{x \in \mathcal{X}_k} f_k(x). \tag{5.2}$$

While $k \leq p$ do

If (5.2) has a unique optimal solution $\hat{x}_k$, STOP, $\hat{x}_k$ is the unique optimal solution of the lexicographic optimization problem.
If (5.2) is unbounded, STOP, the lexicographic optimization problem is unbounded.
If $k = p$, STOP, the set of optimal solutions of the lexicographic optimization problem is

$$\left\{ x \in \mathcal{X}_p : f_p(x) = \min_{x \in \mathcal{X}_p} f_p(x) \right\}.$$

Let $\mathcal{X}_{k+1} := \{x \in \mathcal{X}_k : f_k(x) = \min_{x \in \mathcal{X}_k} f_k(x)\}$ and let $k := k + 1$.
End while.
Output: Set of lexicographically optimal solutions.

In applications of lexicographic optimization, it will often be reasonable to assume that all objectives are bounded over the feasible set $\mathcal{X}$. However, Algorithm 5.1 will also give a correct solution if f_k is unbounded over $\mathcal{X}$, but bounded over $\mathcal{X}_k$. Note that, if a problem $\min_{x \in \mathcal{X}_k} f_k(x)$ is unbounded, it is not possible to define $\mathcal{X}_{k+1}$.

We consider problem (5.2) to justify correctness of Algorithm 5.1.

Proposition 5.3. *If $\hat{x}$ is a unique optimal solution of (5.2) with $k < p$, or if $\hat{x}$ is an optimal solution of (5.2) with $k = p$ then $f(\hat{x}) \leqq_{\text{lex}} f(x)$ for all $x \in \mathcal{X}$.*

Proof. Consider $k < p$ in the first, and $k = p$ in the second case. Suppose there is an $x \in \mathcal{X}$ with $f(x) <_{\text{lex}} f(\hat{x})$. By definition of the problem (5.2) in iteration i and its feasible set $\mathcal{X}_i$ as optimal solutions of (5.2) in iteration $i-1$ we cannot have that $f_i(x) < f_i(\hat{x})$ for any $i \leq k-1$. Therefore $f_i(x) = f_i(\hat{x})$ for $i = 1, \ldots, k-1$. Thus, $f_j(x) < f_j(\hat{x})$ must hold for some $k \leq j \leq p$. If $k < p$ this means that either $\hat{x}$ is not optimal for (5.2) or has has at least two optimal solutions in iteration k, contradicting the assumption. If $k = p$, we must have $f(x) = f(\hat{x})$ contradicting $f(x) <_{\text{lex}} f(\hat{x})$. □

Note also, that if $\hat{x}$ is a unique optimal solution of a problem (5.2) then $\hat{x} \in \mathcal{X}_{sE}$. To see this, consider $x \in \mathcal{X}$ such that $f_k(x) \leq f_k(\hat{x})$ for all $k = 1, \ldots, p$, which due to $\hat{x}$ being efficient can only hold with $f(x) = f(\hat{x})$, and thus by uniqueness of $\hat{x}$ implies $x = \hat{x}$.

Proposition 5.4. *If x is a unique optimal solution of (5.2) for some $k \in \{1, \ldots, p\}$, then $x \in \mathcal{X}_{sE}$.*

Up to now we have used the ranking of the objectives given by the indices, i.e. $f(x) = (f_1(x), f_2(x), \ldots, f_p(x))$. There is no reason to stick with that order. Indeed, we may well choose another ranking of the objectives and apply lexicographic optimization. To deal with this, we look at all possible permutations of the indices $(1, \ldots, p)$. Let $\pi : \{1, \ldots, p\} \to \{1, \ldots, p\}$ be a permutation and consider the permutation $(f_{\pi(1)}, \ldots, f_{\pi(p)})$ of the objective functions. We also use π to denote the model map

$$\pi : \mathbb{R}^p \to \mathbb{R}^p, \quad y \mapsto (y_{\pi(1)}, \ldots, y_{\pi(p)})$$

that defines this permutation of components in the objective function vector. We shall naturally use $\pi(y)$ and $\pi(f)$ to denote the permutation of y and f. As in Lemma 5.2 we can show that optimal solutions of $(\mathcal{X}, f, \mathbb{R}^p)/\pi/(\mathbb{R}^p, <_{\text{lex}})$ are efficient. We denote by Π the set of all permutations of $\{1, \ldots, p\}$ and by

$$\mathcal{X}_\Pi := \bigcup_{\pi \in \Pi} \operatorname{Opt}(\mathcal{X}, f, \mathbb{R}^p)/\pi/(\mathbb{R}^p, <_{\text{lex}})$$

the set of solutions which are lexicographically optimal for any problem $(\mathcal{X}, f, \mathbb{R}^p)/\pi/(\mathbb{R}^p, <_{\text{lex}})$ with $\pi \in \Pi$.

Definition 5.5. *A feasible solution $\hat{x} \in \mathcal{X}$ is a* global lexicographic solution *if there is a $\pi \in \Pi$ such that $\pi(f(\hat{x})) \leq_{\text{lex}} \pi(f(x))$ for all $x \in \mathcal{X}$.*

Then we have .

Proposition 5.6. $\mathcal{X}_\Pi \subset \mathcal{X}_E$.

Example 5.7. It is quite obvious that the inclusion in Proposition 5.6 is strict in general. Let $\mathcal{X} = [0, 1]$ and $f_1(x) = x, \ f_2(x) = 1 - x$.

Clearly $\mathcal{X}_E = \mathcal{X}$. The optimal solution of $([0,1], f, \mathbb{R}^2)/\text{id}/(\mathbb{R}^2, <_{\text{lex}})$ is $\hat{x} = 0$, the optimal solution of $([0,1], f, \mathbb{R}^2)/\pi/(\mathbb{R}^2, <_{\text{lex}})$, where $\pi(y_1, y_2) = (y_2, y_1)$ is $\hat{x} = 1$. Therefore $\mathcal{X}_\Pi = \{0, 1\} \neq \mathcal{X}_E$.

Moreover, because of the uniqueness of both lexicographically optimal solutions in this example $\mathcal{X}_\Pi \subset \mathcal{X}_{sE}$, and again the inclusion is strict, as $\mathcal{X}_E = \mathcal{X}_{sE}$. □

Note that finding $\mathcal{X}_\Pi$ is usually computationally very expensive. It involves solving $|\Pi| = p!$ lexicographic problems, which, using Algorithm 5.1, amounts to $p \cdot p!$ single objective problems. However, if $\mathcal{X}$ is a finite set, finding $\mathcal{X}_\Pi$ can be done in time polynomial in $|\mathcal{X}|$ and p, as we shall see in Section 8.2.

5.2 Max-Ordering Optimality

The second problem class we consider is $\bullet / \max / (\mathbb{R}, \leq)$. Problems of this class can be written as follows

$$\min_{x \in \mathcal{X}} \max_{k=1,\ldots,p} f_k(x), \tag{5.3}$$

and are called max-ordering optimization problems. Let $\mathcal{X}_{MO}$ denote the set of optimal solutions of a max-ordering problem. We encountered such min-max problems in Chapter 4, as special cases of compromise programming problems, when the l_∞-norm is used as distance measure. From these results we easily establish the relationships with efficiency.

Definition 5.8. *A feasible solution* $\hat{x} \in \mathcal{X}$ *is* max-ordering optimal *or a* max-ordering solution *if there is no* $x \in \mathcal{X}$ *such that* $\max_{k=1,\ldots,p} f_k(x) < \max_{k=1,\ldots,p} f_k(\hat{x})$.

Proposition 5.9. *An optimal solution of the max-ordering problem* (5.3) *is weakly efficient but not necessarily efficient.*

Proof. The easy proof and example are left to the reader as Exercise (5.2) □

From Proposition 5.9 we know that $\mathcal{X}_{\text{MO}} \subset \mathcal{X}_{wE}$. By our general assumption that $\min_{x \in \mathcal{X}} f_k(x)$ exists for all $k = 1, \ldots, p$ the max-ordering optimization problem is bounded. Let $y^U < y^I$ be a utopian point. Observe that the efficient set $\mathcal{X}_E$ of the multicriteria optimization problems with objectives $(f_1, \ldots, f_p)$ and $(f_1 - y_1^U, \ldots, f_p - y_p^U)$ is the same. This is true because the subtraction of constants affects only $\mathcal{Y}$ – $\mathcal{Y}$ and $\mathcal{Y}_N$ are translated by y^U – not $\mathcal{X}$. The following result has already been shown as Theorem 4.24:

Proposition 5.10. *A feasible solution* $\hat{x} \in \mathcal{X}$ *is weakly efficient if and only if there is some* $\lambda \in \mathbb{R}^p_>$ *such that* $\hat{x}$ *is an optimal solution of*

$$\min_{x \in \mathcal{X}} \max_{k=1,\ldots,p} \lambda_k (f_k(x) - y_k^U).$$

Therefore $\mathcal{X}_{wE}$ can be determined through the solution of max-ordering problems. Obviously, results concerning efficiency optimality must be weaker. Yet we can show that at least one optimal solution of the max-ordering problem is efficient.

Proposition 5.11. *Suppose that* $\mathcal{Y}_N$ *is externally stable, and that a max-ordering solution exists. Then* $\mathcal{X}_{\text{MO}} \cap \mathcal{X}_E \neq \emptyset$. *If there is* $y \in \mathcal{Y}$ *such that* $f(x) = y$ *for all* $x \in \mathcal{X}_{\text{MO}}$ *then* $\mathcal{X}_{\text{MO}} \subset \mathcal{X}_E$.

Proof. Let $\hat{x} \in \mathcal{X}_{\text{MO}}$ and suppose $\hat{x} \notin \mathcal{X}_E$. Then, because of external stability of $\mathcal{Y}_N$, there is some $x \in \mathcal{X}_E$ such that $f(x) \leq f(\hat{x})$. Thus $\max_{k=1,\ldots,p} f_k(x) \leq \max_{k=1,\ldots,p} f_k(\hat{x})$. Since $\hat{x}$ is max-ordering optimal, equality must hold, and $x \in \mathcal{X}_{\text{MO}}$, too. The second part follows from uniqueness of $f(x)$ for all $x \in \mathcal{X}_{\text{MO}}$. □

We will come back to this result in Section 5.3, where we strengthen max-ordering optimality by combining it with lexicographic optimality in a way that guarantees that all optimal solutions are also efficient. We shall then see how to find part of the intersection $\mathcal{X}_{\text{MO}} \cap \mathcal{X}_E$.

Next, we show that the max-ordering problem (5.3) can be solved as a single objective optimization problem, and that max-ordering solutions have a geometric characterization similar to the one given for efficient solutions in Theorem 2.30.

If we introduce a variable z to stand for $\max_{k=1,\ldots,p} f_k(x)$ we can rewrite (5.3) as

$$\begin{aligned} &\min z \\ \text{subject to } &f_k(x) \leq z \quad k = 1, \ldots, p \\ &x \in \mathcal{X}. \end{aligned} \tag{5.4}$$

Reformulation (5.4) indicates that max-ordering solutions can be characterized through level sets $\mathcal{L}^k_{\leq}(z) = \{x \in \mathcal{X} : f_k(x) \leq z\}$. This geometric characterization is given in Proposition 5.12.

Proposition 5.12. *A feasible solution $\hat{x} \in \mathcal{X}$ is max-ordering optimal, i.e. $\hat{x} \in \mathcal{X}_{\text{MO}}$, if and only if*

$$\bigcap_{k=1}^{p} \mathcal{L}^p_{\leq}\left(\max_{k=1,\ldots,p} f_k(\hat{x})\right) \neq \emptyset \tag{5.5}$$

and for all $z < \max_{k=1,\ldots,p} f_k(\hat{x})$ it holds that $\cap_{k=1}^{p} \mathcal{L}^k_{\leq}(z) = \emptyset$.

In the following we show a case for which the max-ordering problem can be solved easily and give lower and upper bounds for the general case. Let y^I be the ideal point of the multicriteria optimization problem defined by $\mathcal{X}$ and f and let x^k, $k = 1, \ldots, p$ be such that $y^I_k = f_k(x^k)$.

In the max-ordering problem only the worst objective value is considered for each feasible point x. It may happen that there is an objective f_k which is worst for each $x \in \mathcal{X}$: $f_k(x) > f_i(x)$ for all $i \neq k$. In this case the objective function f_k is considerably worse than the others, and the max-ordering problem is "easy" to solve, by simply minimizing that objective.

More precisely, note that for each $k = 1, \ldots, p$ we have

$$y^I_k = f_k(x^k) = \min_{x \in \mathcal{X}} f_k(x) \leq \min_{x \in \mathcal{X}} \max_{i=1,\ldots,p} f_i(x) \leq \max_{i=1,\ldots,p} f_i(x^k). \qquad (5.6)$$

Proposition 5.13. *If for some x^k with $f_k(x^k) = y^I_k$ it holds that that $f_i(x^k) \leq y^I_k$ for all $i = 1, \ldots, p$ then $x^k \in \mathcal{X}_{\mathrm{MO}}$ and the optimal objective value of the max-ordering problem is y^I_k.*

Proof. The assumption $f_i(x^k) \leq y^I_k$ for all $i = 1, \ldots, p$ implies

$$\max_{i=1,\ldots,k} f_i(x^k) \leq y^I_k.$$

Therefore (5.6) holds with equalities, i.e.

$$f_k(x^k) = y^I_k = \min_{x \in \mathcal{X}} \max_{i=1,\ldots,p} f_i(x).$$

□

If the condition of Proposition 5.13 does not apply inequality (5.6) can be used to obtain lower and upper bounds on the optimal value of the max-ordering problem. Taking the maximum over k on the left we obtain

$$\max_{k=1,\ldots,p} f_k(x^k) \leq \min_{x \in \mathcal{X}} \max_{i=1,\ldots,p} f_i(x).$$

Now taking first the minimum over all optimal solutions x^k of the single objective problem $\min_{x \in \mathcal{X}} f_k(x)$ and then the minimum over all $k \in \{1, \ldots, p\}$ on the right we get

$$\max_{k=1,\ldots,p} f_k(x^k) \leq \min_{x \in \mathcal{X}} \max_{i=1,\ldots,p} f_i(x) \leq \min_{k=1,\ldots,p} \min_{x^k \in \mathcal{X}^k} \max_{i=1,\ldots,p} f_i(x^k) \qquad (5.7)$$

where $\mathcal{X}^k = \{x \in \mathcal{X} : f_k(x) = \min_{x \in \mathcal{X}} f_k(x)\}$.

Another lower bound is derived from consideration of minimizing weighted sums of the objectives. Let the set of weights Λ be defined as usual, namely

$$\Lambda = \left\{ \lambda \in \mathbb{R}^p_{\geqq} : \sum_{k=1}^p \lambda_k = 1 \right\}.$$

We obtain the following result.

Proposition 5.14.

$$\max_{\lambda \in \Lambda} \min_{x \in \mathcal{X}} \sum_{k=1}^p \lambda_k f_k(x) \leq \min_{x \in \mathcal{X}} \max_{k=1,\ldots,p} f_k(x).$$

Proof. For each $x \in \mathcal{X}$ and each $\lambda \in \Lambda$ it holds that

$$\sum_{k=1}^{p} \lambda_k f_k(x) \le \sum_{k=1}^{p} \lambda_k \max_{i=1,\ldots,p} f_i(x) \le \max_{i=1,\ldots,p} f_i(x). \tag{5.8}$$

Taking minima over $x \in \mathcal{X}$ on both sides yields that for each $\lambda \in \Lambda$

$$\min_{x\in\mathcal{X}} \sum_{k=1}^{p} \lambda_k f_k(x) \le \min_{x\in\mathcal{X}} \max_{i=1,\ldots,p} f_i(x). \tag{5.9}$$

Since the right hand side in (5.9) is independent of λ the result follows. □

Note that by much the same argument, taking first a maximum over $\lambda \in \Lambda$ on the left and then the minimum over $x \in \mathcal{X}$ on both sides of (5.8), it follows also that

$$\min_{x\in\mathcal{X}} \max_{\lambda\in\Lambda} \sum_{k=1}^{p} \lambda_k f_k(x) \le \min_{x\in\mathcal{X}} \max_{k=1,\ldots,p} f_k(x), \tag{5.10}$$

for a similar lower bound.

We will not go into any further detail of max-ordering optimization here, and continue with a stronger version of it.

5.3 Lexicographic Max-Ordering Optimization

As we have seen, an optimal solution of a max-ordering optimization problem is not necessarily efficient, because the max-ordering optimality concept considers only one of the p objective values at each $x \in \mathcal{X}$, namely the worst. A straightforward idea is to extend this to consider the second worst objective, the third worst objective, etc. in the case that the max-ordering problem has several optimal solutions. This approach is similar to lexicographic optimization and considers a ranking of the objective values $f_1(x), \ldots, f_p(x)$. The difference is that the ranking is from worst to best value and thus depends on x.

We call the result lexicographic max-ordering optimality, because it is a combination of max-ordering and lexicographic optimality, where the lexicographic order is applied to a nonincreasingly ordered sequence of the objectives.

Definition 5.15. *1. For $y \in \mathbb{R}^p$ let* $\text{sort}(y) := (\text{sort}_1(y), \ldots, \text{sort}_p(y))$ *such that* $\text{sort}_1(y) \ge \ldots \ge \text{sort}_p(y)$ *be a function that reorders the components of y in a nonincreasing way.*

2. *A feasible solution* $\hat{x} \in \mathcal{X}$ *is called a* lexicographic max-ordering solution (lex-MO solution) *if*

$$\mathrm{sort}(f(\hat{x})) \leqq_{\mathrm{lex}} \mathrm{sort}(f(x)) \text{ for all } x \in \mathcal{X}. \tag{5.11}$$

A lexicographic max-ordering optimization problem can be written as

$$\min_{x \in \mathcal{X}} \mathrm{sort}(\theta(f(x))). \tag{5.12}$$

According to this definition we apply a mapping sort : $\mathbb{R}^p \to \mathbb{R}^p$ to the objective vectors $f(x)$, which reorders the components, and apply the lexicographic order to compare reordered objective vectors. In case of ties, i.e. equal components of $f(x)$, we assume that the order is given by the index of objective functions.

This means that sort is used as model map and the lexicographic order for comparison. Thus a lexicographic max-ordering problem is denoted, in the classification of Section 1.5, by

$$(\mathcal{X}, f, \mathbb{R}^p) / \mathrm{sort} / (\mathbb{R}^p, <_{\mathrm{lex}}). \tag{5.13}$$

The set of optimal solutions will be denoted by $\mathcal{X}_{\text{lex-MO}}$ and its image in objective space by $\mathcal{Y}_{lexMO} = f(\mathcal{X}_{\text{lex-MO}})$. Because the lexicographic order is total, it is clear that there is only one optimal value, i.e.

$$|\,\mathrm{sort}(\mathcal{Y}_{\text{lex-MO}})| = |\{\mathrm{sort}(f(x)) : x \in \mathcal{X}_{\text{lex-MO}}\}| = 1. \tag{5.14}$$

This unique optimal value may, however, be attained for several $x \in \mathcal{X}$. There might even be several $y \in \mathcal{Y}_{\text{lex-MO}}$, which after resorting are equal to this unique optimal value.

In this section we will show that lexicographic max-ordering can be used to find the efficient set. We will see how lexicographic max-ordering problems can be solved when $\mathcal{X}$ is convex and the objective functions are convex functions. And finally we establish an axiomatic characterization of lexicographic max-ordering, which identifies in which situations a multicriteria problem must be considered as belonging to the lex-MO class.

That lexicographic max-ordering really extends max-ordering is shown next.

Theorem 5.16. *The following relationship between lex-MO solutions, efficient solutions, and max-ordering solutions holds.*

$$\mathcal{X}_{\text{lex-MO}} \subset \mathcal{X}_E \cap \mathcal{X}_{\text{MO}} \tag{5.15}$$

and $\mathcal{X}_{\text{lex-MO}} = \mathcal{X}_E \cap \mathcal{X}_{\text{MO}}$ *if* $f(x)$ *is the same for all* $x \in \mathcal{X}_{\text{MO}}$.

Proof. Let $x \in \mathcal{X}_{\text{lex-MO}}$. First, assume that $x \notin \mathcal{X}_E$. Then we can find $x' \in \mathcal{X}$ such that $f(x') \leq f(x)$. Reordering the components of $f(x')$ and $f(x)$ nonincreasingly it follows that $\text{sort}(f(x')) \leq_{\text{lex}} \text{sort}(f(x))$ and $\text{sort}(f(x')) \neq \text{sort}(f(x))$ because $f(x') \neq f(x)$. This is a contradiction to $x \in \mathcal{X}_{\text{lex-MO}}$.

Second, assume that $x \notin \mathcal{X}_{\text{MO}}$. In this case we can find $x' \in \mathcal{X}$ such that $\max_{k=1,\ldots,p} f_k(x') < \max_{k=1,\ldots,p} f_k(x)$. But with the definition of sort this is the same as $\text{sort}_1(f(x')) < \text{sort}_1(f(x))$, which clearly implies $\text{sort}(f(x')) <_{\text{lex}} \text{sort}(f(x))$ and thus again contradicts $x \in \mathcal{X}_{\text{lex-MO}}$.

The equality follows from (5.15) and from Proposition 5.11. □

Inclusion (5.15) in Theorem 5.16 indicates that the intersection of the efficient set and $\mathcal{X}_{\text{MO}}$ does in general not only contain just the lex-MO solutions. Example 5.17 shows that this is indeed the case.

Example 5.17. Consider problem data with feasible set $\mathcal{X} = \{a, b, c, d, e, f\}$, for which the objective function values are explicitly given as shown in Table 5.1.

Table 5.1. Feasible solutions and objective values in Example 5.17.

x	$f(x)$	sort $f(x)$
a	(1,3,8,2,4)	(8,4,3,2,1)
b	(4,3,8,1,1)	(8,4,3,1,1)
c	(7,5,4,6,1)	(7,6,5,4,1)
d	(3,7,4,6,5)	(7,6,5,4,3)
e	(4,7,5,6,5)	(7,6,5,5,4)
f	(5,6,7,3,8)	(8,7,6,5,3)

The sorted objective vectors are also shown for convenience. It is easily seen that $\mathcal{X}_{\text{MO}} = \{c, d, e\}$, that $\mathcal{X}_E = \{a, b, c, d, f\}$, and that $\mathcal{X}_{\text{lex-MO}} = \{c\}$. Therefore $\mathcal{X}_{\text{lex-MO}} \subset \mathcal{X}_{\text{MO}} \cap X_{\text{Par}}$, but the inclusion is strict. Note that lexicographically optimal solutions with respect to all permutations are $\{a, b, c, d\}$, so that $\mathcal{X}_\Pi \subset X_{\text{Par}}$ and $\mathcal{X}_{\text{lex-MO}} \subset \mathcal{X}_\Pi$ and both of these inclusions are strict. The relationship between these sets is illustrated in Figure 5.1 □

It is important to note that lex-MO solutions are not necessarily lexicographically optimal, because although sort defines a permutation of $f(x)$, it is one which depends on x, see Example 5.7, where $\mathcal{X}_{\text{lex-MO}} = \{0.5\}$ and $\mathcal{X}_\Pi = \{0, 1\}$.

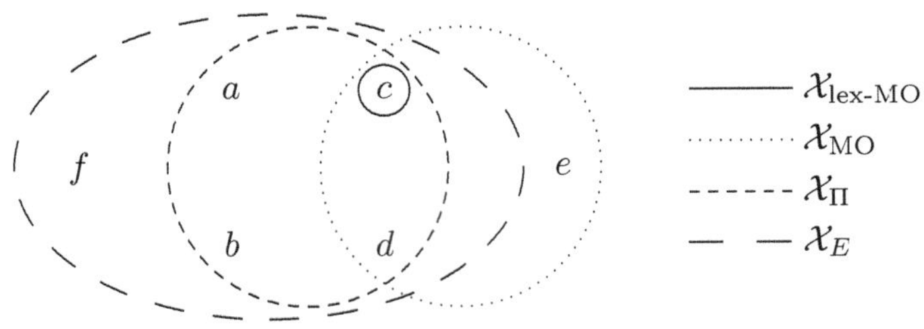

Fig. 5.1. Relationships between optimal solutions according to different MCO classes.

We have shown that lexicographic max-ordering solutions are efficient. If the preimage of some $y \in \mathcal{Y}_{\text{lex-MO}}$ is a singleton, it follows that the lex-MO solution is even strictly efficient.

Corollary 5.18. *All $\hat{x} \in \mathcal{X}_{\text{lex-MO}}$ for which $\{x : f(x) = f(\hat{x})\}$ is a singleton are strictly efficient.*

Next we show that $\mathcal{X}_{\text{lex-MO}}$ is invariant under permutations and strictly increasing mappings. Let $\tau : \mathbb{R} \to \mathbb{R}$ be a function. By abuse of notation let $\tau(f)$ denote $(\tau \circ f_1, \ldots, \tau \circ f_p)$.

Proposition 5.19. *Let $\mathcal{X} \subset \mathbb{R}^n$ and $f : \mathbb{R}^n \to \mathbb{R}^p$.*

1. *Let $\pi \in \Pi$ be a permutation. Then the $\mathcal{X}_{\text{lex-MO}}$ sets of the lex-MO problems $\text{lexmin}_{x \in \mathcal{X}} \text{sort}(\pi(f(x)))$ and $\text{lexmin}_{x \in \mathcal{X}} \text{sort}(f(x))$ are the same.*
2. *Let $\tau : \mathbb{R} \to \mathbb{R}$ be strictly increasing. Then the $\mathcal{X}_{\text{lex-MO}}$ sets of the lex-MO problems $\text{lexmin}_{x \in \mathcal{X}} \text{sort}(\tau(f(x)))$ and $\text{lexmin}_{x \in X} \text{sort}(f(x))$ are the same.*

Proof. 1. The first statement is obvious because $\text{sort}(f_1(x), \ldots, f_p(x)) = \text{sort}(f_{\pi(1)}(x), \ldots, f_{\pi(p)}(x))$ for all $\pi \in \Pi$.
2. By the strict monotonicity of τ we know that

$$f_i(x) < f_i(x') \iff \tau(f_i(x)) < \tau(f_i(x'))$$

and therefore

$$\text{sort}(f(x)) <_{\text{lex}} \text{sort}(f(x')) \iff \text{sort}(\tau(f(x))) <_{\text{lex}} \text{sort}(\tau(f(x'))).$$

□

We will now prove that $\mathcal{X}_E$ can be identified by solving lex-MO problems with positive weights λ for the objective functions. Theorem 5.20 strengthens the result of Proposition 5.10, and the corresponding results of Section 4.5, which allow only a characterization of weakly efficient solutions. Here the lexicographic extension of max-ordering guarantees that only efficient solutions will be found – and not just weakly efficient ones. Let y^U be a utopia point.

Theorem 5.20. *A feasible solution $\hat{x} \in \mathcal{X}$ is efficient if and only if there exists some $\lambda \in \mathbb{R}^p_>$ such that $\hat{x}$ is an optimal solution of the lex-MO optimization problem*

$$\operatorname*{lexmin}_{x \in \mathcal{X}} \operatorname{sort}(\lambda \odot (f(x) - y^U)).$$

Proof. "$\Longleftarrow$" Let $\hat{x} \in \mathcal{X}$ be an optimal solution of the lex-MO optimization problem

$$\operatorname*{lexmin}_{x \in \mathcal{X}} \operatorname{sort}(\lambda \odot (f(x) - y^U))$$

and assume that $\hat{x} \notin \mathcal{X}_E$. For any $x \in \mathcal{X}$ with $f(x) \leq f(\hat{x})$ we also have

$$\lambda \odot (f(x) - y^U) \leq \lambda \odot (f(\hat{x}) - y^U).$$

Note that all λ_k are positive. Therefore

$$\operatorname{sort}(\lambda \odot (f(x) - y^U)) <_{\text{lex}} \operatorname{sort}(\lambda \odot (f(\hat{x}) - y^U)),$$

a contradiction.

"$\Longrightarrow$" Let $\hat{x} \in \mathcal{X}_E$. As in Theorem 4.24 define $\lambda_k := 1/(f_k(\hat{x}) - y^U_k)$. Thus, $\lambda_k(f_k(\hat{x}) - y^U_k) = 1$ for all $k = 1, \ldots, p$. Now let $x \in \mathcal{X}$ be such that $f(x) \neq f(\hat{x})$. Because $\hat{x} \in \mathcal{X}_E$ we must have $f_k(x) > f_k(\hat{x})$ for at least one objective f_k. This implies $\lambda_k(f_k(x) - y^U_k) > 1$ and

$$\operatorname{sort}(\lambda \odot (f(x) - y^U)) >_{\text{lex}} (1, \ldots, 1) = \operatorname{sort}(\lambda \odot (f(\hat{x}) - y^U)).$$

□

Let us discuss the solution of lex-MO problems now. Could we apply a procedure like the lexicographic method? First we would have to solve the max-ordering problem. Then fix the value of the worst objective, solve the max-ordering problem for the remaining $p - 1$ objectives and so on. Unfortunately, we do not know which objective will be the worst, and there may be several max-ordering solutions x with the worst value obtained for different objectives. In Example 5.17 we have $f_1(c) = 7$, $f_2(d) = f_2(e) = 7$ for the three max-ordering solutions $\{c, d, e\}$, yet only c is a lex-MO solution. Taking into account all possible combinations would mean $p!$ sequences of the objectives, which would be computationally prohibitive in general.

There are exceptions, however. In Chapter 8.2, we shall see that lex-MO problems are easily solved, when $\mathcal{X}$ is a finite set. The other exception is convexity. Under this additional assumption on $\mathcal{X}$ and f, we can show that there is *one* objective f_k such that

$$f_k(x) = \min_{x \in \mathcal{X}} \max_{i=1,\ldots,p} f_i(x) \tag{5.16}$$

for all $x \in \mathcal{X}_{\text{MO}}$. We now present some results that have been proved by Behringer (1977a).

Let $\mathcal{X}$ be a convex set and let $f_k : \mathbb{R}^n \to \mathbb{R}, k = 1, \ldots, p$ be convex functions. We use $\mathcal{X}_{\text{MO}}$ to denote the set of all optimal solutions of the max-ordering problem and $\mathcal{X}_{\text{lex-MO}}$ for the optimal solutions of the lex-MO problem, and some further notation to facilitate readability of proofs. Let

$$\begin{aligned} z_{\text{MO}} &:= \min_{x \in \mathcal{X}} \max_{k=1,\ldots,p} f_k(x), \\ \mathcal{A}_i &:= \left\{ x \in \mathcal{X} : f_i(x) = \max_{k=1,\ldots,p} f_k(x) \right\}, \\ \mathcal{L}_i &:= \left\{ x \in \mathcal{A}_i : f_i(x) = \min_{x \in \mathcal{A}_i} f_i(x) \right\}. \end{aligned}$$

Example 5.21. In Figure 5.2 the maximum of three functions f_1, f_2, f_3 of one variable is shown as a bold line. The sets $\mathcal{A}_1$ and $\mathcal{A}_2$ are indicated by bold lines on the x-axis, $\mathcal{A}_3$ is in between $\mathcal{A}_1$ and $\mathcal{A}_2$.

In Figure 5.2 all three sets $\mathcal{A}_i$ are nonempty. Minimizing f_i over $\mathcal{A}_i$, we get $\mathcal{L}_1 = \{3\}, \mathcal{L}_2 = \{1 - \sqrt{3}\}$, and $\mathcal{L}_3 = [1 - \sqrt{3}, 1]$. □

Note that $\max_{k=1,\ldots,p} f_k(x)$ is a convex function and therefore continuous. Hence if $\mathcal{X}$ is compact, $\mathcal{X}_{\text{MO}} \neq \emptyset$ and compact again. Then, iteratively, we get that $\mathcal{X}_{\text{lex-MO}} \neq \emptyset$ and compact.

Lemma 5.22. *If all f_k are convex functions and $\mathcal{X}$ is a convex set then $\mathcal{X}_{\text{MO}}$ is convex.*

Proof. Assume that $\mathcal{X}_{\text{MO}} \neq \emptyset$. Because all f_k are convex, the function $\text{sort}_1(x) := \max_{k=1,\ldots,p} f_k(x)$ is convex. Thus,

$$\begin{aligned} \mathcal{X}_{\text{MO}} &= \{x \in \mathcal{X} : \text{sort}_1(f(x)) = z_{\text{MO}}\} \\ &= \{x \in \mathcal{X} : \text{sort}_1(f(x)) \leq z_{\text{MO}}\} \\ &= \bigcap_{k=1}^{p} \{x \in \mathcal{X} : f_i(x) \leq z_{\text{MO}}\} \\ &= \bigcap_{k=1}^{p} \mathcal{L}^k_{\leq}(z_{\text{MO}}) \end{aligned}$$

is convex as an intersection of p convex level sets. □

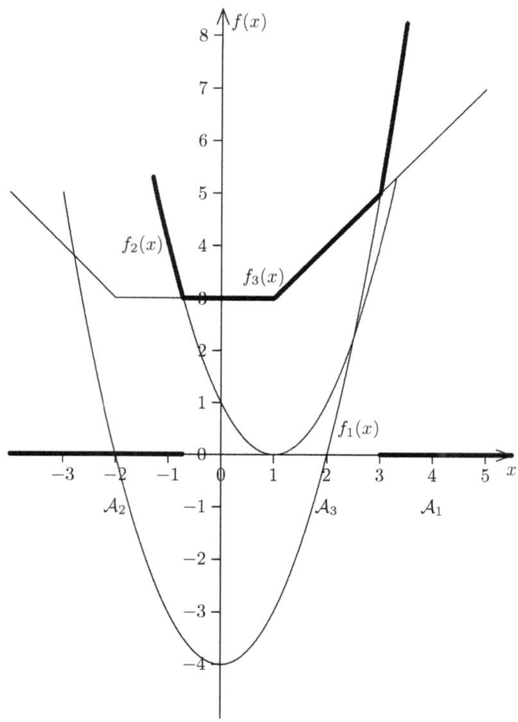

Fig. 5.2. The sets $\mathcal{A}_i$ for three convex functions.

Theorem 5.23 (Behringer (1977a)). *Let $\mathcal{X} \subset \mathbb{R}^n$ be a convex set and let f_k be convex functions. Furthermore, suppose $\mathcal{X}_{\text{MO}} \neq \emptyset$. Then there is an index $k \in \{1, \ldots, p\}$ such that $f_k(x) = z_{\text{MO}}$ for all $x \in \mathcal{X}_{\text{MO}}$.*

Proof. Let $\hat{x} \in \mathcal{X}_{\text{MO}}$. Then for some $j \in \{1, \ldots, p\}$, $f_j(\hat{x}) = z_{\text{MO}}$ and in particular $f_j(\hat{x}) \geq f_i(\hat{x})$ for all $i = 1, \ldots, p$.

Suppose there is no $k \in \{1, \ldots, p\}$ with $f_k(x) = f_j(\hat{x})$ for all $x \in \mathcal{X}_{\text{MO}}$. Then for each $k \in \{1, \ldots, p\}$ we must have some $x^k \in \mathcal{X}_{\text{MO}}$ such that $f_k(x^k) < f_j(\hat{x})$ and $f_i(x^k) \leq f_j(\hat{x})$ for $i = 1, \ldots, p$. Note that $x^k \in \mathcal{X}_{\text{MO}}$ does not allow $f_i(x^k) > f_j(\hat{x})$.

Let $x^* := \sum_{k=1}^p \alpha_k x^k$ with $\alpha_k > 0, \sum_{k=1}^p \alpha_k = 1$ be a strict convex combination of these x^k. Then $x^* \in \mathcal{X}_{\text{MO}}$, because of convexity of $\mathcal{X}_{\text{MO}}$ (Lemma 5.22) but

$$f_i(x^*) \leq \sum_{k=1}^p \alpha_k f_k(x^k) < f_j(\hat{x}), \tag{5.17}$$

because $f_k(x^k) < f_j(\hat{x})$ holds for all f_k. (5.17) contradicts $\hat{x} \in \mathcal{X}_{\text{MO}}$. □

Theorem 5.23 says that z_{MO} is attained for all $x \in \mathcal{X}_{\text{MO}}$ for at least one objective. The index k in 5.23 is called a *common index.* Having established the existence of a common index, we address the problem of finding one. The answer is given by Theorems 5.24 and 5.25.

Theorem 5.24. *Under the assumptions of Theorem 5.23, k is a common index if and only if $\mathcal{X}_{\text{MO}} = \mathcal{L}_k$.*

Proof. "$\Longrightarrow$" Let k be a common index. To show $\mathcal{X}_{\text{MO}} = \mathcal{L}_k$, we prove both inclusions. First let $x \in \mathcal{X}_{\text{MO}}$. Then $f_k(x) = z_{\text{MO}}$, because k is a common index. Thus $x \in \mathcal{L}_k$ and consequently $\mathcal{X}_{\text{MO}} \subset \mathcal{L}_k$.
Let $x \in \mathcal{L}_k$. Then by definition of $\mathcal{L}_k$

$$f_k(x') \geq f_k(x) \text{ for all } x' \in \mathcal{A}_k. \tag{5.18}$$

Assume that $x \notin \mathcal{X}_{\text{MO}}$. Then $\max_{i=1,\ldots,p} f_i(x) > z_{\text{MO}}$. Since we assume $\mathcal{X}_{\text{MO}}$ to be nonempty, there is some $\hat{x} \in \mathcal{X}_{\text{MO}}$ and since k is a common index, $f_k(\hat{x}) = \max_{i=1,\ldots,p} f_i(\hat{x}) = z_{\text{MO}}$ and in particular $\hat{x} \in \mathcal{A}_k$. Because $x \notin \mathcal{X}_{\text{MO}}$ it must hold that $\max_{i=1,\ldots,p} f_i(x) > z_{\text{MO}} = f_k(\hat{x})$. Applying (5.18) to $x' = \hat{x}$, and using that $\mathcal{L}_k \subset \mathcal{A}_k$, we get

$$f_k(\hat{x}) \geq f_k(x) = \max_{i=1,\ldots,p} f_i(x) > f_k(\hat{x}). \tag{5.19}$$

As this is impossible, we conclude $x \in \mathcal{X}_{\text{MO}}$ and therefore $\mathcal{L}_k \subset \mathcal{X}_{\text{MO}}$.
"$\Longleftarrow$" Let $x \in \mathcal{L}_k = \mathcal{X}_{\text{MO}}$. Then $f_k(x) = \max_{i=1,\ldots,p} f_i(x)$ by definition of $\mathcal{L}_k$ and $\max_{i=1,\ldots,p} f_i(x) = \min_{x \in \mathcal{X}} \max_{i=1,\ldots,p} f_i(x)$ by definition of $\mathcal{X}_{\text{MO}}$. Therefore k is a common index. □

The following theorem gives criteria for k to be a common index. These criteria use the sets $\mathcal{L}_i$. First observe that, if $\mathcal{L}_i$ is empty, i cannot be a common index, as f_i has no minimum over $\mathcal{A}_i$. This happens in particular if $\mathcal{A}_i$ is empty, i.e. if there is no feasible solution x for which $f_i(x) \geq f_j(x)$ for all $j \neq i$. Then, among all nonempty $\mathcal{L}_i$ only those with the smallest value of $\min_{x \in \mathcal{A}_i}$ need to be considered for common indices. The main part of Theorem 5.25 below shows how to identify the common indices among those.

Theorem 5.25 (Behringer (1977a)). *Suppose the assumptions of Theorem 5.23 are satisfied. Then the following statements hold.*

1. *If $\mathcal{L}_i = \emptyset$ then i is not a common index.*
2. *Let $\mathcal{J} := \{i \in \{1, \ldots, p\} : \mathcal{L}_i \neq \emptyset\}$ and $m_i := \min_{x \in \mathcal{A}_i} f_i(x)$. Define $\overline{m} := \min_{i \in \mathcal{J}} m_i$. If $m_i > \overline{m}$ then i is not a common index.*
3. *Let $\mathcal{J}^* := \{i \in \mathcal{J} : m_i = \overline{m}\}$. Then $\mathcal{L}_k = \cup_{j \in \mathcal{J}^*} \mathcal{L}_j$ if and only if $k \in \mathcal{J}^*$ is a common index.*

Proof. 1. From Theorem 5.24 i is a common index if and only if $\mathcal{X}_{\text{MO}} = \mathcal{L}_i$. But $\mathcal{X}_{\text{MO}}$ is nonempty, whereas $\mathcal{L}_i$ is empty.

2. Suppose that $m_i > m_j$ and that i is a common index. Then $\mathcal{L}_i = \mathcal{X}_{\text{MO}} \neq \emptyset$. Let $x^0 \in \mathcal{X}_{\text{MO}}$ and $\hat{x} \in \mathcal{L}_j \neq \emptyset$. Then

$$z_{\text{MO}} = \max_{l=1,\ldots,p} f_l(x^0) = f_i(x^0) = m_i > m_j = f_j(\hat{x}) = \max_{i=1,\ldots,p} f_i(\hat{x}),$$

an impossibility.

3. We prove necessity and sufficiency of the condition separately.

"$\Longleftarrow$" Let $k \in \mathcal{J}^*$ be a common index. Obviously $\mathcal{L}_k \subseteq \cup_{j\in\mathcal{J}^*}\mathcal{L}_j$, so it remains to show that $\cup^*_{j\in\mathcal{J}}\mathcal{L}_j \subseteq \mathcal{L}_k$. So let $x \in \cup_{j\in\mathcal{J}^*}\mathcal{L}_j$. Then $x \in \mathcal{L}_j$ for some $j \in \mathcal{J}^*$ and

$$f_j(x) = \max_{i=1,\ldots,p} f_i(x) = \min_{x\in\mathcal{A}_j} f_j(x) = m_j = \overline{m}. \tag{5.20}$$

From Theorem 5.24 $\mathcal{L}_k = \mathcal{X}_{\text{MO}}$, i.e.

$$f_k(\hat{x}) = m_k = \overline{m} = \max_{i=1,\ldots,p} f_i(\hat{x}) = z_{\text{MO}} \tag{5.21}$$

for all $\hat{x} \in \mathcal{L}_k$. Putting (5.20) and (5.21) together, we see that $f_j(x) = z_{\text{MO}}$ and therefore $x \in \mathcal{X}_{\text{MO}}$ and $x \in \mathcal{L}_k$.

"$\Longrightarrow$" Now suppose we have $\mathcal{L}_k = \cup_{i\in\mathcal{J}^*}\mathcal{L}_i$ for some $k \in \mathcal{J}^*$. Since $\mathcal{X}_{\text{MO}} \neq \emptyset$ we know that a common index $\hat{k}$ exists and that $\mathcal{X}_{\text{MO}} = \mathcal{L}_{\hat{k}}$. From the first and second part of this Theorem, we know that $\hat{k} \in \mathcal{J}^*$. Finally from necessity $\mathcal{L}_{\hat{k}} = \cup_{i\in\mathcal{J}^*}\mathcal{L}_i$. Altogether we have

$$\mathcal{X}_{\text{MO}} = \mathcal{L}_{\hat{k}} = \bigcup_{i\in\mathcal{J}*} \mathcal{L}_i = \mathcal{L}_k \tag{5.22}$$

and by Theorem 5.24 k is a common index. □

Part 3 actually says that common indices are defined by maximal sets $\mathcal{L}_k, k \in \mathcal{J}$. In Example 5.21, none of the sets $\mathcal{L}_i$ is empty. But 1 is not a common index, because of the second statement. While $m_2 = m_3$ the second statement confirms that 3 is the only common index.

With Theorem 5.25 we have a method to find a common index, which will be used as a subroutine in the algorithm to solve lexicographic max-ordering problems for convex data.

Algorithm 5.2 (Finding a common index.)

Input: Feasible set $\mathcal{X}$ *and objective function* f.

Find $\mathcal{J} := \{i \in \{1,\ldots,p\} : \mathcal{L}_i \neq \emptyset\}$ *by subdividing the feasible set into sets* $\mathcal{A}_i$*, and solving the single objective optimization problems* $\min_{x\in\mathcal{A}_i} f_i(x)$.

Choose one optimal solution $x^i \in \mathcal{L}_i$ *for each* $i \in \mathcal{J}$ *and determine* $\mathcal{J}^* := \{i \in \mathcal{J} : f_i(x^i) \leq f_j(x^j)$ *for all* $j \in \mathcal{J}\}$.
$\hat{\mathcal{J}} := \{i \in \mathcal{J}^* : \mathcal{L}_j \subset \mathcal{L}_i$ *for all* $j \in calJ^*\}$ *is the set of all common indices.*
Output: $\hat{\mathcal{J}}$.

The idea of solving lexicographic max-ordering problems that we outlined above can now be formalized. It consists of repeatedly solving max-ordering problems, identifying common indices, and reducing the set of objectives that still have to be considered.

Algorithm 5.3 (Lexicographic max-ordering.)
Input: *Feasible set* $\mathcal{X}$ *and objective function* f.
Initialization: *Set* $\mathcal{X}' := \mathcal{X}$, $\mathcal{Q} := \{1, \ldots, p\}$, *and* $f' := f$.
While $|\mathcal{Q}| > 1$ *do*
Solve the max-ordering problem $\min_{x \in \mathcal{X}'} \max_{k \in \mathcal{Q}} f'(x)$ *and find the set* $\mathcal{X}'_{\text{MO}}$ *of all max-ordering solutions.*
Apply Algorithm 5.2 to find a common index k.
Let $\mathcal{X}' := \mathcal{X}'_{\text{MO}}$, $\mathcal{Q} := \mathcal{Q} \setminus \{k\}$, *and* $f' := f \setminus f_k$.
End while.
If $|\mathcal{Q}| = 1$ *let* $\mathcal{X}_{\text{lex-MO}} = \mathcal{X}'_{\text{MO}}$ *and STOP.*
Output: $\mathcal{X}_{\text{lex-MO}}$.

At the end of this chapter we study some properties of lex-MO solutions in the framework of multicriteria optimization classes. Recall that a multicriteria optimization class is the set of all multicriteria optimization problems with the same model map and ordered set. The properties are that, if only one objective is present, the problem should reduce to a single objective optimization problem, that optimal solutions should be max-ordering optimal, and a reduction property. The reduction property states that, if the values of some objective functions at some optimal solution are known, then the set of optimal solutions of the original problem which attain these values should be equal to the set of optimal solutions of a problem with a restricted feasible set where the known objective values are included as constraints. These results are from Ehrgott (1997) and Ehrgott (1998).

Recall that we denote the set of optimal solutions of a multicriteria optimization problem $(\mathcal{X}, f, \mathbb{R}^p)/\theta/(\mathbb{R}^P, \preceq)$ by

$$\text{Opt}((\mathcal{X}, f, \mathbb{R}^p)/\theta/(\mathbb{R}^P, \preceq)).$$

Definition 5.26. *1. An MCO class* $\bullet/\theta/(\mathbb{R}^P, \preceq)$ *satisfies the normalization property if* $\theta = \text{id}$ *and* $(\mathbb{R}^P, \preceq) = (\mathbb{R}, <)$ *whenever* $f : \mathbb{R}^n \mathbb{R}$.
2. An MCO class $\bullet/\theta/(\mathbb{R}^P, \preceq)$ *satisfies the regularity property if* $\text{Opt}((\mathcal{X}, f, \mathbb{R}^p)/\theta/(\mathbb{R}^P, \preceq)) \subset \mathcal{X}_{\text{MO}}$ *for all* $\mathcal{X}$ *and* f.

The normalization property means that for optimization problems with a single objective function the optimal solutions according to MCO class $\bullet/\theta/(\mathbb{R}^P, \preceq)$ are exactly the optimal solutions of the single objective optimization problem $\min_{x \in \mathcal{X}} f(x)$. All MCO classes discussed in this book have this property.

The regularity property means that an optimal solution according to MCO class $\bullet/\theta/(\mathbb{R}^P, \preceq)$ must also be an optimal solution of the max-ordering problem $\min_{x \in \mathcal{X}} \max_{k=1,\dots,p} f_k(x)$.

The third property requires some more preparation. Let $\mathcal{X} \subset \mathbb{R}^n$ be a feasible set and $f : \mathbb{R}^n \to \mathbb{R}^p$ be a vector valued objective function and let $\mathcal{Q} \subset \{1, \dots, p\}$. Furthermore, let $\{y_1, \dots, y_p\} \subset \mathbb{R}$ be such that there is at least one $x \in \text{Opt}((\mathcal{X}, f, \mathbb{R}^p)/\theta/(\mathbb{R}^P, \preceq))$ such that $\{f_k(x) : k = 1, \dots, p\} = \{y_k : k = 1, \dots, p\}$ (these sets are understood as multisets and may contain multiple copies of some elements). The *reduced problem* $\text{RP}(\mathcal{Q})$

$$(\mathcal{X}^{\mathcal{Q}}, f^{\mathcal{Q}}, \mathbb{R}^{|\mathcal{Q}|})/\theta/(\mathbb{R}^P, \preceq)$$

is defined by the feasible set

$$\mathcal{X}^{\mathcal{Q}} := \{x \in \mathcal{X} : \{f_k(x) : k \in \{1, \dots, p\} \setminus \mathcal{Q}\} = \{y_k : k \in \{1, \dots, p\} \setminus \mathcal{Q}\}\}.$$

and the objective function

$$f^{\mathcal{Q}} = (f_k : k \in \mathcal{Q}).$$

We denote the complement of $\mathcal{Q}$ by $\bar{\mathcal{Q}} := \{1, \dots, p\} \setminus \mathcal{Q}$.

Definition 5.27. *An MCO class satisfies the reduction property, if for all data* $(\mathcal{X}, f, \mathbb{R}^p)$*, for all* $q \leq p$*, and for all* y *as above*

$$\begin{aligned}&\text{Opt}\left((\mathcal{X}^{\mathcal{Q}}, f, \mathbb{R}^{|\mathcal{Q}|})/\theta/(\mathbb{R}^P, \preceq)\right) = \\ &= \left\{x \in \text{Opt}\left((\mathcal{X}, f, \mathbb{R}^p)/\theta/(\mathbb{R}^p, \preceq)\right) : \{f_k(x) : k \in \bar{\mathcal{Q}}\} = \{y_k : k \in \bar{\mathcal{Q}}\}\right\}.\end{aligned}$$

Proposition 5.28. *The* lex-MO *class satisfies the normalization, regularity and reduction properties.*

Proof. 1. The normalization property is obvious because $\text{sort}(f(x)) = f(x)$ and $<_{\text{lex}}$ is $<$ if $p = 1$.

2. The regularity property follows from Theorem 5.16 which states that $\mathcal{X}_{\text{lex-MO}} \subset \mathcal{X}_{\text{MO}}$.
3. We write Opt and Opt(RP($\mathcal{Q}$)) for the sets of lex-MO solutions of the original and reduced problems, respectively. Let

$$\text{Opt}^* := \{x \in \text{Opt} : \{f_k(x) : k \in \bar{\mathcal{Q}}\} = \{y_k : k \in \bar{\mathcal{Q}}\}\}.$$

We have to show Opt(RP($\mathcal{Q}$)) $=$ Opt*. By the choice of y, Opt* is nonempty.
First note that for all $x \in \text{Opt}^*$ and for all $x' \in \mathcal{X}^{\mathcal{Q}}$ we have

$$\text{sort}(f(x)) \leqq_{\text{lex}} \text{sort}(f(x')). \tag{5.23}$$

since $\mathcal{X}^{\mathcal{Q}} \subset \mathcal{X}$. Moreover, $\{f_k(x) : k \in \bar{\mathcal{Q}}\} = \{f_k(x') : k \in \bar{\mathcal{Q}}\}$ by the definition of $\mathcal{X}^{\mathcal{Q}}$ and Opt*, and therefore

$$\text{sort}(f^{\bar{\mathcal{Q}}}(x)) = \text{sort}(f^{\bar{\mathcal{Q}}}(x')) = \text{sort}(y^{\bar{\mathcal{Q}}}). \tag{5.24}$$

Let $x \in \text{Opt}^*$ and assume that x is not optimal for RP($\mathcal{Q}$). Then there is some $\hat{x} \in \mathcal{X}^{\mathcal{Q}}$ such that sort $f^{\mathcal{Q}}(\hat{x}) <_{\text{lex}}$ sort $f^{\mathcal{Q}}(x)$. Together with (5.24) this implies $\text{sort}(f(\hat{x})) <_{\text{lex}} \text{sort}(f(x))$, a contradiction to (5.23).
Let $x' \in$ Opt(RP($\mathcal{Q}$)). Since Opt$^* \subset$ Opt(RP($\mathcal{Q}$)) and $\{\text{sort}(f^{\mathcal{Q}}(x))\} : x \in \text{Opt(RP}(\mathcal{Q}))$ is a singleton we must have that $\{f_k(x') : k \in \mathcal{Q}\} = \{y_k : k \in \mathcal{Q}\}$ and thus (5.24) implies $\text{sort}(f(x')) = \text{sort}(y)$ and therefore $x' \in Opt^*$. □

It will be left to the reader as Exercise 5.5 to determine which of the other MCO classes considered so far satisfy the regularity and reduction property.

5.4 Notes

Lexicographic optimization plays an important role in goal programming, see e.g. Romero (2001). It is more often encountered in linear and combinatorial optimization than in nonlinear programming.

Max-ordering optimization models can be found in many areas of optimization. Usually (5.3) is reformulated as the single objective optimization problem 5.4.

For lex-MO optimization we have seen that convexity is important to obtain a reasonable algorithm to solve (5.12). In this context, we remark that for Lemma 5.22 to Theorem 5.25 to be true it is sufficient that f_k are lower semicontinuous and strictly quasiconvex. Naturally, if further assumptions hold, better results can be expected. An algorithm for the linear case is given in

Marchi and Oviedo (1992) and location problems in the plane (where $\mathcal{X} \subset \mathbb{R}^2$) are dealt with in Ehrgott *et al.* (1999).

Lex-MO optimality implies the absence of any preference between the objectives: Behringer (1977b) calls it "optimality under complete ignorance". It is therefore an approach that treats all objectives in an equitable way, and indeed if an efficient solution x exists such that $f_1(x) = \cdots = f_p(x)$ then such an $x \in \mathcal{X}_{\text{lex-MO}}$. This leads to the idea of equitable solutions of multicriteria optimization problems, investigated in more detail in Kostreva and Ogryczak (1999). Lex-MO optimality can be considered as an "extreme" case of equitable optimality.

Galperin (1992) considers yet another multicriteria optimization class, introducing the balance space approach. While Ehrgott *et al.* (1997) show that it is in some sense equivalent to the efficiency approach because it has led to some interesting research. The balance space approach is also closely related to max-ordering optimality, see Ehrgott and Galperin (2002).

Exercises

5.1. Solve the following lexicographic optimization problem with linear objectives and linear constraints

$$\begin{aligned} \min f_1(x) &= -x_1 + x_2 - x_3 \\ \min f_2(x) &= x_2 \\ \min f_3(x) &= -x_1 - 2x_2x \\ \text{subject to } x_1 + x_2 &\leq 1 \\ x_1 - x_2 + x_3 &\leq 4 \\ x_1, x_2, x_3 &\geq 0. \end{aligned}$$

What happens if you reverse the order of objective functions?

5.2. Prove that an optimal solution $\hat{x}$ of the max-ordering problem

$$\min_{x \in \mathcal{X}} \max_{k=1,\ldots,p} f_k(x)$$

is weakly efficient. Give an example that shows that $\hat{x}$ is not necessarily efficient.

5.3. Find an optimal solution of the max-ordering optimization problem

$$\begin{aligned} \min \max \left(x_1 + x_2 + x_3, -x_1 + x_2, -x_2 + 2x_3\right) & \\ \text{subject to} \quad x_1 + x_2 &\geq 1 \\ x_1 - x_2 + x_3 &\geq 4 \\ x_1, x_2, x_3 &\geq 0. \end{aligned}$$

Is the optimal solution you found efficient?

5.4. Let $\mathcal{Y} \subset \mathbb{R}^p$. Show that $y^1 \leq y^2$ implies $y^1 <_{\text{lex}} y^2, \max_{k=1,\ldots,p} y^1_k \leq y^2_k$, and $sort(y^1) <_{\text{lex}} sort(y^2)$. What about the converse?

5.5. Find out which of the multicriteria optimization classes $\bullet/\,\text{id}\,/(\mathbb{R}^p, \leq)$, $\bullet/\lambda/(\mathbb{R}, <)$, $\bullet/\pi/(\mathbb{R}^p, <_{\text{lex}})$, and $\bullet/\max/(\mathbb{R}, <)$ satisfy the regularity and reduction property.

5.6. Construct an example of an MCO class that does not have the normalization property. I.e. define a model map $\theta : \mathbb{R}^p \to \mathbb{R}^P$ and an ordering such that the problem of finding optimal solutions according to this MCO class is not the same as solving the single objective optimization problem $\min_{x \in \mathcal{X}} f(x)$.

5.7. Find a lexicographic max-ordering solution of the optimization problems in Exercises 6.2 and 6.7.

5.8. Show that if there exists an $x \in \mathcal{X}_E$ with $f_1(x) = f_2(x) = \ldots = f_p(x)$ then $x \in \mathcal{X}_{\text{lex-MO}}$.

6

Introdcution to Multicriteria Linear Programming

This chapter commences the second part of this book, in which we focus on multicriteria problems with linear and combinatorial structures, i.e. multiobjective linear programming and multiobjective combinatorial optimization.

We give an example from the design of radiotherapy treatment plans to show that multiobjective linear programming has important applications. We repeat the main definitions of multicriteria optimization and summarize the main results from linear programming to make this part of the book self-contained. We apply some of the general results proved in Chapters 2 and 3 and show how to use parametric linear programming to solve linear programs with two objectives. We also prove the main theorem of linear programming, which states that all efficient solutions are properly efficient. Adding the convexity of linear problems this means that all efficient solutions can be characterized by weighted sum scalarization.

Example 6.1. The goal of radiation therapy in the treatment of cancer is to destroy a tumour by damaging the DNA of cancerous cells, thereby rendering them incapable of reproduction. This is done by focusing intensity modulated beams on the patient from a number of beam directions. Intensity modulation is achieved by a mechanical device called multileaf collimator. It essentially allows subdividing beams into sub-beams in a rectangular grid pattern so that intensity of each individual sub-beam can be decided separately. Given the beam directions, an intensity map defines the intensity of radiation of each sub-beam of all beam directions. The intensity map has to be determined according to a treatment prescription, which can take the form of lower and upper bounds on the radiation dose delivered to the tumour as well as upper bounds on the radiation dose delivered to critical structures (such as healthy organs) and normal tissue.

Radiation dose distribution in the body depends on intensity of radiation beams in a linear fashion. Let $x \in \mathbb{R}^n$ be a vector describing an intensity map, where n is the total number of sub-beams. The patient body is discretized into m dose points according to magnetic resonance imaging (MRI) or computed tomography (CT) scans. The dose delivered to the dose points is then Ax, where A is a $m \times n$ matrix. Assuming that we have l critical structures, we can partition the rows of A according to the set of dose points in the tumour $\mathcal{T}$, in a critical structure $\mathcal{S}_i, i = 1, \ldots, l$, or in normal tissue $\mathcal{N}$ and form submatrices $A_\mathcal{T}$, $A_{\mathcal{S}_i}$, $A_\mathcal{N}$ accordingly. Let $l_\mathcal{T}$ denote the prescribed tumouricidal dose, $u_\mathcal{T}$ be an upper bound on the dose in the tumour, $u_{\mathcal{S}_i}$ be upper bounds on the dose in critical structure i, and $u_\mathcal{N}$ be an upper bound on the dose in normal tissue. We assume that these bounds apply to every dose point in the tumour, critical structure, and normal tissue, respectively.

Ideally, we would like to design a treatment that delivers a uniform dose of $l_\mathcal{T}$ to the tumour and no dose at all to critical structures and normal tissue. Since this is usually physically impossible we have to accept some underdosing $z_\mathcal{T}$ in the tumour or overdosing $z_{\mathcal{C}_i}, i = 1, \ldots, l$ and $z_\mathcal{N}$ in critical structures and normal tissue. Naturally, the values of $z_\mathcal{T}, z_{\mathcal{S}_1}, \ldots, z_{\mathcal{C}_l}$ should be kept as small as possible.

We can therefore describe the problem via the following multiobjective optimization problem Holder (2004), where e is a vector of ones of appropriate dimension.

$$
\begin{array}{lrll}
\min & (z_\mathcal{T}, z_{\mathcal{S}_1}, \ldots, z_{\mathcal{S}_l}, z_\mathcal{N}) & & \\
\text{subject to} & A_\mathcal{T} x + z_\mathcal{T} e & \geq l_\mathcal{T} & \\
& A_\mathcal{T} x & \leq u_\mathcal{T} & \\
& A_{\mathcal{S}_i} x - z_{\mathcal{S}_i} e & \leq u_{\mathcal{S}_i} & i = 1, \ldots, l \\
& A_\mathcal{N} x - z_\mathcal{N} e & \leq u_\mathcal{N} & \\
& z_{\mathcal{S}_i} & \geq -u_{\mathcal{S}_i} & i = 1, \ldots, l \\
& z_\mathcal{N} & \geq 0 & \\
& x & \geqq 0. &
\end{array}
$$

In this model, the goal is to find efficient solutions $(x, z) \in \mathbb{R}^{n+l+2}$ such that the maximal underdosing of any tumour dose point and the maximal overdosing of any critical structure and any normal tissue dose point is simultaneously minimized. □

6.1 Notation and Definitions

A multiobjective linear program (MOLP) is a special case of the multiobjective program

$$\begin{array}{ll} \min & (f_1(x), \ldots, f_k(x)) \\ \text{subject to} & g_j(x) \leq 0 \quad j = 1, \ldots, m \end{array}$$

that arises if all objective functions and constraints are linear. Thus, the objective functions are

$$f_k(x) = c_k^T x \quad k = 1, \ldots, p,$$

where $c_k \in \mathbb{R}^n$. The constraints $g_j(x) \leq 0$ are summarily written in matrix form and as equality constraints

$$Ax = b.$$

As usual in linear programming we restrict the variables to the nonnegative orthant of $\mathbb{R}^n : x \geqq 0$. Recall that we use the notation

$$\begin{aligned} &y^1 < y^2 \text{ if } y_k^1 < y_k^2 \quad k = 1, \ldots, p \\ &y^1 \leqq y^2 \text{ if } y_k^1 \leq y_k^2 \quad k = 1, \ldots, p \\ &y^1 \leq y^2 \text{ if } y^1 \leqq y^2, y^1 \neq y^2 \end{aligned}$$

and

$$\begin{aligned} \mathbb{R}^p_> &:= \{y \in \mathbb{R}^p : y > 0\} \\ \mathbb{R}^p_\geq &:= \{y \in \mathbb{R}^p : y \geq 0\} \\ \mathbb{R}^p_\geqq &:= \{y \in \mathbb{R}^p : y \geqq 0\}. \end{aligned}$$

A multiobjective linear program is then the following optimization problem

$$\begin{array}{ll} \min & Cx \\ \text{subject to} & Ax = b \\ & x \geqq 0 \end{array} \tag{6.1}$$

with a $p \times n$ objective or *criteria matrix* C consisting of the rows $c_k^T, k = 1, \ldots, p$. The feasible set in decision space is $\mathcal{X} = \{x \in \mathbb{R}^n : Ax = b, x \geqq 0\}$ defined by the $m \times n$ *constraint matrix* A and the *right hand side vector* $b \in \mathbb{R}^m$. The *feasible set* in *decision space* is $\mathcal{X} = \{x \in \mathbb{R}^n : Ax = b, x \geqq 0\}$. The feasible set in *objective space* is $\mathcal{Y} = C\mathcal{X} = \{Cx : x \in \mathcal{X}\}$.

In terms of the classification of Section 1.5 we can write the MOLP (6.1) as $(\mathcal{X}, C, \mathbb{R}^p)/\text{id}/(\mathbb{R}^p, \leq)$. We shall make the following basic assumption. Let

$$\mathcal{X}_k := \left\{\hat{x} \in \mathcal{X} : c_k^T \hat{x} \leq c_k^T x \text{ for all } x \in \mathcal{X}\right\}$$

be the set of optimal solutions of the LP with the k-th objective function. We assume that

$$\cap_{k=1}^p \mathcal{X}_k = \emptyset. \tag{6.2}$$

Assumption (6.2) guarantees that there is no feasible solution that minimizes all p objectives at the same time, i.e. the MOLP (6.1) is a true multiobjective problem. In other words $y^I \notin \mathcal{Y}$.

Definition 6.2. *Let $\hat{x} \in \mathcal{X}$ be a feasible solution of the MOLP* (6.1) *and let $\hat{y} = C\hat{x}$.*

1. *$\hat{x}$ is called* weakly efficient *if there is no $x \in \mathcal{X}$ such that $Cx < C\hat{x}$; $\hat{y} = C\hat{x}$ is called* weakly nondominated.
2. *$\hat{x}$ is called* efficient *if there is no $x \in \mathcal{X}$ such that $Cx \leq C\hat{x}$; $\hat{y} = C\hat{x}$ is called* nondominated.
3. *$\hat{x}$ is called* properly efficient *if it is efficient and if there exists a real number $M > 0$ such that for all i and x with $c_i^T x < c_i^T \hat{x}$ there is an index j and $M > 0$ such that $c_j^T x > c_j^T \hat{x}$ and*

$$\frac{c_i^T \hat{x} - c_i^T x}{c_j^T x - c_j^T \hat{x}} \leq M.$$

Let us consider an example to illustrate efficient solutions and nondominated points.

Example 6.3 (Steuer (1985)). This MOLP has two objectives, two constraints, and two variables so that we can graphically illustrate it in decision and objective space. It is

$$\begin{array}{llrcl}
\min & 3x_1 + & x_2 & & \\
\min & -x_1 - & 2x_2 & & \\
\text{subject to} & & x_2 & \leq & 3 \\
& 3x_1 - & x_2 & \leq & 6 \\
& & x & \geqq & 0.
\end{array}$$

Figures 6.1 and 6.2 show the feasible sets $\mathcal{X}$ and $\mathcal{Y}$. The extreme points of $\mathcal{X}$ and of $\mathcal{Y}$ are labeled.

In Figure 6.2, point y^1 is dominated: All points in $(y^1 - \mathbb{R}^p_{\geq}) \cap \mathcal{Y}$, illustrated by the right angle attached to y^1, dominate it. On the other hand, y^2 is nondominated, as the right angle attached to it does not contain any point of $\mathcal{Y}$ except y^2 itself. □

Let us now summarize some consequences of results proved in Chapters 2 and 3.

Lemma 6.4. *The feasible sets $\mathcal{X}$ in decision space and $\mathcal{Y}$ in objective space of the MOLP 6.1 are convex and closed.*

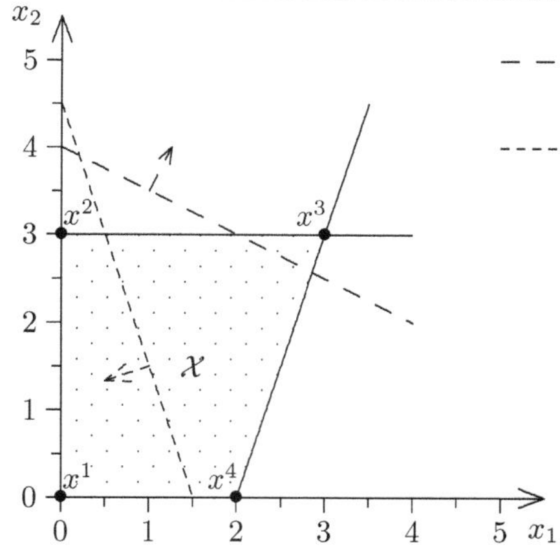

Fig. 6.1. Feasible set and objective contours in Example 6.3.

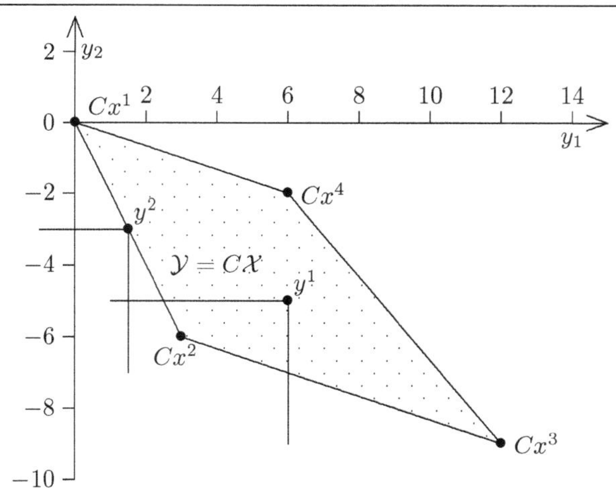

Fig. 6.2. Feasible set in objective space in Example 6.3.

Theorem 6.5. *1. If $\mathcal{Y} \neq \emptyset$ and there is some $y \in \mathbb{R}^p$ such that $\mathcal{Y} \subset y + \mathbb{R}^p_{\geqq}$ (i.e. $\mathcal{Y}$ is bounded from below) then $\mathcal{Y}_N \neq \emptyset$.*

2. $\mathcal{S}(\mathcal{Y}) = \mathcal{Y}_{pN} \subset \mathcal{Y}_N \subset \operatorname{cl}\mathcal{S}(\mathcal{Y})$, where $\mathcal{S}(\mathcal{Y}) = \{\hat{y} \in \mathcal{Y} :$ There is $\lambda > 0$ such that $\lambda^T\hat{y} < \lambda^T y$ for all $y \in \mathcal{Y}\}$.

3. If there is some $y \in \mathbb{R}^p$ such that $\mathcal{Y} \subset y + \mathbb{R}^p_{\geqq}$ then $\mathcal{Y}_N$ is connected.

4. If $\mathcal{X}$ is bounded then $\mathcal{X}_{wE}$ and $\mathcal{X}_E$ are connected.

Proof. 1. This follows from Theorem 2.10. Boundedness of $\mathcal{Y}$ implies compactness of all sections $(y^0 - \mathbb{R}^p_{\geqq}) \cap \mathcal{Y}, y^0 \in \mathcal{Y}$, because $\mathcal{Y}$ is closed.

2. This follows from Theorem 3.17 and convexity of $\mathcal{Y}$.

3. This follows from the first observation, noting that with convexity the assumptions of Theorem 3.35 are satisfied.
4. Boundedness of $\mathcal{X}$ implies that $\mathcal{X}$ is compact. With convexity of $\mathcal{X}$ and the objective functions, Theorems 3.38 and 3.40 apply. □

A strengthened version of the second statement is of fundamental importance for multiobjective linear programming and we shall prove it shortly. To that end let us consider the solution of the weighted sum scalarization of MOLP.

Let $\lambda \in \mathbb{R}^p_{\geq}$. The weighted sum linear program, which we often refer to as LP(λ), is

$$\begin{aligned} \min \quad & \lambda^T Cx \\ \text{subject to} \quad & Ax = b \\ & x \geqq 0. \end{aligned}$$

Theorem 6.6. *Let $\hat{x} \in \mathcal{X}$ be an optimal solution of the weighted sum LP (6.3).*

1. If $\lambda \geq 0$ then $\hat{x}$ is weakly efficient.
2. If $\lambda > 0$ then $\hat{x}$ is efficient.

Proof. 1. Suppose that $x \in \mathcal{X}$ strictly dominates $\hat{x}$, i.e.

$$c_k^T x < c_k^T x \quad k = 1, \ldots, p. \tag{6.3}$$

Thus,

$$\lambda_k c_k^T x \leq \lambda_k c_k^T x \quad k = 1, \ldots, p \tag{6.4}$$

with strict inequality holding at least once since $\lambda \neq 0$. Summing over k we have $\lambda^T Cx < \lambda^T C\hat{x}$, a contradiction.

2. In this case we have "$\leq$" instead of "$<$" in (6.3), with one strict inequality. Then, because $\lambda > 0$, (6.4) holds, too, and $\lambda^T Cx < \lambda^T C\hat{x}$ gives a contradiction once again. □

Theorem 6.6 gives a way to find efficient solutions of (6.1). Graphically, this is very similar to the graphic method of solving LPs with two variables, we just apply it in criterion space of an MOLP with two objectives. In fact, we find nondominated points graphically and the efficient solutions of the MOLP are the preimages of $\mathcal{Y}_N$ under the linear mapping C.

Example 6.7. Figure 6.3 shows the same set $\mathcal{Y}$ as Figure 6.2. Some level curves of weighted sum objectives

$$\{y \in \mathbb{R}^p : \lambda^T y = \gamma\}$$

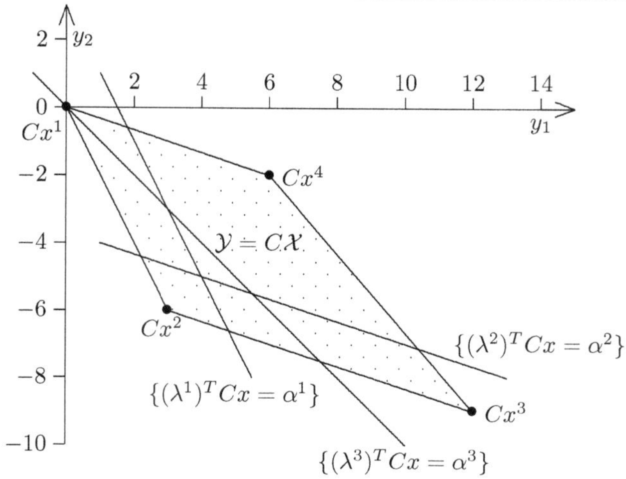

Fig. 6.3. Level Sets of the objective of a weighted sum LP.

are shown. Nondominated points are identified by moving these lines in parallel to the left and downward, as this decreases the level γ in accordance with minimizing $\lambda^T Cx$ (note that $y = Cx$).

Clearly, with λ^1 all points on the line between Cx^1 and Cx^2 are identified as nondominated, with λ^2 all points on the line connecting Cx^2 and Cx^3, and with λ^3 the single point Cx^3. The two line segments together constitute the whole set $\mathcal{Y}_N$. □

The following observations about Example 6.7 turn out to be important.

- A single nondominated point can be identified by many different weighting vectors λ.
- A single weighting vector λ can identify many nondominated points.
- The linearity of the constraints and objectives appears to make it possible to find all nondominated points with (only a finite number of) weighting vectors, because $\mathcal{X}$ and $\mathcal{Y}$ are polyhedra.

The reader should keep these in mind for what follows. We shall elaborate on the last point now. In order to do so, we need duality of linear programming. Let

$$\begin{array}{ll} \min & c^T x \\ \text{subject to} & Ax = b \\ & x \geqq 0 \end{array} \qquad (6.5)$$

be a single objective linear program (LP). For every LP (6.5) a dual linear program is defined as

$$
\begin{array}{ll}
\max & b^T u \\
\text{subject to} & A^T u \leqq c \\
& u \in \mathbb{R}^m.
\end{array}
\tag{6.6}
$$

Let us denote by $\mathcal{U} := \{u \in \mathbb{R}^m : A^T u \leq c\}$ the feasible set of the dual linear program (6.6). The relationship between the primal and dual linear programs are stated in Theorem 6.8, which can be found in every textbook on linear programming, e.g. Dantzig and Thapa (2003).

Theorem 6.8 (Linear Programming Duality).

1. *(Weak duality). Let $x \in \mathcal{X}$ and $u \in \mathcal{U}$ be feasible solutions of (6.5) and (6.6), respectively. Then*
$$b^T u \leq c^T x.$$
2. *If (6.5) is unbounded then (6.6) is infeasible and vice versa.*
3. *It is possible that both (6.5) and (6.6) are infeasible.*
4. *(Strong duality). If both (6.5) and (6.6) are feasible, i.e. $\mathcal{X} \neq \emptyset$ and $\mathcal{U} \neq \emptyset$, then*
$$\min_{x \in \mathcal{X}} c^T x = \max_{u \in \mathcal{U}} b^T u$$
and $b^T \hat{u} = c^T \hat{x}$ for any optimal solution $\hat{x} \in \mathcal{X}$ of (6.5) and any optimal solution $\hat{u} \in \mathcal{U}$ of (6.6).

We are now ready to prove the fundamental result of multiobjective linear programming, which we do in several steps.

Lemma 6.9. *A feasible solution $x^0 \in \mathcal{X}$ is efficient if and only if the linear program*

$$
\begin{array}{lr}
\max & e^T z \\
\text{subject to} & Ax = b \\
& Cx + Iz = Cx^0 \\
& x, z \geqq 0,
\end{array}
\tag{6.7}
$$

where $e^T = (1, \ldots, 1) \in \mathbb{R}^p$ and I is the $p \times p$ identity matrix, has an optimal solution $(\hat{x}, \hat{z})$ with $\hat{z} = 0$.

Proof. This is in fact Theorem 4.14 applied to the MOLP (6.1). Let $(x, z) \in \mathcal{X} \times \mathbb{R}^p_{\geqq}$ be a feasible solution of (6.7). Then $Cx + Iy = Cx^0$ and therefore $z = Cx^0 - Cx \geqq 0$ by the nonnegativity of z. If $\hat{x}$ in an optimal solution $(\hat{x}, \hat{z})$ is efficient there is no $x \in \mathcal{X}$ such that $Cx \leqq C\hat{x}$, so we must have $\hat{z} = 0$. On the other hand, if $\hat{x}$ is not efficient there must be $x \in \mathcal{X}$ such that $Cx \leq Cx^0$. But then there is a z with $z_k > 0$ for at least one k, contradicting optimality of $(\hat{x}, 0)$. Note that (6.7) is always feasible. □

Lemma 6.10. *A feasible solution $x^0 \in \mathcal{X}$ is efficient if and only if the linear program*

$$\begin{array}{rl} \min & u^T b + w^T C x^0 \\ \text{subject to} & u^T A + w^T C \geqq 0 \\ & w \geqq e \\ & u \in \mathbb{R}^m \end{array} \tag{6.8}$$

has an optimal solution $(\hat{u}, \hat{w})$ with $\hat{u}^T b + \hat{w}^T C x^0 = 0$.

Proof. Note that (6.8) is the dual of (6.7). Therefore $(\hat{x}, \hat{z})$ is an optimal solution of the LP (6.7) if and only if the LP (6.8) has an optimal solution $(\hat{u}, \hat{w})$ such that $e^T \hat{z} = \hat{u}^T b + \hat{w}^T C x^0 = 0$. □

With Lemma 6.10 we can now prove that all efficient solutions of an MOLP (6.1) can be found by solving a weighted sum LP (6.3). In the proof, we consider an efficient solution x^0 and construct an appropriate weight $\lambda \in \mathbb{R}^p_>$ such that x^0 is an efficient solution of the weighted sum LP(λ) (6.3).

Theorem 6.11 (Isermann (1974)). *A feasible solution $x^0 \in \mathcal{X}$ is an efficient solution of the MOLP (6.1) if and only if there exists a $\lambda \in \mathbb{R}^p_>$ such that*

$$\lambda^T C x^0 \leq \lambda^T C x \tag{6.9}$$

for all $x \in \mathcal{X}$.

Proof. "$\Longleftarrow$" We know from Theorem 6.6 that an optimal solution of a weighted sum LP with positive weights is efficient.

"$\Longrightarrow$" Let $x^0 \in \mathcal{X}_E$. From Lemma 6.10 it follows that the LP (6.8) has an optimal solution $(\hat{u}, \hat{w})$ such that

$$\hat{u}^T b = -\hat{w}^T C x^0. \tag{6.10}$$

It is easy to see that this same $\hat{u}$ is also an optimal solution of the LP

$$\min \left\{ u^T b : u^T A \geqq -\hat{w}^T C \right\}, \tag{6.11}$$

which is just (6.8) with $w = \hat{w}$ fixed. Therefore, an optimal solution of the dual of (6.11)

$$\max \left\{ -\hat{w}^T C x : Ax = b,\ x \geqq 0 \right\} \tag{6.12}$$

exists. Since by weak duality $u^T b \geq -\hat{w}^T C x$ for all feasible solutions u of (6.11) and for all feasible solutions x of (6.12) and we already know that $\hat{u}^T b = -\hat{w}^T C x^0$ from (6.10), it follows that x^0 is an optimal solution of (6.12). Finally, we note that (6.12) is equivalent to

$$\min \left\{ \hat{w}^T C x : Ax = b,\ x \geqq 0 \right\}$$

and that, from the constraints in (6.8), $\hat{w} \geqq e > 0$. Therefore x^0 is an optimal solution of the weighted sum LP (6.3) with $\lambda = \hat{w}$ as weighting vector. □

Applying the second statement of Theorem 6.5 we have just proved that

$$\mathcal{X}_E = \mathcal{X}_{pE}$$

and

$$\mathcal{S}(\mathcal{Y}) = \mathcal{Y}_N = \mathcal{Y}_{pN}$$

hold for multiobjective linear programs. Therefore, every efficient solution is properly efficient and we can find all efficient solutions by weighted sum scalarization.

Regarding the first statement of Theorem 6.5 we have the following condition for the existence of efficient solutions, respectively nondominated points.

Proposition 6.12. *Let $x^0 \in \mathcal{X}$. Then the LP* (6.7) *is feasible and the following statements hold.*

1. *If $(\hat{x}, \hat{z})$ is an optimal solution of* (6.7) *then $\hat{x}$ is an efficient solution of the MOLP 6.1.*
2. *If* (6.7) *is unbounded then $\mathcal{X}_E = \emptyset$.*

The proof is left to the reader, see Exercise 6.3.

6.2 The Simplex Method and Biobjective Linear Programs

The purpose of this section is to review the Simplex method for linear programming and to extend it to the case of multiobjective linear programs with two objective functions. For more details and proofs we refer once more to textbooks on linear programming such as Dantzig and Thapa (2003), Dantzig (1998), or Padberg (1999). We repeat the formulation of a linear program from (6.5):

$$\min \left\{ c^T x : Ax = b, x \geqq 0 \right\}, \tag{6.13}$$

where $c \in \mathbb{R}^n$ and A is an $m \times n$ matrix. We will always assume that rank $A = m$ and that $b \geqq 0$.

A nonsingular $m \times m$ submatrix $A_\mathcal{B}$ of A is called *basis matrix*, where $\mathcal{B}$ is the set of indices of the columns of A defining $A_\mathcal{B}$. $\mathcal{B}$ is called a basis. Let

$\mathcal{N} := \{1, \ldots, n\} \setminus \mathcal{B}$ be the set of nonbasic column indices. A variable x_i and an index i are called *basic* if $i \in \mathcal{B}$, nonbasic otherwise.

With the notion of a basis it is possible to split A, c, and x into a basic and nonbasic part, using $\mathcal{B}$ and $\mathcal{N}$ as index sets, i.e. $A = (A_\mathcal{B}, A_\mathcal{N}), c^T = (c_\mathcal{B}^T, c_\mathcal{N}^T)$, and $x = (x_\mathcal{B}^T, x_\mathcal{N}^T)^T$. This allows us to rewrite the constraints $Ax = b$ as

$$(A_\mathcal{B}, A_\mathcal{N})(x_\mathcal{B}^T, x_\mathcal{N}^T)^T = b.$$

Since $A_\mathcal{B}$ is invertible we get

$$x_\mathcal{B} = A_\mathcal{B}^{-1}(b - A_\mathcal{N} x_\mathcal{N}). \tag{6.14}$$

Setting $x_\mathcal{N} = 0$ in (6.14) we obtain $x_\mathcal{B} = A_\mathcal{B}^{-1} b$. $(x_\mathcal{B}, 0)$ is called a *basic solution* of the LP (6.13). If in addition $x_\mathcal{B} \geqq 0$ it is called a *basic feasible solution* or BFS for short. The basis $\mathcal{B}$ is also called feasible.

We can compute the objective function value of $x = (x_\mathcal{B}^T, x_\mathcal{N}^T)^T$ as follows

$$\begin{aligned}(c_\mathcal{B}^T, c_\mathcal{N}^T)(x_\mathcal{B}^T, x_\mathcal{N}^T)^T &= c_\mathcal{B}^T x_\mathcal{B} + c_\mathcal{N}^T x_\mathcal{N} \\ &= c_\mathcal{B}^T A_\mathcal{B}^{-1} b + \left(c_\mathcal{N}^T - c_\mathcal{B}^T A_\mathcal{B}^{-1} A_\mathcal{N}\right) x_\mathcal{N}.\end{aligned} \tag{6.15}$$

The vector $\bar{c}^T = c^T - c_\mathcal{B}^T A_\mathcal{B}^{-1} A$ is called vector of *reduced costs*. Note that writing $\bar{c} = (\bar{c}_\mathcal{B}, \bar{c}_\mathcal{N})$ we always have $\bar{c}_\mathcal{B} = 0$.

Let $(x_\mathcal{B}, 0)$ be a basic feasible solution. From (6.15) it is clear that if there is some $s \in \mathcal{N}$ such that $\bar{c}_s < 0$ the value of $c^T x$ decreases if x_s increases from 0. Recomputing $x_\mathcal{B}$ as in (6.14) ensures that the constraints $Ax = b$ will still be satisfied. The increase of x_s must therefore be limited by the nonnegativity $x_\mathcal{B} \geqq 0$. Let $\tilde{A} := A_\mathcal{B}^{-1} A$ and $\tilde{b} := A_\mathcal{B}^{-1} b$.

Consider (6.14) for a basic variable $x_j, j \in \mathcal{B}$, i.e.

$$x_j = \tilde{b}_j - \tilde{A}_{js} x_s \geq 0, \tag{6.16}$$

where $\tilde{A}_{js}$ is the element of $\tilde{A}$ in row j and column s. If $\tilde{A}_{js} \leq 0$ then (6.16) is true for any $x_s \geq 0$, i.e. the objective value is unbounded. Otherwise x_s must be chosen so that $x_s \leq \tilde{b}_j / \tilde{A}_{js}$ for all $j \in \mathcal{B}$. The largest feasible value of x_s to retain a feasible solution is then

$$x_s = \min \left\{ \frac{\tilde{b}_j}{\tilde{A}_{js}} : j \in \mathcal{B}, \tilde{A}_{js} > 0 \right\}. \tag{6.17}$$

At this value, one basic variable $x_j, j \in \mathcal{B}$, will become zero, blocking any further increase of x_s. Let $r \in \mathcal{B}$ be an index for which the minimum in (6.17) is attained. Variable x_s is called *entering variable*, x_r is called *leaving variable*. The new basis $\mathcal{B}' = (\mathcal{B} \setminus \{r\}) \cup \{s\}$ defines a basic feasible solution $(x_{\mathcal{B}'}, 0)$ with a better objective value than $(x_\mathcal{B}, 0)$ as long as $\tilde{b}_j > 0$, which we shall assume for the moment. We discuss this after Algorithm 6.1.

Theorem 6.13 below justifies the Simplex method of linear programming.

Theorem 6.13. *1. If the LP (6.13) is feasible, i.e. if $\mathcal{X} \neq \emptyset$, then a basic feasible solution exists.*

2. If, furthermore, the objective function $c^T x$ is bounded from below on $\mathcal{X}$, then an optimal basic feasible solution exists.

3. A basic feasible solution $(x_{\mathcal{B}}, 0)$ is optimal if $\bar{c}_{\mathcal{N}} \geqq 0$.

If $(x_{\mathcal{B}}, 0)$ is an optimal BFS then $\mathcal{B}$ is called optimal basis.

Let $\mathcal{B}$ be a basis and $(x_{\mathcal{B}}, 0)$ be a basic feasible solution of the LP (6.13). Starting from this basis and BFS the Simplex algorithm finds an optimal basis and an optimal BFS.

Algorithm 6.1 (Simplex algorithm for linear programming.)

Input: *Basis $\mathcal{B}$ and BFS $(x_{\mathcal{B}}, 0)$.*

While $\{i \in \mathcal{N} : \bar{c}_i < 0\} \neq \emptyset$

Choose $s \in \{i \in \mathcal{N} : \bar{c}_i < 0\}$.

If $\tilde{A}_{js} \leq 0$ *for all* $j \in \mathcal{B}$, *STOP, the LP (6.13) is unbounded.*

Otherwise choose $r \in \operatorname{argmin}\left\{j \in \mathcal{B} : \frac{\tilde{b}_j}{\tilde{A}_{sj}}, \tilde{A}_{sj} > 0\right\}$.

Let $\mathcal{B} := (\mathcal{B} \setminus \{r\}) \cup \{s\}$ *and update* $\tilde{A} := A_{\mathcal{B}}^{-1} A$ *and* $\tilde{b} := A_{\mathcal{B}}^{-1} b$.

End while.

Output: *Optimal basis $\mathcal{B}$ and optimal basic feasible solution $(x_{\mathcal{B}}, 0)$.*

Assuming that $\tilde{b}_r > 0$ in every iteration, the algorithm terminates after finitely many iterations with an optimal solution or the conclusion that the LP is unbounded. This is because the value of x_s in the new basis is positive and so the objective value has decreased by $\bar{c}_s x_s$ and there are at most $\frac{n!}{m!(n-m)!}$ feasible bases.

On the other hand, if $\tilde{b}_r = 0$, i.e. $x_r = 0$, the new basis will have $x_s = 0$. In fact, both bases define the same BFS. Bases containing a variable at value 0 are called degenerate. It is then possible that the Simplex algorithm iterates between a sequence of degenerate bases without terminating. Rules to avoid this can be found in linear programming textbooks. From now on we shall assume that the LPs we consider are nondegenerate, but in Example 7.11 we will demonstrate why degeneracy is problematic.

Lemma 6.14. *Let the LP be nondegenerate and let $\mathcal{B}$ be an optimal basis. Then $\bar{c}_{\mathcal{N}} \geqq 0$.*

The proof of this result is left as Exercise 6.4.

In what follows we will use the tableau notation for the Simplex algorithm. A Simplex tableau summarizes the information in any iteration of the algorithm in a tabular form:

	$\bar{c}$	$-c_{\mathcal{B}}^T x_{\mathcal{B}}$
$\mathcal{B}$	$\tilde{A}$	$\tilde{b}$

The first row contains the reduced cost vector $\bar{c}$ and the negative value of the objective at the current basis. The second "row" contains the basic indices as well as the constraint matrix and right hand side at the current basis. One iteration consists in the determination of an entering and a leaving variable, x_s and x_r, and the update of $\tilde{A}$ and $\tilde{b}$. This is called a pivot and is done by Gaussian elimination to convert column s of $\tilde{A}$ into a unit column with $\tilde{A}_{rs} = 1$. $\tilde{A}_{rs}$ is called the *pivot element.*

Algorithm 6.1 is initialized with a feasible basis. To find one, an auxiliary linear program is solved:

$$\begin{aligned} &\min e^T z \\ &\text{subject to } Ax + z = b \\ &\qquad\qquad\quad x, z \geqq 0. \end{aligned} \tag{6.18}$$

The LP (6.18) is always feasible and $(x, z) = (0, b)$ is a basic feasible solution, because $b \geqq 0$ by general assumption.

Proposition 6.15. *The LP* (6.13) *is feasible, i.e.* $\mathcal{X} \neq \emptyset$, *if and only if the auxiliary LP* (6.18) *has an optimal solution* $(\hat{x}, \hat{z})$ *with* $\hat{z} = 0$.

If $\hat{x}$ in an optimal solution of (6.18) is not a BFS of the original LP, it can always be easily converted into one.

This proposition concludes our summary of the algebra of linear programming, which we use to construct the multicriteria Simplex algorithm. Let us now consider the geometry.

Let $a \in \mathbb{R}^n$ and $b \in \mathbb{R}$. The set

$$\mathcal{H}_{a,b} := \{x \in \mathbb{R}^n : a^T x = b\} \tag{6.19}$$

is a *hyperplane.* A hyperplane defines a halfspace

$$\mathcal{H}_{a,b} := \{x \in \mathbb{R}^n : a^T x \leq b\}.$$

Let $\mathcal{X} \subset \mathbb{R}^n$ be a nonempty set. A hyperplane $\mathcal{H}_{a,b}$ is called *supporting hyperplane* of $\mathcal{X}$ at $\hat{x}$ if $\hat{x} \in \mathcal{X} \cap \mathcal{H}_{a,b}$ and $\mathcal{X} \subset \mathcal{H}_{a,b}$. We also say $\mathcal{H}_{a,b}$ supports $\mathcal{X}$ at $\hat{x}$.

Let $\mathcal{X}$ be the intersection of finitely many halfspaces. Then $\mathcal{X}$ is called *polyhedron.* For example, the feasible set $\mathcal{X} = \{x \in \mathbb{R}^n : Ax = b,\ x \geqq 0\}$ of an LP is a polyhedron. A polyhedron $\mathcal{X}$ is called a *polytope* if $\mathcal{X}$ is bounded.

$x \in \mathcal{X}$ is an *extreme point* of $\mathcal{X}$ if $x = \alpha x^1 + (1-\alpha)x^2$ with $x^1, x^2 \in \mathcal{X}$ and $0 \leq \alpha \leq 1$ implies that $x^1 = x^2 = x$.

Let $\mathcal{X} \neq \emptyset$ be a polyhedron written $\mathcal{X} = \{x : Ax \leqq b\}$. If $d \in \mathbb{R}^n$ is such that $Ad \leqq 0$, then d is called a *ray* of $\mathcal{X}$. A ray is called an *extreme ray* if there are no rays d^1, d^2, $d^1 \neq \alpha d^2$ for all $\alpha \in \mathbb{R}_>$, such that $r = (1/2)(r^1 + r^2)$.

Lemma 6.16. *Let $\mathcal{X}$ be a polyhedron and $d \in \mathbb{R}^n$ be a ray of $\mathcal{X}$. Then $x + \alpha d \in \mathcal{X}$ for all $\alpha \geq 0$. The set $\{d : Ad \leqq 0\}$ is a convex cone.*

The *dimension* of a polyhedron $\mathcal{X}$, $\dim \mathcal{X}$, is the maximal number of affinely independent points of $\mathcal{X}$, minus 1. Let $\mathcal{H}_{a,b}$ be a supporting hyperplane of polyhedron $\mathcal{X}$. Then $\mathcal{F} = \mathcal{X} \cap \mathcal{H}$ is called a *face* of $\mathcal{X}$. A face $\mathcal{F}$ of polyhedron $\mathcal{X}$ is itself a polyhedron, thus $\dim \mathcal{F}$ is defined. We consider only faces $\mathcal{F}$ with $\emptyset \neq \mathcal{F} \neq \mathcal{X}$. These are called *proper faces.* With the definitions given so far an *extreme point* is a face of dimension 0. A face of dimension 1 is called an *edge* of $\mathcal{X}$. A *facet* is a face of dimension $\dim \mathcal{X} - 1$. Note that all proper faces of $\mathcal{X}$ belong to the boundary of $\mathcal{X}$.

The algebra and geometry of linear programming are related through the following theorem.

Theorem 6.17.
1. *A basic feasible solution $(x_\mathcal{B}, 0)$ of a linear program (6.13) is an an extreme point of the feasible set $\mathcal{X}$. However, several feasible bases may define the same basic feasible solution and therefore the same extreme point (in case of degeneracy).*
2. *If $\mathcal{X} \neq \emptyset$ and the LP (6.13) is bounded, the set of all optimal solutions of the LP is either $\mathcal{X}$ itself or a face of $\mathcal{X}$.*
3. *For each extreme point $\hat{x}$ of $\mathcal{X}$ there exists a cost vector $c \in \mathbb{R}^n$ such that $\hat{x}$ is an optimal solution of* $\min\{c^T x : \ x \in \mathcal{X}\}$.

This concludes our review of single objective linear programming and we are now in a position to extend the Simplex algorithm 6.1 to deal with LPs with two objective functions. This extension is objective row parametric linear programming.

Let us consider a biobjective linear program

$$\begin{array}{lr} \min & ((c^1)^T x, (c^2)^T x) \\ \text{subject to} & Ax = b \\ & x \geqq 0. \end{array} \tag{6.20}$$

From Theorem 6.11 we know that finding the efficient solutions of (6.20) is equivalent to solving the LP

$$\min \left\{ \lambda_1 (c^1)^T x + \lambda_2 (c^2)^T x : Ax = b, x \geqq 0 \right\}$$

for all $\lambda \in \mathbb{R}^2_>$. Without loss of generality (dividing the objective function by $\lambda_1 + \lambda_2$) we can assume that $(\lambda_1, \lambda_2) = (\lambda, 1 - \lambda)$. We define the parametric objective function

$$c(\lambda) := \lambda c^1 + (1 - \lambda)c^2.$$

Thus we need to solve

$$\min \left\{ c(\lambda)^T x : Ax = b, x \geqq 0 \right\}, \tag{6.21}$$

which is a parametric linear program. We need to determine which values of λ are relevant. To this end consider a feasible basis $\mathcal{B}$ and the reduced cost vector $c(\lambda)$ of the parametric objective,

$$\bar{c}(\lambda) = \lambda \bar{c}^1 + (1 - \lambda)\bar{c}^2. \tag{6.22}$$

Suppose $\hat{\mathcal{B}}$ is an optimal basis for the LP (6.21) for some $\lambda = \hat{\lambda}$. The optimality criterion of Theorem 6.13 with Lemma 6.14 implies that $\bar{c}(\lambda) \geqq 0$. Two cases need to be distinguished.

The first case is $\bar{c}^2 \geq 0$. Then from (6.22) $\bar{c}(\lambda) \geqq 0$ for all $\lambda < \hat{\lambda}$ and $\hat{\mathcal{B}}$ is an optimal basis for all $0 \leq \lambda \leq \hat{\lambda}$. Otherwise there is at least one $i \in \mathcal{N}$ such that $\bar{c}^2_i < 0$. Therefore there is a value of $\lambda < \hat{\lambda}$ such that $\bar{c}(\lambda)_i = 0$, i.e.

$$\begin{aligned} \lambda \bar{c}^1_i + (1 - \lambda)\bar{c}^2_i &= 0 \\ \lambda(\bar{c}^1_i - \bar{c}^2_i) + \bar{c}^2_i &= 0 \\ \lambda &= \frac{-\bar{c}^2_i}{\bar{c}^1_i - \bar{c}^2_i}. \end{aligned}$$

Define $\mathcal{I} = \{i \in \mathcal{N} : \bar{c}^2_i < 0, \bar{c}^1_i \geq 0\}$ and

$$\lambda' := \max_{i \in \mathcal{I}} \frac{-\bar{c}^2_i}{\bar{c}^1_i - \bar{c}^2_i}. \tag{6.23}$$

Basis $\hat{\mathcal{B}}$ is optimal for the parametric LP (6.21) for all $\lambda \in [\lambda', \hat{\lambda}]$. As soon as $\lambda \leq \lambda'$ new bases become optimal. Therefore an entering variable x_s has to be chosen, where the maximum in (6.23) is attained for $i = s$.

The above means that we can solve (6.21) by first solving it with $\lambda = 1$ (we assume that LP(1) has an optimal solution, what to do otherwise is discussed in the next chapter) and then iteratively finding entering variables and new λ values according to (6.23) until $\bar{c}^2 \geqq 0$. This procedure is stated as Algorithm 6.2. Note that if $\mathcal{B}$ is an optimal basis of LP(1) then it is not an optimal basis of LP(0), due to our basic assumption. Thus $\mathcal{I} \neq \emptyset$ initially.

Algorithm 6.2 (Parametric Simplex for biobjective LPs.)

Input: *Data A, b, C for a biobjective LP.*

Phase I: *Solve the auxiliary LP* (6.18) *using the Simplex algorithm 6.1. If the optimal value is positive, STOP, $\mathcal{X} = \emptyset$. Otherwise let $\mathcal{B}$ be an optimal basis.*

Phase II: *Solve the LP* (6.21) *for $\lambda = 1$ starting from basis $\mathcal{B}$ found in Phase I yielding an optimal basis $\hat{\mathcal{B}}$. Compute $\tilde{A}$ and $\tilde{b}$.*

Phase III: *While $\mathcal{I} = \{i \in \mathcal{N} : \bar{c}_i^2 < 0, \bar{c}_i^1 \geq 0\} \neq \emptyset$.*

$\lambda := \max_{i \in \mathcal{I}} \frac{-\bar{c}_i^2}{\bar{c}_i^1 - \bar{c}_i^2}.$

$s \in \operatorname{argmin} \left\{ i \in \mathcal{I} : \frac{-\bar{c}_i^2}{\bar{c}_i^1 - \bar{c}_i^2} \right\}.$

$r \in \operatorname{argmin} \left\{ j \in \mathcal{B} : \frac{\tilde{b}_j}{\tilde{A}_{sj}}, \tilde{A}_{sj} > 0 \right\}.$

Let $\mathcal{B} := (\mathcal{B} \setminus \{r\}) \cup \{s\}$ and update $\tilde{A}$ and $\tilde{b}$.

End while.

Output: *Sequence of λ-values and sequence of optimal BFSs.*

In every iteration, to determine a new optimal basis, an index s, at which the critical value λ' of (6.23) is attained, is chosen as pivot column (entering variable x_s). The pivot row (leaving variable x_r) is chosen by the usual quotient rule. We pivot x_s into the basis. Proceeding in this way, we generate a sequence of critical λ values $1 = \lambda^1 > \ldots > \lambda^l = 0$ and optimal bases $\mathcal{B}^1, \ldots, \mathcal{B}^{l-1}$ which define optimal BFSs of (6.21) for all λ: $\mathcal{B}^i$ is an optimal basis of (6.21) for all $\lambda \in [\lambda^i, \lambda^{i+1}], i = 1, \ldots, l$.

Example 6.18 (Steuer (1985)). We solve the parametric linear program resulting from the biobjective LP of Example 6.3 with slack variables x_3, x_4.

$$\begin{array}{llll}
\min & (4\lambda - 1)x_1 + (3\lambda - 2)x_2 & & \\
\text{subject to} & \qquad\qquad x_2 + x_3 & & = 3 \\
& 3x_1 - \qquad x_2 & + x_4 & = 6 \\
& & & x \geqq 0.
\end{array}$$

We show the Simplex tableaus with both reduced cost vectors $\bar{c}^1$ and $\bar{c}^2$. An optimal basis for $\lambda = 1$ is obviously given by $\mathcal{B} = \{3, 4\}$ and an optimal basic feasible solution is $x = (0, 0, 3, 6)$. We can therefore start with Phase III. Apart from the tableau we show the computation of $\mathcal{I}$ and λ' in every iteration. Pivot elements in the tableaus are indicated by a square frame.

Iteration 1:

$\bar{c}^1$	3	1	0	0	0
$\bar{c}^2$	-1	-2	0	0	0
x_3	0	[1]	1	0	3
x_4	3	-1	0	1	6

$\lambda = 1, \bar{c}(\lambda) = (3, 1, 0, 0)$, $\mathcal{B}^1 = \{3, 4\}$, $x^1 = (0, 0, 3, 6)$
$\mathcal{I} = \{1, 2\}$, $\lambda' = \max\left\{\frac{1}{3-1}, \frac{2}{1+2}\right\} = \frac{2}{3}$
$s = 2$, $r = 3$

Iteration 2:

$\bar{c}^1$	3	0	-1	0	-3
$\bar{c}^2$	-1	0	2	0	6
x_2	0	1	1	0	3
x_4	[3]	0	1	1	9

$\lambda = 2/3, \bar{c}(\lambda) = (5/3, 0, 0, 0)$, $\mathcal{B}^2 = \{2, 4\}$, $x^2 = (0, 3, 0, 9)$
$\mathcal{I} = \{1\}$, $\lambda' = \max\left\{\frac{1}{3+1}\right\} = \frac{1}{4}$
$s = 1$, $r = 4$

Iteration 3:

$\bar{c}^1$	0	0	-2	-1	-12
$\bar{c}^2$	0	0	7/3	1/3	9
x_2	0	1	1	0	3
x_1	1	0	1/3	1/3	3

$\lambda = 1/4, \bar{c}(\lambda) = (0, 0, 5/4, 0)$, $\mathcal{B}^3 = \{1, 2\}$, $x^3 = (3, 3, 0, 0)$
$\mathcal{I} = \emptyset$.
The algorithm STOPS and returns the values $\lambda^1 = 1, \lambda^2 = 2/3, \lambda^3 = 1/4, \lambda^4 = 0$ and the basic feasible solutions x^1, x^2, x^3.

Note that in each iteration $\bar{c}(\lambda)$ can be calculated with the previous value λ and the new λ' and both indicate optimality, as predicted by (6.23). To summarize, we have the following results.

- Basis $\mathcal{B}^1 = (3, 4)$ and BFS $x^1 = (0, 0, 3, 6)$ are optimal for $\lambda \in [2/3, 1]$.
- Basis $\mathcal{B}^2 = (2, 4)$ and BFS $x^2 = (0, 3, 0, 9)$ are optimal for $\lambda \in [1/4, 2/3]$, and
- Basis $\mathcal{B}^3 = (1, 2)$ and BFS $x^3 = (3, 3, 0, 0)$ are optimal for $\lambda \in [0, 1/4]$.

The (negative of the) objective vectors of the three basic feasible solutions are also shown in the Simplex tableaus, they are $Cx^1 = (0, 0)$, $Cx^2 = (3, -6)$, and $Cx^3 = (12, -9)$.

The two critical values $\lambda = 2/3$ and $\lambda = 1/4$ that induce basis changes correspond to weighting vectors $(2/3, 1/3)$ and $(1/4, 3/4)$. Note that the contour lines in decision space defined by these vectors are parallel to the edges between x^1 and x^2 respectively x^2 and x^3 in Figure 6.4 because

$$\frac{2}{3}(3x_1 + x_2) + \frac{1}{3}(-x_1 - 2x_2) = \frac{5}{3}x_1$$
$$\frac{1}{4}(3x_1 + x_2) + \frac{3}{4}(-x_1 - 2x_2) = -\frac{5}{4}x_2.$$

Thus these two vertical and horizontal lines are also geometrically identified as optimal solutions of the weighted sum problems.

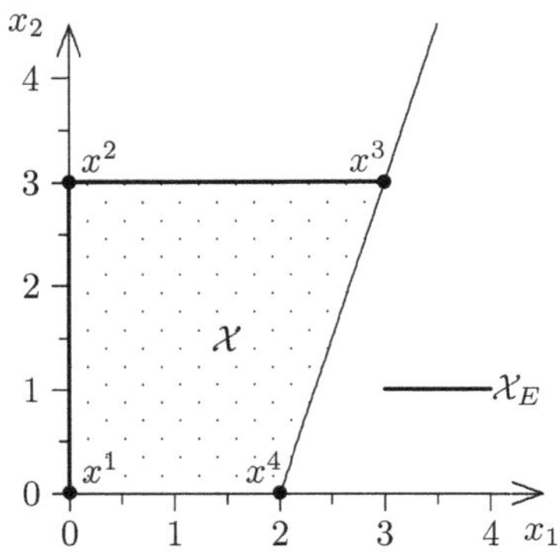

Fig. 6.4. Feasible set and efficient set in Example 6.18.

□

In the sequence $1 = \lambda^1 > \lambda^2 > \ldots > \lambda^l = 0$ an optimal basic feasible solution x^i is always an optimal solution of (6.21) for all $\lambda \in [\lambda^{i+1}, \lambda^i]$. Therefore, for each λ^i, $2 \leq i \leq l-1$ we have *two* optimal basic feasible solutions x^i and x^{i-1}. In Figure 6.4 this means that the level curves $c(\lambda)^T x = \gamma$ are parallel to the face of $\mathcal{X}$ defined by x^i and x^{i-1}. Therefore, linearity implies that every feasible solution on the edge between these extreme points is an optimal solution of (6.21) with $\lambda = \lambda^i$. This edge is $\text{conv}(x^i, x^{i-1})$, a face of $\mathcal{X}$. In particular, it follows that the subset of $\mathcal{X}_E$, which consists of the efficient BFSs found by Algorithm 6.2 and the edges connecting them is connected. Of course the image of this set is also a connected subset of $\mathcal{Y}_N$ due to the continuity of the linear map C.

Now reconsidering what we have actually done by solving (6.21), Theorem 6.11 shows that we have solved a bicriteria linear program in some sense: For

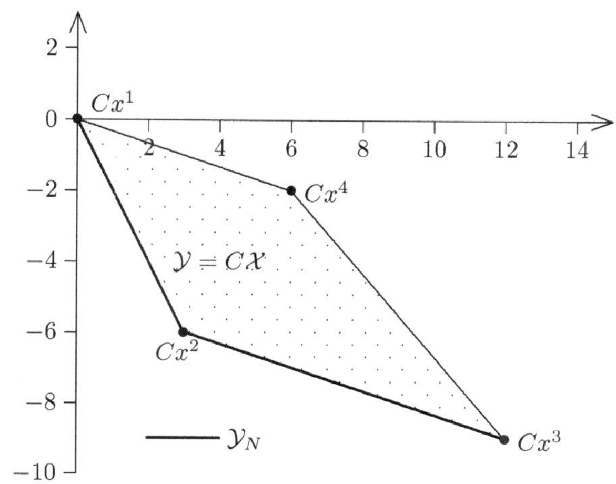

Fig. 6.5. Objective space and nondominated set in Example 6.18.

all $\lambda \in (0, 1)$ we have found an optimal solution of LP(λ). However, there is no guarantee that all efficient solutions can be reconstructed from the bases $\mathcal{B}^1, \ldots, \mathcal{B}^{l-1}$. Imagine an extension of Example 6.3 with a third variable x_3 and constraints $0 \leq x_3 \leq 1$, but otherwise unchanged. Thus, $\mathcal{X}$ is now three-dimensional and $\mathcal{X}_E$ consists of two two-dimensional faces of $\mathcal{X}$, but $\mathcal{Y}$ and $\mathcal{Y}_N$ are unchanged. Algorithm 6.2 may then yield the same bases as before, i.e. it finds a path through $\mathcal{X}_E$ that traces $\mathcal{Y}_N$ but not all bases defining efficient solutions. This issue will be discussed in the next chapter. Thus, Algorithm 6.2 finds, for each $y \in \mathcal{Y}_N$ an $x \in \mathcal{X}_\mathcal{E}$ with $Cx = y$.

It may also be the case that for the initial LP with objective $c(1)$ an optimal BFS has been found that is a weakly efficient rather than an efficient solution of the biobjective LP. This problem can be avoided if a second LP is solved initially. In this LP the second objective is minimized under a constraint that the first retains its optimal value (that is, a lexicographic LP is solved):

$$\begin{aligned} &\min (c^2)^T x \\ \text{subject to } &(c^1)^T x = \hat{c} \\ &x \in \mathcal{X}, \end{aligned}$$

where $\hat{c} = \min\{(c^1)^T x : Ax = b, x \geqq 0\}$. For the subsequent iterations of Algorithm 6.2 the additional constraint is dropped again.

In the next chapter we generalize these observations to linear programs with p objectives. We devise a method to identify an initial efficient BFS, if one exists, and a method to pivot among efficient bases.

Exercises

6.1. Prove that for the MOLP $\min\{Cx : Ax = b, x \geqq 0\}$ both $\mathcal{X}$ and $\mathcal{Y}$ are closed and convex.

6.2. Give an example of an MOLP with $p = 2$ objectives such that an optimal solution of the weighted sum LP(λ) with $\lambda \geq 0$ but $\lambda_i = 0$ for $i = 1$ or $i = 2$ is weakly efficient, but not efficient.

6.3. Prove Proposition 6.12. Let $x^0 \in \mathcal{X}$. Then the LP (6.7) is feasible and the following statements hold.

1. If $(\hat{x}, \hat{z})$ is an optimal solution of (6.7) then $\hat{x}$ is an efficient solution of the MOLP 6.1.
2. If (6.7) is unbounded then $\mathcal{X}_E = \emptyset$.

6.4. Let the LP $\{\min c^T x : Ax = b, x \geqq 0\}$ be nondegenerate and let $\mathcal{B}$ a feasible basis. Show that $\bar{c}_N \geqq 0$ if $\mathcal{B}$ is an optimal basis.

6.5. Prove Proposition 6.15, i.e. show that $\mathcal{X} \neq \emptyset$ if and only if the auxiliary LP (6.18) has optimal value zero.

6.6. Consider the parametric linear program

$$\begin{array}{lr} \min & \lambda(-2x_1 + x_2) + (1-\lambda)(-4x_1 - 3x_2) \\ \text{subject to} & x_1 + 2x_2 \leq 10 \\ x_1 & \leq 5 \\ x_1, x_2 & \geq 0. \end{array}$$

Solve the problem with Algorithm 6.2. Determine $\mathcal{X}_E$ as well as $\mathcal{Y}_E$ and illustrate the results graphically.

6.7. This exercise is about the structure of $\mathcal{X}_E$. Give examples of MOLPs with the following properties.

1. $\mathcal{X}_E$ is a singleton, although $\mathcal{X}$ is full dimensional, i.e $\dim \mathcal{X} = n$.
2. $\mathcal{X} \neq \emptyset$ but $\mathcal{X}_E = \emptyset$.
3. It is possible that some objectives are unbounded, yet $\mathcal{X}_E \neq \emptyset$. Show this behaviour for the MOLP

$$\begin{array}{lrcl} \min & x_1 + 2x_2 & & \\ \min & - 2x_2 & & \\ \text{subject to} & -x_1 + x_2 & \leq & 3 \\ & x_1 + x_2 & \geq & 3 \\ & x_1, \quad x_2 & \geq & 0. \end{array}$$

What can you say about $\mathcal{X}_E$ in this case?

7

A Multiobjective Simplex Method

An MOLP with two objectives can be conveniently solved using the parametric Simplex method presented in Algorithm 6.2. With three or more objectives, however, this is no longer possible because we deal with at least two parameters in the objective function $c(\lambda)$.

7.1 Algebra of Multiobjective Linear Programming

In this section we consider the general MOLP

$$\begin{array}{ll} \min & Cx \\ \text{subject to} & Ax = b \\ & x \geqq 0. \end{array} \tag{7.1}$$

For $\lambda \in \mathbb{R}^p_>$ we denote by LP(λ) the weighted sum linear program

$$\min \left\{ \lambda^T C x : Ax = b, x \geqq 0 \right\}. \tag{7.2}$$

We use the notation $\bar{C} = C - C_{\mathcal{B}} A_{\mathcal{B}}^{-1} A$ for the reduced cost matrix with respect to basis $\mathcal{B}$ and $R := \bar{C}_{\mathcal{N}}$ for the nonbasic part of the reduced cost matrix. Note that $\bar{C}_{\mathcal{B}} = 0$ according to (6.15) and is therefore uninteresting. Proofs in this section will make use of Theorem 6.11. These results are multicriteria analogies of well known linear programming results, or necessary extensions to cope with the increased complexity of multiobjective compared to single objective linear programming.

Lemma 7.1. *If $\mathcal{X}_E \neq \emptyset$ then $\mathcal{X}$ has an efficient basic feasible solution.*

Proof. By Theorem 6.11 there is some $\lambda \in \mathbb{R}^p_>$ such that $\min_{x \in \mathcal{X}} \lambda^T C x$ has an optimal solution. But by Theorem 6.13 the LP(λ) $\min_{x \in \mathcal{X}} \lambda^T C x$ has a optimal basic feasible solution solution, which is an efficient solution of the MOLP by Theorem 6.6. □

Lemma 7.1 justifies the definition of an efficient basis.

Definition 7.2. *A feasible basis $\mathcal{B}$ is called* efficient basis *if $\mathcal{B}$ is an optimal basis of LP(λ) for some $\lambda \in \mathbb{R}^p_>$.*

We now look at pivoting among efficient bases. We say that a pivot is a *feasible pivot* if the solution obtained after the pivot step is feasible, even if the pivot element $\tilde{A}_{rs} < 0$.

Definition 7.3. *Two bases $\mathcal{B}$ and $\hat{\mathcal{B}}$ are called* adjacent *if one can be obtained from the other by a single pivot step.*

Definition 7.4.
1. *Let $\mathcal{B}$ be an efficient basis. Variable $x_j, j \in \mathcal{N}$ is called* efficient nonbasic variable *at $\mathcal{B}$ if there exists a $\lambda \in \mathbb{R}^p_>$ such that $\lambda^T R \geqq 0$ and $\lambda^T r^j = 0$, where r^j is the column of R corresponding to variable x_j.*
2. *Let $\mathcal{B}$ be an efficient basis and let x_j be an efficient nonbasic variable. Then a feasible pivot from $\mathcal{B}$ with x_j entering the basis is called an* efficient pivot *with respect to $\mathcal{B}$ and x_j.*

The system $\lambda^T R \geqq 0, \lambda^T r^j = 0$ is the general form of the equations we used to compute the critical λ values in parametric linear programming that were used to derive (6.23): We chose s such that $\bar{c}(\lambda) \geq 0, \bar{c}(\lambda)_s = 0$.

Proposition 7.5. *Let $\mathcal{B}$ be an efficient basis. There exists an efficient nonbasic variable at $\mathcal{B}$.*

Proof. Because $\mathcal{B}$ is an efficient basis there exists $\lambda > 0$ such that $\lambda^T R \geqq 0$. Thus the set $\mathcal{L} := \{\lambda > 0 : \lambda^T R \geqq 0\}$ is not empty. We have to show that there is $\lambda \in \mathcal{L}$ and $j \in \mathcal{N}$ such that $\lambda^T r^j = 0$.

First we observe that there is no column r of R such that $r \leq 0$. There also must be at least one column with positive and negative elements, because of the general assumption (6.2). Now let $\lambda^* \in \mathcal{L}$. In particular $\lambda^{*T} \geqq 0$. Let $\lambda' \in \mathbb{R}^p_>$ be such that $\mathcal{I} := \{i \in \mathcal{N} : \lambda'^T r^j < 0\} \neq \emptyset$. Such a λ must exist, because R contains at least one negative entry.

We define $\phi : \mathbb{R} \to \mathbb{R}^{|\mathcal{N}|}$ by

$$\phi_i(t) := (t\lambda^{*T} + (1-t)\lambda'^T)r^i, i \in \mathcal{N}.$$

Thus, $\phi(0) = \lambda'^T R$ and $\phi(1) = \lambda^{*T} R \geqq 0$. For each $i \in \mathcal{N} \setminus \mathcal{I}$ we have that $\phi_i(t) \geq 0$ for all $t \in [0,1]$. For all $i \in \mathcal{I}$ there exists some $t_i \in [0,1]$ such that

$$\phi_i(t) \begin{cases} < 0, \, t \in [0, t_i) \\ = 0, \, t = t_i \\ \geq 0, \, t \in [t_i, 1]. \end{cases}$$

With $t^* := \max\{t_i : i \in \mathcal{I}\}$ we have that $\phi_i(t^*) \geqq 0$ and $\phi_i(t^*) = 0$ for some $i \in \mathcal{I}$. Thus $\hat{\lambda} := t\lambda^* + (1-t)\lambda' \in \mathcal{L}$ and the proof is complete. □

Example 7.6. It might appear that any nonbasic variable such that r^j contains positive and negative entries is an efficient nonbasic variable. This is not the case, as the following example shows. Let

$$R = \begin{pmatrix} -4 & 1 \\ -2 & 1 \end{pmatrix}.$$

Then there is no $\lambda \in \mathbb{R}^2_>$ such that $\lambda^T R \geqq 0$ and $\lambda^T r^2 = 0$. The latter equation means $\lambda_2 = 2\lambda_1$. Then $\lambda^T r^1 \geq 0$ would require $-2\lambda_2 \geq 0$, an impossibility.

□

Lemma 7.7. *Let $\mathcal{B}$ be an efficient basis and x_j be an efficient nonbasic variable. Then any efficient pivot from $\mathcal{B}$ leads to an adjacent efficient basis $\hat{\mathcal{B}}$.*

Proof. Let x_j be the entering variable at basis $\mathcal{B}$. Because x_j is an efficient nonbasic variable, we have $\lambda \in \mathbb{R}^p_>$ with $\lambda^T R \geqq 0$ and $\lambda^T r^j = 0$. Thus x_j is a nonbasic variable with reduced cost 0 in LP(λ). This means that the reduced costs of LP(λ) do not change after a pivot with x_j entering. Let $\hat{\mathcal{B}}$ be the resulting basis with any feasible pivot and entering variable x_j. Then $\lambda^T R \geqq 0$ and $\lambda^T r^j = 0$ at $\hat{\mathcal{B}}$, i.e. $\hat{\mathcal{B}}$ is an optimal basis for LP(λ) and therefore an adjacent efficient basis. □

We need a method to check whether a nonbasic variable x_j at an efficient basis $\mathcal{B}$ is efficient. This can be done by performing a test that consists in solving an LP.

Theorem 7.8 (Evans and Steuer (1973)). *Let $\mathcal{B}$ be an efficient basis and let x_j be a nonbasic variable. Variable x_j is an efficient nonbasic variable if and only if the LP*

$$\begin{array}{ll} \max & e^t v \\ \text{subject to } & Rz - r^j\delta + Iv = 0 \\ & z, \delta, v \geqq 0 \end{array} \tag{7.3}$$

has an optimal value of 0.

Proof. By Definition 7.4 x_j is an efficient nonbasic variable if the LP

$$\begin{array}{lr} \min & 0^T\lambda = 0 \\ \text{subject to} & R^T\lambda \geqq 0 \\ & (r^j)^T\lambda = 0 \\ & I\lambda \geqq e \\ & \lambda \geqq 0 \end{array} \tag{7.4}$$

has an optimal objective value of 0, i.e. if it is feasible. The first two constraints of (7.4) together are equivalent to $R^T\lambda \geqq 0, (r^j)^T\lambda \leq 0$, or $R^T\lambda \geqq 0, (-r^j)^T\lambda \geq 0$, which gives the LP

$$\begin{array}{lr} \min & 0^T\lambda = 0 \\ \text{subject to} & R^T\lambda \geqq 0 \\ & -(r^j)^T\lambda \geq 0 \\ & I\lambda \geqq e \\ & \lambda \geqq 0. \end{array} \tag{7.5}$$

The dual of (7.5) is

$$\begin{array}{lr} \max & e^Tv \\ \text{subject to} & Rz - r^j\delta + Iv + It = 0 \\ & z, \delta, v, t \geqq 0. \end{array} \tag{7.6}$$

Since an optimal solution of (7.6) will always contain t at value zero, this is equivalent to

$$\begin{array}{lr} \max & e^Tv \\ \text{subject to} & Rz - r^j\delta + Iv = 0 \\ & z, \delta, v \geqq 0, \end{array}$$

which is (7.3). □

It is important to note that the test problem (7.3) is always feasible since $(z, \delta, v) = 0$ can be chosen. The proof also 7.8 also shows that (7.3) can only have either an optimal solution with $v = 0$ (the objective value of (7.4) is zero), or be unbounded. With this observation we conclude that

- x_j is an efficient nonbasic variable if and only if (7.3) is bounded and has optimal value 0,
- x_j is an "inefficient" nonbasic variable if and only if (7.3) is unbounded.

The Simplex algorithm works by moving along adjacent bases until an optimal one is found. We want to make use of this principle to identify all efficient bases, i.e. we want to move from efficient basis to efficient basis. Therefore we must prove that it is indeed possible to restrict ourselves to adjacent bases only, i.e. that the efficient bases are connected in terms of adjacency.

Definition 7.9. *Two efficient bases $\mathcal{B}$ and $\hat{\mathcal{B}}$ are called* connected *if one can be obtained from the other by performing only efficient pivots.*

We prove that all efficient bases are connected using parametric programming. Note that single objective optimal pivots (i.e. the entering variable is

x_s with $\bar{c}_s = 0$) as well as parametric pivots are efficient pivots (one of the two reduced costs is negative, the other positive) according to (6.23). These cases are also covered by Proposition 7.5. Theorem 7.10 is the foundation for the multicriteria Simplex algorithm. We present a proof by Steuer (1985).

Theorem 7.10 (Steuer (1985)). *All efficient bases are connected.*

Proof. Let $\mathcal{B}$ and $\hat{\mathcal{B}}$ be two efficient bases. Let $\lambda, \hat{\lambda} \in \mathbb{R}^p_>$ be the positive weighting vectors for which $\mathcal{B}$ and $\hat{\mathcal{B}}$ are optimal bases for LP(λ) and LP($\hat{\lambda}$), respectively. We consider the parametric LP with objective function

$$c(\Phi) = \Phi\hat{\lambda}^T C + (1-\Phi)\lambda^T C \tag{7.7}$$

with $\Phi \in [0,1]$.

Let $\hat{B}$ be the first basis (for $\Phi = 1$). After several parametric programming or optimal pivots we get a basis $\tilde{\mathcal{B}}$ which is optimal for LP(λ). Since $\lambda^* = \Phi\hat{\lambda} + (1-\Phi)\lambda \in \mathbb{R}^p_>$ for all $\Phi \in [0,1]$ all intermediate bases are optimal for LP(λ^*) for some $\lambda^* \in \mathbb{R}^p_>$, i.e. they are efficient bases. All parametric and optimal pivots are efficient pivots as explained above. If $\tilde{\mathcal{B}} = \mathcal{B}$ we are done. Otherwise $\mathcal{B}$ can be obtained from $\tilde{\mathcal{B}}$ by efficient pivots (i.e. optimal pivots for LP(λ)), because both $\mathcal{B}$ and $\tilde{\mathcal{B}}$ are optimal bases for this LP. □

It is now possible to explain why the nontriviality assumption is necessary. Without it, the existence of efficient nonbasic variables is not guaranteed, and therefore Theorem 7.10 may fail. Example 7.11 also demonstrates a problem with degenerate MOLPs.

Example 7.11 (Steuer (2002)). We want to solve the following MOLP

$$\begin{array}{lrcrcrl} \min & & - & 2x_2 & + & x_3 & \\ \min & -x_1 & + & 2x_2 & + & x_3 & \\ \text{subject to} & & & x_2 & + & 4x_3 & \leq 8 \\ & x_1 & + & x_2 & & & \leq 8 \\ x_1, & & & x_2, & & x_3 & \geq 0. \end{array}$$

We introduce slack variables x_4, x_5 to write the LP in equality form. It is clear that both objective functions are minimized at the same solution, $\hat{x} = (0,4,0,0,0)$. Thus $\mathcal{X}_E = \{\hat{x}\}$. Because the only nonzero variable at $\hat{x}$ is $\hat{x}_2 = 2$, there are four different bases that define the same efficient basic feasible solution, namely $\{1,2\}, \{2,3\}, \{2,4\}$, and $\{2,5\}$ (the problem is degenerate). Below we show the Simplex tableaus for these four bases.

$\bar{c}^1$	0	0	7	2	0	16
$\bar{c}^2$	0	0	8	$\frac{9}{4}$	$-\frac{1}{4}$	16
x_1	1	0	-1	$-\frac{1}{4}$	$\frac{1}{4}$	0
x_2	0	1	4	1	0	8

$\bar{c}^1$	7	0	0	$\frac{1}{4}$	$\frac{7}{4}$	16
$\bar{c}^2$	8	0	0	$\frac{1}{4}$	$\frac{7}{4}$	16
x_3	-1	0	1	$\frac{1}{4}$	-$\frac{1}{4}$	0
x_2	4	1	0	0	1	8

$\bar{c}^1$	8	0	-1	0	2	16
$\bar{c}^2$	9	0	-1	0	2	16
x_4	-4	0	4	1	-1	0
x_2	4	1	0	0	1	8

$\bar{c}^1$	0	0	7	2	0	16
$\bar{c}^2$	1	0	7	2	0	16
x_2	0	1	4	1	0	8
x_5	4	0	-4	-1	1	0

Bases $\{1,2\}$ and $\{2,4\}$ are not efficient according to Definition 7.2 because R contains columns that do not have positive entries. This is due to degeneracy, which makes the negative reduced cost values possible, despite the BFS being efficient/optimal.

Furthermore, bases $\{2,3\}$ and $\{2,5\}$ are efficient. The definition is satisfied for all $\lambda \in \mathbb{R}^2_>$. However, for these bases R has no negative entries at all, hence no efficient nonbasic variable according to Definition 7.3 exist. The example therefore shows, that the assumption (6.2) is necessary to guarantee existence of efficient nonbasic variables, and the validity of Theorem 7.10. □

From Theorem 7.8 we know that we must consider negative pivot elements, i.e. $\tilde{A}_{rj} < 0$. What happens if nonbasic variable x_j is efficient and column j of $\tilde{A}$ contains no positive elements at all? Then the increase of x_j is unbounded, a fact that indicated an unbounded LP in the single objective case. However, since $\lambda^T r^j = 0$ this is not the case in the multiobjective LP. Rather, unboundedness of $\mathcal{X}_E$ is detected in direction d given by the vector with components $-\tilde{b}_i/\tilde{A}_{ij}, i \in \mathcal{B}, x_j = 1$. Of course, this is not a feasible pivot, as it does not lead to another basis.

The results so far allow us to move from efficient basis to efficient basis. To formulate a multiobjective Simplex algorithm we now need an efficient basis to start with.

For the MOLP

$$\min\{Cx : Ax = b, x \geqq 0\}$$

one and only one of the following cases can occur:

- The MOLP is infeasible, i.e. $\mathcal{X} = \emptyset$,
- it is feasible ($\mathcal{X} \neq \emptyset$) but has no efficient solutions ($\mathcal{X}_E = \emptyset$), or
- it is feasible and has efficient solutions, i.e. $\mathcal{X}_E \neq \emptyset$.

The multicriteria Simplex algorithm deals with these situations in three phases as follows.

Phase I: Determine an initial basic feasible solution or stop with the conclusion that $\mathcal{X} = \emptyset$. This phase does not involve the objective function matrix C, and the usual auxiliary LP (6.18) can be used.

Phase II: Determine an initial efficient efficient basis or stop with the conclusion that $\mathcal{X}_E = \emptyset$.

Phase III: Pivot among efficient bases to determine all efficient bases and directions of unboundedness of $\mathcal{X}_E$.

In Phase II, the solution of a weighted sum LP(λ) with $\lambda > 0$ will yield an efficient basis, provided LP(λ) is bounded. If we do not know that in advance it is necessary to come up with a procedure that either concludes that $\mathcal{X}_E = \emptyset$ or returns an appropriate λ for which LP(λ) has an optimal solution. Assuming that $\mathcal{X} \neq \emptyset$ Phase I returns a basic feasible solution $x^0 \in \mathcal{X}$, which may or may not be efficient. We proceed in two steps: First, the auxiliary LP (6.8) is solved to check whether $\mathcal{X}_E = \emptyset$. Proposition 6.12 and duality imply that $\mathcal{X}_\mathcal{E} \neq \emptyset$ if and only if (6.8) has an optimal solution. In this case the optimal solution of (6.8) returns an appropriate weighting vector $\hat{w}$, analogously to the argument we have used in the proof of Theorem 6.11.

From Proposition 6.12 the MOLP $\min\{Cx : Ax = b, x \geqq 0\}$ has an efficient solution if and only if the LP 7.8

$$\max \left\{e^T z : Ax = b, Cx + Iz = Cx^0, x, z \geqq 0\right\} \tag{7.8}$$

has an optimal solution. Moreover $\hat{x}$ in an optimal solution of (6.7) is efficient. However, we do not know if $\hat{x}$ is a basic feasible solution of the MOLP and we can in general not choose $\hat{x}$ to start Phase III of the algorithm.

Instead we apply linear programming duality (Theorem 6.8): (7.8) has an optimal solution if and only if its dual (7.9)

$$\min \left\{u^T b + w^T C x^0 : u^T A + w^T C \geqq 0, w \geqq e\right\} \tag{7.9}$$

has an optimal solution $(\hat{u}, \hat{w})$ with $\hat{u}^T b + \hat{w}^T C x^0 = e^T \hat{z}$. Then $\hat{u}$ is also an optimal solution of the LP (7.10)

$$\min \left\{u^T b : u^T A, \geqq -\hat{w}TC\right\} \tag{7.10}$$

which is just (7.9) for $w = \hat{w}$ fixed. As in the proof of Theorem 6.11 the dual of (7.10) has an optimal solution, and therefore an optimal basic feasible solution, which is efficient. The dual of (7.10) is equivalent to the weighted sum LP($\hat{w}$)

$$\min\left\{\hat{w}^T Cx : Ax = b, x \geqq 0.\right\}$$

It follows that the LPs (7.9) and LP($\hat{w}$) are the necessary tools in Phase II. If (7.9) is infeasible, $\mathcal{X}_E = \emptyset$. Otherwise an optimal solution of (7.9) yields an appropriate weighting vector $\lambda = \hat{w}$ for which LP(λ) has an optimal basic feasible solution, which is an initial efficient basic feasible solution of the MOLP.

In the following description of the multiobjective Simplex algorithm, which finds all efficient bases and all efficient basic feasible solutions, we need to store a list $\mathcal{L}_1$ of efficient bases to be processed and a list of efficient bases $\mathcal{L}_2$ for output, as well as a list $\mathcal{EN}$ of efficient nonbasic variables.

Algorithm 7.1 (Multicriteria Simplex algorithm.)

Input: *Data* A, b, C *of an MOLP.*

Initialization: *Set* $\mathcal{L}_1 := \emptyset$, $\mathcal{L}_2 := \emptyset$.

Phase I: *Solve the LP* $\min\{e^T z : Ax + Iz = b, x, z \geqq 0\}$. *If the optimal value of this LP is nonzero, STOP,* $\mathcal{X} = \emptyset$. *Otherwise let* x^0 *be a basic feasible solution* x^0 *of the MOLP.*

Phase II: *Solve the LP* $\min\{u^T b + w^T Cx^0 : u^T A + w^T C \geqq 0, w \geqq e\}$. *If this problem is infeasible, STOP,* $\mathcal{X}_E = \emptyset$. *Otherwise let* $(\hat{u}, \hat{w})$ *be an optimal solution.*

Find an optimal basis $\mathcal{B}$ *of the LP* $\min\{\hat{w}^T Cx : Ax = b, x \geqq 0\}$.

$\mathcal{L}_1 := \{\mathcal{B}\}$, $\mathcal{L}_2 := \emptyset$.

Phase III:

While $\mathcal{L}_1 \neq \emptyset$

Choose $\mathcal{B}$ *in* $\mathcal{L}_1$, *set* $\mathcal{L}_1 := \mathcal{L}_1 \setminus \{\mathcal{B}\}$, $\mathcal{L}_2 := \mathcal{L}_2 \cup \{\mathcal{B}\}$.

Compute $\tilde{A}$, $\tilde{b}$, *and* R *according to* $\mathcal{B}$.

$\mathcal{EN} := \mathcal{N}$.

For all $j \in \mathcal{N}$.

Solve the LP $\max\{e^T v : Ry - r^j\delta + Iv = 0; y, \delta, v \geqq 0\}$.

If this LP is unbounded$\mathcal{EN} := \mathcal{EN} \setminus \{j\}$.

End for

For all $j \in \mathcal{EN}$.

For all $i \in \mathcal{B}$.

If $\mathcal{B}' = (\mathcal{B} \setminus \{i\}) \cup \{j\}$ *is feasible and* $\mathcal{B}' \notin \mathcal{L}_1 \cup \mathcal{L}_2$

then $\mathcal{L}_1 := \mathcal{L}_1 \cup \mathcal{B}'$.

End for.

End for.

End while.
Output: $\mathcal{L}_2$.

We have formulated the algorithm only using bases. It is clear that efficient basic feasible solutions can be computed form the list $\mathcal{L}_2$ after completion of the algorithm, or during the algorithm. It is of course necessary to update $\tilde{A}$ and $\tilde{b}$ when moving from one basis to the next. Since this has been described in Algorithm 6.1, we omitted details. It is also possible to get directions in which $\mathcal{X}_E$ is unbounded. As mentioned before, these are characterized by columns of $\tilde{A}$ that do not contain positive entries.

Is Algorithm 7.1 an efficient algorithm? While we only introduce computational complexity in Section 8.1 we comment on the performance of multicriteria Simplex algorithms here. Because the (single objective) Simplex algorithm may require an exponential number of pivot steps (in terms of problem size m, n, p, see e.g. Dantzig and Thapa (1997) for a famous example), the same is true for our multicriteria Simplex algorithm.

The question, whether a polynomial time algorithm for multicriteria linear programming (e.g. a generalization of Karmarkar's interior point algorithm Karmarkar (1984)) is possible depends on the number of efficient extreme points. Unfortunately, it is easy to construct examples with exponentially many.

Example 7.12. Consider a multicriteria linear program, the feasible set of which is a hypercube in $\mathbb{R}^n$, i.e. $\mathcal{X} = [0,1]^n$ and which has objectives to minimize x_i as well as $-x_i$. Formally,

$$\begin{array}{lrl} \min & x_i & i = 1, \ldots, n \\ \min & -x_i & i = 1, \ldots, n \\ \text{subject to} & x_i \leq 1 & i = 1, \ldots, n \\ & -x_i \leq 1 & i = 1, \ldots, n. \end{array}$$

This problem has n variables, $m = 2n$ constraints and $p = 2n$ objective functions. It is obvious, that all 2^n extreme points of the feasible set are efficient. □

Some investigations show that the average number of efficient extreme points can be huge. Benson (1998c) reports on such numerical tests. Results on three problem classes (with inequality constraints) with 10 random examples each are summarized in Table 7.1.

However, Küfer (1998) did a probabilistic analysis and found that the expected number of efficient extreme points for a certain family of randomly generated MOLPs is polynomial in n, m, and p.

Table 7.1. Number of efficient extreme points.

n	m	Q	Number of efficient extreme points
30	25	4	7,245.9 on average
50	50	4	83,780.6 on average
60	50	4	more than 200,000 in each problem

We close this section with an example for the multicriteria Simplex algorithm.

Example 7.13 (Wiecek (1995)). We solve an MOLP with three objectives, three variables, and three constraints:

$$\begin{array}{llllll} \min & -x_1 & -2x_2 & & \\ \min & -x_1 & & +2x_3 & \\ \min & x_1 & & -x_3 & \\ \text{subject to} & x_1 & +x_2 & & \leq 1 \\ & & x_2 & & \leq 2 \\ & x_1 & -x_2 & +x_3 & \leq 4. \end{array}$$

Slack variables x_4, x_5, x_6 are introduced to write the constraints in equality form $Ax = b$.

Phase I: It is clear that $\mathcal{B} = \{4, 5, 6\}$ is a feasible basis and $x^0 = (0, 0, 0, 1, 2, 4)$ is a basic feasible solution.

Phase II: We solve (7.9) with x^0 from Phase I:

$$\begin{array}{ll} \min & u_1 + 2u_2 + 4u_3 \\ \text{subject to } & u^T \begin{pmatrix} 1 & 1 & 0 & 1 & 0 & 0 \\ 0 & 1 & 0 & 0 & 1 & 0 \\ 1 & -1 & 1 & 0 & 0 & 1 \end{pmatrix} + w^T \begin{pmatrix} -1 & -2 & 0 & 0 & 0 & 0 \\ -1 & 0 & 2 & 0 & 0 & 0 \\ 1 & 0 & -1 & 0 & 0 & 0 \end{pmatrix} \geqq 0 \\ & \hfill w \geqq e \end{array}$$

The w component of the optimal solution is $\hat{w} = (1, 1, 1)$.
We now solve $\min\{\hat{w}^T Cx : x \in \mathcal{X}\}$. x^0 is an initial basic feasible solution for this problem. An optimal basis is $\mathcal{B}^1 = \{2, 5, 6\}$ with optimal basic feasible solution $x^1 = (0, 1, 0, 0, 1, 5)$. Therefore we initialize $\mathcal{L}_1 = \{\{2, 5, 6\}\}$ and move to Phase III.

Phase III:

Iteration 1: We choose basis $\mathcal{B}^1 = \{2, 5, 6\}$ and set $\mathcal{L}_1 = \emptyset$, $\mathcal{L}_2 = \{\{2, 5, 6\}\}$. The tableau for this basis is given below.

$\bar{c}^1$	1	0	0	2	0	0	2
$\bar{c}^2$	-1	0	2	0	0	0	0
$\bar{c}^3$	1	0	-1	0	0	0	0
x_2	1	1	0	1	0	0	1
x_5	-1	0	0	-1	1	0	1
x_6	2	0	1	1	0	1	5

$\mathcal{EN} := \{1, 3, 4\}$.

The LP to check if x_1 is an efficient nonbasic variable is given in tableu form, where the objective coefficients of 1 for variables v have been eliminated by subtracting all constraint rows from the objective row to obtain a basic feasible solution with basic variables $v = 0$. This LP does have an optimal solution that is found after only one pivot. Pivot elements are highlighted by square frames.

1	1	2	-1	0	0	0	0
1	0	2	-1	1	0	0	0
-1	2	0	[1]	0	1	0	0
1	-1	0	-1	0	0	1	0

The LP to check if variable x_3 is an efficient nonbasic variable is shown below. The problem has an optimal solution, proved by performing the indicated pivot.

1	1	2	-1	0	0	0	0
1	0	2	0	1	0	0	0
-1	2	0	-2	0	1	0	0
1	-1	0	[1]	0	0	1	0

Finally, we check nonbasic variable x_4. In the tableau displayed below, column three indicates that the LP is unbounded, and x_4 is not efficient.

1	1	2	-2	0	0	0	0
1	0	2	-2	1	0	0	0
-1	2	0	0	0	1	0	0
1	-1	0	0	0	0	1	0

As a result of these checks we have that $\mathcal{EN} = \{1, 3\}$. Checking in the tableau for $\mathcal{B}^1 = \{2, 5, 6\}$, we find that the feasible pivots are 1) x_1 enters and x_2 leaves, giving basis $\mathcal{B}^2 = \{1, 5, 6\}$ and 2) x_3 enters and x_6 leaves, yielding basis $\mathcal{B}^3 = \{2, 3, 5\}$.

$\mathcal{L}_1 := \{\{1, 5, 6\}, \{2, 3, 5\}\}$.

Iteration 2: Choose $\mathcal{B}^2 = \{1,5,6\}$ with BFS $x^2 = (1,0,0,0,2,3)$.

$\mathcal{L}_1 = \{\{2,3,5\}\}$, $\mathcal{L}_2 = \{\{2,5,6\},\{2,3,5\}\}$.

The tableau for the basis is as follows.

$\bar{c}^1$	0	-1	0	1	0	0	1
$\bar{c}^2$	0	1	2	1	0	0	1
$\bar{c}^3$	0	-1	-1	-1	0	0	-1
x_2	1	1	0	1	0	0	1
x_5	0	1	0	0	1	0	2
x_6	0	-2	1	-1	0	1	3

$\mathcal{EN} = \{2,3,4\}$.

If x_2 enters the basis, x_1 leaves, which leads to basis $(2,5,6)$ which is the previous one. Therefore x_2 need not be checked.

The tableau for checking x_3 is displayed below. After one pivot column 3 shows that the LP is unbounded and x_3 is not efficient.

-1	1	1	-1	0	0	0	0
-1	0	1	0	1	0	0	0
[1]	2	1	-2	0	1	0	0
-1	-1	-1	1	0	0	1	0

We check nonbasic variable x_4. One iteration is again enough to exhibit unboundedness, and x_4, too, is not efficient.

-1	1	1	-1	0	0	0	0
-1	0	1	-1	1	0	0	0
[1]	2	1	-1	0	1	0	0
-1	-1	-1	1	0	0	1	0

These checks show that there are no new bases and BFSs to add, therefore $\mathcal{EN} = \emptyset$ and we proceed to the next iteration.

Iteration 3: We choose $\mathcal{B}^3 = \{2,3,5\}$ with BFS $x^3 = (0,1,5,0,1,0)$.

$\mathcal{L}_1 = \emptyset$, $\mathcal{L}_2 = \{\{2,5,6\},\{1,5,6\},\{2,3,5\}\}$.

The tableau for the basis is shown below.

$\bar{c}^1$	1	0	0	2	0	0	2
$\bar{c}^2$	-5	0	0	-2	0	-2	-10
$\bar{c}^3$	3	0	0	1	0	1	5
x_2	1	1	0	1	0	0	1
x_5	-1	0	0	-1	1	0	1
x_3	2	0	1	1	0	1	5

$\mathcal{EN} = \{1,4,6\}$.

We test nonbasic variable x_1. After one pivot column 4 in the tableau shows that the LP is unbounded.

-1	1	-1	1	0	0	0	0
$\boxed{1}$	2	0	-1	1	0	0	0
-5	-2	-2	5	0	1	0	0
3	1	1	-3	0	0	1	0

The test of nonbasic variable x_4 yields the following tableau, and again one pivot is enough to determine unboundedness.

-1	1	-1	-1	0	0	0	0
1	2	0	-2	1	0	0	0
-5	-2	-2	2	0	1	0	0
3	1	$\boxed{1}$	-1	0	0	1	0

Since $\mathcal{EN} = \emptyset$ the iteration is finished.

Iteration 4: Since $\mathcal{L}_1 = \emptyset$ the algorithm terminates.

Output: List of efficient bases $\mathcal{B}^1 = \{2, 5, 6\}, \mathcal{B}^2 = \{1, 5, 6\}, \mathcal{B}^3 = \{2, 3, 5\}$.

During the course of the algorithm, we identified three efficient bases and three corresponding efficient basic feasible solutions. Their adjacency structure is shown in Figure 7.1. A line indicates that bases are adjacent. Note that bases $\{1, 5, 6\}$ and $\{2, 3, 5\}$ are not adjacent, because at least two pivots are needed to obtain one from the other. They are, however, connected via basis $\{2, 5, 6\}$

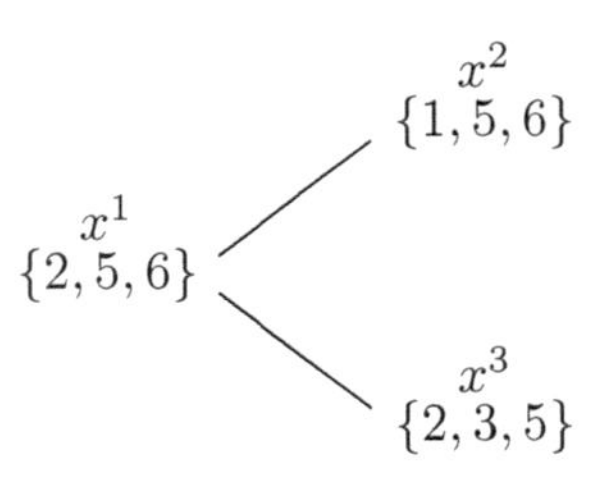

Fig. 7.1. Adjacency of efficient bases and corresponding BFSs.

The problem is displayed in decision space in Figure 7.2. The efficient set consists of the edges connecting x^1 and x^2 and x^1 and x^3.

□

In the following section we study the geometry of multiobjective linear programming. Amongst other things we shall see, how the results from the

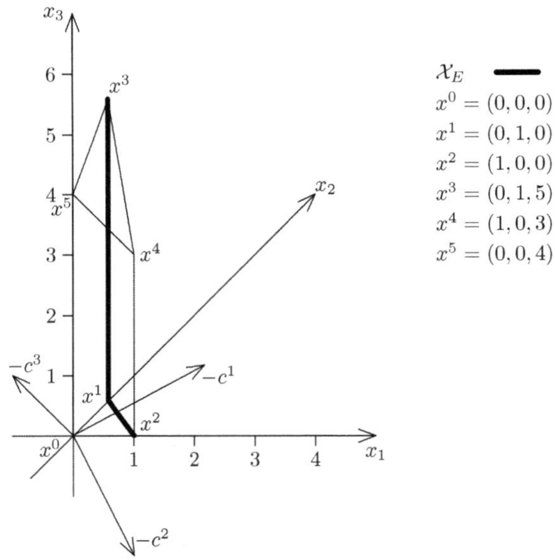

Fig. 7.2. Feasible and efficient set in Example 7.13.

multicriteria Simplex algorithm (i.e. the list of efficient bases and their adjacency structure) is exploited to identify the maximal efficient faces.

7.2 The Geometry of Multiobjective Linear Programming

First we observe that efficient BFS correspond to extreme points of $\mathcal{X}_E$.

Lemma 7.14. *1. Let $\mathcal{B}$ be an efficient basis and $(x_\mathcal{B}, 0)$ be the corresponding basic feasible solution. Then $(x_\mathcal{B}, 0)$ is an extreme point of $\mathcal{X}_E$.*
2. Let $x \in \mathcal{X}_E$ be an extreme point. Then there is an efficient basis $\mathcal{B}$ such that $x = (x_\mathcal{B}, 0)$.

Proof. This result follows from Theorem 6.11, the definition of an efficient basis and the single objective counterpart in Theorem 6.17. □

Note that, as in the single objective case, several efficient bases might identify the same efficient extreme point, if the MOLP is degenerate.

If $(x_\mathcal{B}, 0)$ and $(x_{\hat{\mathcal{B}}}, 0)$ are the efficient basic feasible solutions defined by adjacent efficient bases $\mathcal{B}$ and $\hat{\mathcal{B}}$, we see from the proof of Lemma 7.7 that both $(x_{\mathcal{B}}, 0)$ and $(x_{\hat{\mathcal{B}}}, 0)$ are optimal solutions of the same LP(λ). Therefore, due to linearity, the edge $\mathrm{conv}((x_{\mathcal{B}}, 0), (x_{\hat{\mathcal{B}}}, 0))$ is contained in $\mathcal{X}_E$.

Lemma 7.15. *Let $\mathcal{B}$ and $\hat{\mathcal{B}}$ be optimal bases for LP(λ). Then the edge* $\mathrm{conv}((x_{\mathcal{B}}, 0), (x_{\hat{\mathcal{B}}}, 0))$ *is contained in $\mathcal{X}_E$.*

We also have to take care of efficient unbounded edges: $\mathcal{X}_E$ may contain some unbounded edges $\mathcal{E} = \{x : x = x^i + \mu d^j,\ \mu \geq 0\}$, where d^j is an extreme ray and x^i is an extreme point of $\mathcal{X}$. This can happen even if the LP(λ) is bounded if $c(\lambda)$ is parallel to d^j. An unbounded edge always starts at an extreme point, which must therefore be efficient.

Let $\mathcal{B}$ be an efficient basis associated with that extreme point. Then the unbounded efficient edge is detected by an efficient nonbasic variable, in which the column $\tilde{A}^j$ contains only nonpositive elements, showing that $\mathcal{X}$ is unbounded in that direction. Because $\lambda^T r^j = 0$ this does not constitute unboundedness of the objective function.

Definition 7.16. *Let $\mathcal{F} \subset \mathcal{X}$ be a face of $\mathcal{X}$. $\mathcal{F}$ is called* efficient face, *if $\mathcal{F} \subset \mathcal{X}_E$. It is called* maximal efficient face, *if there is no efficient face $\mathcal{F}'$ of higher dimension with $\mathcal{F} \subset \mathcal{F}'$.*

Lemma 7.17. *If there is a $\lambda \in \mathbb{R}^p_>$ such that $\lambda^T Cx = \gamma$ is constant for all $x \in \mathcal{X}$ then $\mathcal{X}_E = \mathcal{X}$. Otherwise*

$$\mathcal{X}_E \subset \bigcup_{t=1}^{T} \mathcal{F}_t, \tag{7.11}$$

where $\{\mathcal{F}_t : t = 1, \ldots, T\}$ is the set of all proper faces of $\mathcal{X}$ and T is the number of proper faces of $\mathcal{X}$.

Proof. The first case is obvious, because if $\lambda^T Cx = \gamma$ for all $x \in \mathcal{X}$ then the whole feasible set is optimal for this particular LP(λ). Then from Theorem 6.6 $\mathcal{X} \subset \mathcal{X}_E$.

The second part follows from the fact that optimal solutions of LP(λ) are on the boundary of $\mathcal{X}$ (Theorem 6.17) and, once more, Theorem 6.11. Of course $\mathrm{bd}\,\mathcal{X} = \cup_{t=1}^T \mathcal{F}_t$. □

Thus, in order to describe the complete efficient set $\mathcal{X}_E$, we need to identify the maximally efficient faces of $\mathcal{X}$. We will need the representation of a point x in a face $\mathcal{F}$ as a convex combination of the extreme points and a nonnegative combination of the extreme rays of $\mathcal{F}$. This result is know as Minkowski's theorem. A proof can be found in Nemhauser and Wolsey (1999, Chapter I.4, Theorem 4.8).

Theorem 7.18 (Minkowski's Theorem). *Let $\mathcal{X}$ be a polyhedron and $x \in \mathcal{X}$. Let $x^1, \ldots, x^k$ be the extreme points and let $d^1, \ldots, d^l$ be the extreme rays of $\mathcal{X}$, then there are nonnegative real numbers $\alpha_i, i = 1, \ldots, k$ and $\mu_j, j = 1, \ldots, l$ such that $0 \leq \alpha_i \leq 1, i = 1, \ldots, k$, $\sum_{i=1}^k \alpha_i = 1$, and*

$$x = \sum_{i=1}^{k} \alpha_i x^i + \sum_{j=1}^{l} \mu_j d^j. \tag{7.12}$$

Furthermore, if $x \in \operatorname{ri} \mathcal{X}$ the numbers α_i and μ_j can be chosen to be positive.

Example 7.19. Consider the polyhedron $\mathcal{X}$ defined as follows:

$$\mathcal{X} := \{x \in \mathbb{R}^2 : x \geqq 0, 2x_1 + x_2 \geq 2, -x_1 + x_2 \leq 2\}$$

shown in Figure 7.3. Clearly, $\mathcal{X}$ has two extreme points $x^1 = (0, 2)$ and $x^2 = (1, 0)$. The two extreme rays are $d^1 = (1, 1)$ and $d^2 = (1, 0)$.

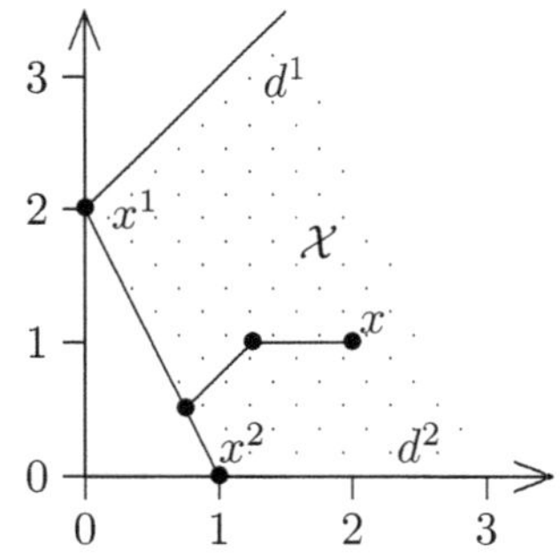

Fig. 7.3. A polyhedron with extreme points and extreme rays.

The point $x = (2, 1) \in \operatorname{ri} \mathcal{X}$ can be written as

$$x = \frac{1}{4}\begin{pmatrix} 0 \\ 2 \end{pmatrix} + \frac{3}{4}\begin{pmatrix} 1 \\ 0 \end{pmatrix} + \frac{1}{2}\begin{pmatrix} 1 \\ 1 \end{pmatrix} + \frac{3}{4}\begin{pmatrix} 1 \\ 0 \end{pmatrix} = \begin{pmatrix} 2 \\ 1 \end{pmatrix}.$$

□

Suppose that $\emptyset \neq \mathcal{X}_E \neq \mathcal{X}$. Using Minkowski's theorem, applied to a face $\mathcal{F}$, we prove that the whole face is efficient if and only if it contains an efficient solution in its relative interior.

Theorem 7.20. *A face $\mathcal{F} \subset \mathcal{X}$ is an efficient face if and only if it has an efficient solution $\hat{x}$ in its relative interior.*

Proof. "$\Longrightarrow$" If $\mathcal{F}$ is an efficient face all its relative interior points are efficient by definition.

"$\Longleftarrow$" Let $\hat{x} \in \mathcal{X}_E$ belong to the relative interior of $\mathcal{F}$. We show that there is a $\hat{\lambda} \in \mathbb{R}^p_>$ such that the whole face $\mathcal{F}$ is optimal for LP($\hat{\lambda}$).
First, by Theorem 6.11 we can find a $\hat{\lambda} \in \mathbb{R}^p_>$ such that $\hat{x}$ is an optimal solution of LP($\hat{\lambda}$). In particular LP($\hat{\lambda}$) is bounded. Therefore

$$\hat{\lambda}^T C x^i \geq \hat{\lambda}^T C \hat{x} \tag{7.13}$$

for all extreme points $x^i, i = 1, \ldots, k$ of $\mathcal{F}$ and

$$\hat{\lambda}^T C d^j \geq 0 \tag{7.14}$$

for all extreme rays $d^j, j = 1, \ldots, l$ of $\mathcal{F}$. Note that whenever $\hat{\lambda}^T C d^j < 0$ for some extreme ray d^j LP($\hat{\lambda}$) will be unbounded. Assume there is an extreme point x^i, $i \in \{1, \ldots, k\}$ which is not optimal for LP($\hat{\lambda}$), i.e.

$$\hat{\lambda}^T C x^i > \hat{\lambda}^T C x^0. \tag{7.15}$$

Then from Theorem 7.18 there are positive α_i and μ_j such that with (7.12) $\hat{x} = \sum_{i=1}^k \alpha_i x^i + \sum_{j=1}^l \mu_j d^j$ and

$$\begin{aligned} \hat{\lambda}^T C \hat{x} &= \sum_{i=1}^{k} \alpha_i \hat{\lambda}^T C x^i + \sum_{j=1}^{l} \mu_j \hat{\lambda}^T C d^j \\ &> \sum_{i=1}^{k} \alpha_i \hat{\lambda}^T C x^0 = \hat{\lambda}^T C \hat{x}. \end{aligned} \tag{7.16}$$

We have used positivity of α_i, nonnegativity of μ_i, (7.13), (7.14), and (7.15) for the inequality, and $\sum_{i=1}^k \alpha_i = 1$ for the second equality. The impossibility (7.16) means that

$$\hat{\lambda}^T C x^i = {\lambda^0}^T C x^0. \tag{7.17}$$

for all extreme points x^i, which are thus optimal solutions of LP($\hat{\lambda}$). To complete the proof, consider (7.16) again, using (7.17) this time to get that ${\lambda^0}^T C d^j = 0$ for all extreme rays d^j, because $\mu_j > 0$ since $\hat{x}$ is a relative interior point of $\mathcal{F}$. □

We state to further results about efficient edges and efficient faces, which we leave as exercises for the reader, see Exercises 7.5 and 7.6.

Proposition 7.21. *Assume that the MOLP is not degenerate. Let x^1 and x^2 be efficient extreme points of $\mathcal{X}$ and assume that the corresponding bases are adjacent (i.e. one can be obtained from the other by an efficient pivot). Then* $\operatorname{conv}(x^1, x^2)$ *is an efficient edge.*

Theorem 7.22. *A face $\mathcal{F}$ of $\mathcal{X}$ is efficient if and only if there is a $\lambda > 0$ such that all extreme points of $\mathcal{F}$ are optimal solutions of LP(λ) and $\lambda^T d = 0$ for all extreme rays of $\mathcal{F}$.*

With Theorem 7.20 and Lemma 7.17 we know that $\mathcal{X}_E$ is the union of maximally efficient faces, each of which is the set of optimal solutions of LP(λ), for some $\lambda \in \mathbb{R}^p_>$. If we combine this with the fact that the set of efficient extreme points is connected by efficient edges, (as follows again from Theorem 7.10 and Theorem 7.20, see also page 185) we get the connectedness result for the efficient set of multicriteria linear programs.

Theorem 7.23. *$\mathcal{X}_E$ is connected and, therefore, $\mathcal{Y}_N$ is connected.*

Proof. The result for $\mathcal{X}_E$ follows from Theorem 7.10 and Lemma 7.17 together with Theorem 7.20. Thus, $\mathcal{Y}_N$ is connected because $\mathcal{X}_E$ is and C is linear, i.e. continuous. □

Example 7.24. Figure 7.4 shows the feasible set of a biobjective LP, with the two maximal efficient faces indicated by bold lines.

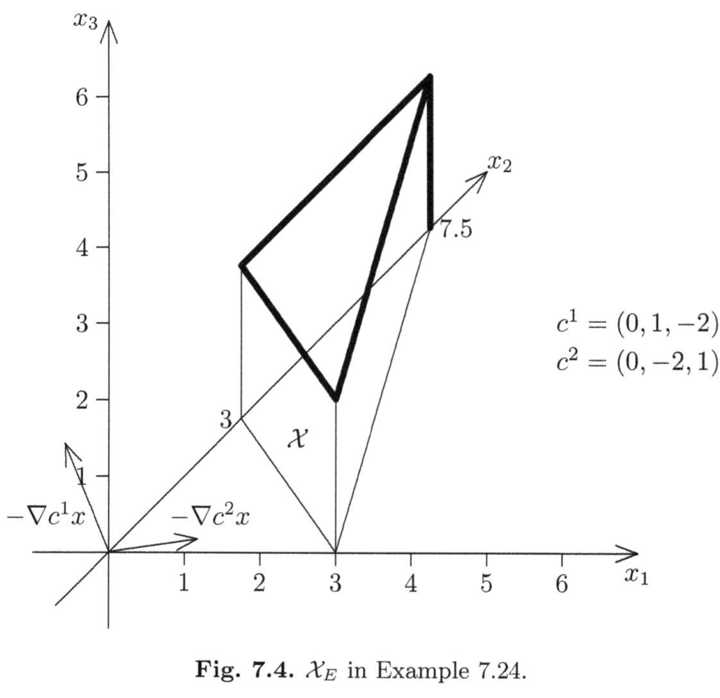

Fig. 7.4. $\mathcal{X}_E$ in Example 7.24.

To check that $\mathcal{X}_E$ is correct we can use Theorem 6.11, i.e $x \in \mathcal{X}_E$ if and only if there is $c(\lambda) = \lambda c^1 + (1-\lambda)c^2$ such that x is an optimal solution of LP(λ), and apply it graphically in this case. The negative gradient of the objective $c(\lambda)$ for different values of λ can be used to graphically determine the optimal faces. In this example, $\mathcal{X}_E$ has a 2-dimensional face and a 1-dimensional face as the only maximal efficient faces. However, the three edges of the efficient triangle and the four efficient extreme points are not maximal efficient faces. The example clearly shows that – even for linear multicriteria optimization problems – the efficient set is in general not convex. □

In the proof of Theorem 7.20 we have seen that for each efficient face $\mathcal{F}$ there exists a $\lambda \in \mathbb{R}^p_>$ such that $\mathcal{F}$ is the set of optimal solutions of LP(λ). Suppose we know efficient face $\mathcal{F}$, how can we find all λ with that property?

Essentially, we want to subdivide the set $\Lambda = \{\lambda \in \mathbb{R}^p_> : \sum_{k=1}^p \lambda_k = 1\}$ into regions that correspond to those weighting vectors λ, which make a certain face efficient. That is, for each efficient face $\mathcal{F}$ we want to find $\Lambda_{\mathcal{F}} \subset \Lambda$ such that $\mathcal{F}$ is optimal for LP(λ) for all $\lambda \in \Lambda_{\mathcal{F}}$.

Let us first assume that $\mathcal{X}$ is nonempty and bounded, so that in particular $\mathcal{X}_E$ is nonempty. Let $\mathcal{F}$ be an efficient face, and x^i, $i = 1, \dots, k$ be the set of all extreme points of $\mathcal{F}$. Because $\mathcal{F}$ is an efficient face, from the proof of Theorem 7.20 there is some $\lambda_{\mathcal{F}} \in \Lambda$ such that $\mathcal{F} = \operatorname{conv}(x^1, \dots, x^k)$ is optimal for LP($\lambda_{\mathcal{F}}$). In particular, $x^1, \dots, x^k$ are optimal solutions of LP($\lambda_{\mathcal{F}}$).

Hence we can apply the optimality condition for linear programs. Let R^i be the reduced cost matrix of a basis associated with x^i. Then x^i is optimal if and only if $\lambda^T R^i \geqq 0$ (note that we assume nondegeneracy here, see Lemma 6.14). Therefore, the face $\mathcal{F}$ is optimal if and only if $\lambda^T R^i \geqq 0,\ i = 1, \dots, k$.

Proposition 7.25. *The set of all λ for which efficient face $\mathcal{F}$ is the optimal solution set of LP(λ) is defined by the linear system*

$$\begin{aligned} \lambda^T e &= 1 \\ \lambda^T R^i &\geqq 0 \quad i = 1, \dots, k \\ \lambda &\geqq 0, \end{aligned}$$

where R^i is the reduced cost matrix of a basis associated with extreme point x^i of $\mathcal{F}$.

Example 7.26. Let us consider the efficient face $\operatorname{conv}(x^1, x^2)$ in Example 7.13. Extreme point x^1 corresponds to basis $\{2, 5, 6\}$ with

$$R^1 = \begin{pmatrix} 1 & 0 & 2 \\ -1 & 2 & 0 \\ 1 & -1 & 0 \end{pmatrix}$$

and extreme point x^2 corresponds to basis $\{1, 5, 6\}$

$$R^2 = \begin{pmatrix} -1 & 0 & 1 \\ 1 & 2 & 1 \\ -1 & -1 & -1 \end{pmatrix}.$$

The linear system of Proposition 7.25 is $\lambda^T R^1 \geqq 0,\ \lambda^T R^2 \geqq 0, \lambda^T e = 1, \lambda \geq 0$, which we write as

$$\begin{aligned} \lambda_1 - \lambda_2 + \lambda_3 &\geq 0 \\ 2\lambda_2 - \lambda_3 &\geq 0 \\ 2\lambda_1 &\geq 0 \\ -\lambda_1 + \lambda_2 - \lambda_3 &\geq 0 \\ 2\lambda_2 - \lambda_3 &\geq 0 \\ \lambda_1 + \lambda_2 - \lambda_3 &\geq 0 \\ \lambda_1 + \lambda_2 + \lambda_3 &= 1 \\ \lambda_1, \lambda_2, \lambda_3 &\geq 0 \end{aligned} \qquad \text{or} \qquad \begin{aligned} \lambda_1 - \lambda_2 + \lambda_3 &= 0 \\ 2\lambda_2 - \lambda_3 &\geq 0 \\ \lambda_1 + \lambda_2 - \lambda_3 &\geq 0 \\ \lambda_1 + \lambda_2 + \lambda_3 &= 1 \\ \lambda_1, \lambda_2, \lambda_3 &\geq 0. \end{aligned}$$

Eliminating λ_3 we obtain $\lambda_2 = 0.5$, $0 \leq \lambda_1 \leq 0.5$. Proceeding in the same way for the efficient face $\text{conv}(x^1, x^2)$ and the efficient extreme points, we obtain the subdivision of Λ depicted in Figure 7.5. For the efficient extreme points x^i there are two-dimensional regions, for the edges, there are line segments that yield the respective face as optimal solutions of LP(λ).

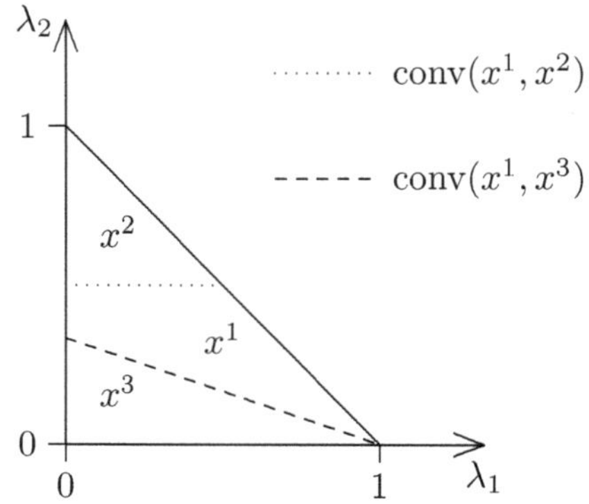

Fig. 7.5. Weights to obtain efficient faces in Example 7.13.

□

If $\mathcal{X}$ is unbounded, it may happen that $\mathcal{X}_E$ contains unbounded efficient faces . In this case an efficient face $\mathcal{F}$ contains unbounded edges, i.e. we must take care of extreme rays in the linear system of Proposition 7.25. We extend it by $\lambda^t C d^j = 0$ for the extreme rays $d^1, \ldots, d^l$ of face $\mathcal{F}$. The proof of Theorem 7.20 shows that this condition has to be satisfied.

If there is some $\lambda \in \Lambda$ such that LP(λ) is unbounded, there is, in addition to the sets $\Lambda_{\mathcal{F}} \subset \Lambda$ for all efficient faces $\mathcal{F}$, a subset $\Lambda_0 \subset= \{\lambda \in \Lambda :$ LP(λ) is unbounded$\}$. This set is the remainder of Λ, which is not associated with any of the efficient faces. Note that this case can only occur if there is a $\lambda > 0$ and an extreme ray d of $\mathcal{X}$ such that $\lambda^T C d < 0$.

Let us finally turn to the determination of maximal efficient faces. The method we present is from Isermann (1977). Let $\mathcal{B}$ be an efficient basis and let $\mathcal{N}^f \subset \mathcal{N}$ be the set of nonbasic variables, which allow feasible pivots. Let $\mathcal{J} \subset \mathcal{N}^f$. Then we have the following proposition.

Proposition 7.27. *All variables in $\mathcal{J}$ are efficient nonbasic variables if and only if the LP*

$$\begin{aligned} \max \quad & e^T v \\ \text{subject to } & Rz - R^{\mathcal{J}}\delta + Iv = e \\ & z, \delta, v \geqq 0 \end{aligned} \tag{7.18}$$

has an optimal solution. Here $R^{\mathcal{J}}$ denotes the columns of R pertaining to variables in $\mathcal{J}$.

Proof. The proof is similar to the proof of Theorem 7.8 and is left to the reader, see Exercise 7.1. □

Let us call $\mathcal{J} \subset \mathcal{N}^f$ a maximal set of efficient nonbasic variables , if there is no $\mathcal{J}' \subset \mathcal{N}^f$ such that $\mathcal{J} \subset J'$ and (7.18) has an optimal solution for $\mathcal{J}'$. Now let $\mathcal{B}^\tau, \tau = 1, \ldots, t$ be the efficient bases and $\mathcal{J}^{\tau,\rho}, \tau = 1, \ldots, t, \rho = 1, \ldots, r$ be all maximal index sets of efficient nonbasic variables at efficient basis $\mathcal{B}^\tau$. Furthermore, let $\mathcal{E}^\nu = (\mathcal{B}^\tau, d^\nu), \nu = 1, \ldots, v$ denote unbounded efficient edges, where d^ν is an extreme ray of $\mathcal{X}$.

We define $\mathcal{Q}^{\tau,\rho} := \mathcal{B}^\tau \cup \mathcal{J}^{\tau,\rho}$. $\mathcal{Q}^{\tau,\rho}$ contains bases adjacent to $\mathcal{B}^\tau$, and the convex hull of the extreme points associated with all bases found in $\mathcal{Q}^{\tau,\rho}$ plus the conical hull of any unbounded edges attached to any of these bases constitutes a candidate for an efficient face.

As we are only interested in identifying maximal efficient faces, we select a minimal number of index sets representing all $\mathcal{Q}^{\tau,\rho}$, i.e. we choose index sets $\mathcal{U}^1, \ldots, \mathcal{U}^o$ with the following properties:

1. For each $\mathcal{Q}^{\tau,\rho}$ there is a set $\mathcal{U}^s$ such that $\mathcal{Q}^{\tau,\rho} \subset \mathcal{U}^s$.
2. For each $\mathcal{U}^s$ there is a set $\mathcal{Q}^{\tau,\rho}$ such that $\mathcal{U}^s = \mathcal{Q}^{\tau,\rho}$.
3. There are no two sets $\mathcal{U}^s, \mathcal{U}^{s'}$ with $s \neq s'$ and $\mathcal{U}^s \subset \mathcal{U}^{s'}$.

Now we determine which extreme points and which unbounded edges are associated with bases in the sets $\mathcal{U}^s$. For $s \in \{1, \ldots, o\}$ let

$$\mathcal{I}_b^s := \{\tau \in \{1, \ldots, t\} : \mathcal{B}^\tau \subset \mathcal{U}^s\},$$
$$\mathcal{I}_u^s := \{\nu \in \{1, \ldots, v\} : \mathcal{B}^\tau \subset U^s\}$$

and define

$$\mathcal{X}_s = \left\{ x \in \mathcal{X} : x = \sum_{\tau \in I_b^s} \alpha_\tau x^\tau + \sum_{\nu \in I_u^s} \mu_\nu d^\nu, \sum_{\tau \in I_b^s} \alpha_\tau = 1, \alpha_\tau \geq 0, \mu_\nu \geq 0 \right\}. \tag{7.19}$$

The sets $\mathcal{X}_s$ are faces of $\mathcal{X}$ and efficient (Theorem 7.28) and in fact they are the maximal efficient faces (Theorem 7.29), if the MOLP is not degenerate.

Theorem 7.28 (Isermann (1977)). $\mathcal{X}_s \subset \mathcal{X}_E$ *for* $s = 1, \ldots, o$.

Proof. By definition of $\mathcal{U}^s$ there is a set $\mathcal{Q}^{\tau,\rho}$ such that $\mathcal{Q}^{\tau,\rho} = \mathcal{U}^s$. Therefore the linear program (7.18) with $\mathcal{J} = \mathcal{Q}^{\tau,\rho} \setminus \mathcal{B}^\tau$ in Proposition 7.27 has an optimal solution. Thus, the dual of this LP

$$\begin{array}{lr} \min & e^T\lambda \\ \text{subject to} & R^T\lambda \geqq 0 \\ & (-R^J)^T\lambda \geqq 0 \\ & \lambda \geqq e \end{array}$$

has an optimal solution $\hat{\lambda}$. But the constraints of the LP above are the optimality conditions for $LP(\lambda)$, where in particular $(R^J)^T\lambda = 0$. Therefore all $x \in \mathcal{X}_s$ are optimal solutions of LP($\hat{\lambda}$) and $\mathcal{X}_s \subset \mathcal{X}_E$. □

Theorem 7.29 (Isermann (1977)). *If* $x \in \mathcal{X}_E$ *there is an* $s \in \{1, \ldots, o\}$ *such that* $x \in \mathcal{X}_s$.

Proof. Let $x \in \mathcal{X}_E$. Then x is contained in a maximal efficient face $\mathcal{F}$, which is optimal for some LP(λ). Let $\mathcal{I}_b$ be the index set of efficient bases corresponding to the extreme points of $\mathcal{F}$ and $\mathcal{I}_u$ be the index set of extreme rays of face $\mathcal{F}$. Then, according to (7.12), x can be written as

$$x = \sum_{i \in \mathcal{I}_b} \alpha_i x^i + \sum_{j \in I_u} \mu_j d^j.$$

We choose any extreme point x^i of $\mathcal{F}$ and let $\mathcal{B}^i$ be a corresponding basis. Furthermore, we let $\mathcal{J}^0 := \{\cup_{\tau \in \mathcal{I}_b} \mathcal{B}^\tau\} \setminus \mathcal{B}^i$. Because all $\mathcal{B}^\tau$ are efficient, $\mathcal{J}^0$ is a set of efficient nonbasic variables at $\mathcal{B}^i$.

Therefore (7.18) has an optimal solution and there exists a maximal index set of efficient nonbasic variables $\mathcal{J}$ with $\mathcal{J}^0 \subset \mathcal{J}$. During the further construction of index sets, none of the indices of extreme points in $\mathcal{J}^0$ is lost, and $\mathcal{B}^i \cup \mathcal{J}^0 \subset \mathcal{U}^s$ for some s. Therefore $x \in \mathcal{X}_s$ for some $s \in \{1, \ldots, o\}$. □

The proofs show that if all efficient bases are nondegenerate, $\mathcal{X}_s$ are exactly the maximal efficient faces of $\mathcal{X}$. Otherwise some $\mathcal{X}_s$ may not be maximal, because there is a choice of bases representing an efficient extreme point, and the maximal sets of efficient nonbasic variables need not be the same for all of them.

Example 7.30. We apply this method to Example 7.13. $\mathcal{X}$ does not contain unbounded edges. The computation of the index sets is summarized in Table 7.2.

Table 7.2. Criteria and alternatives in Example 7.30.

Efficient basis $\mathcal{B}^\tau$	Maximal index set $\mathcal{J}^{\tau,\rho}$	$\mathcal{Q}^{\tau,\rho}$
$\mathcal{B}^1 = \{2,5,6\}$	$\mathcal{J}^{1,1} = \{1\}$	$\mathcal{Q}^{1,1} = \{1,2,5,6\}$
	$\mathcal{J}^{1,2} = \{3\}$	$\mathcal{Q}^{1,2} = \{3,2,5,6\}$
$\mathcal{B}^2 = \{1,5,6\}$	$\mathcal{J}^{2,1} = \{2\}$	$\mathcal{Q}^{2,1} = \{1,2,5,6\}$
$\mathcal{B}^3 = \{2,3,5\}$	$\mathcal{J}^{3,1} = \{6\}$	$\mathcal{Q}^{3,1} = \{2,3,5,6\}$

The sets $\mathcal{U}^s$ are $\mathcal{U}^1 = \{1,2,5,6\}$ and $\mathcal{U}^2 = \{2,3,5,6\}$ and checking, which bases are contained in these sets, we get $\mathcal{I}_b^1 = \{1,2\}$ and $\mathcal{I}_b^2 = \{1,3\}$. From (7.19) we get

$$\mathcal{X}_1 = \{x = \alpha_1 x^1 + \alpha_2 x^2 : \alpha_1 + \alpha_2 = 1,\ \alpha_i \geq 0\} = \text{conv}(x^1, x^2),$$
$$\mathcal{X}_2 = \{x = \alpha_1 x^1 + \alpha_2 x^3 : \alpha_1 + \alpha_2 = 1,\ \alpha_i \geq 0\} = \text{conv}(x^1, x^3),$$

and confirm $\mathcal{X}_E = \mathcal{X}_1 \cup \mathcal{X}_2$, as expected. □

7.3 Notes

A number of multicriteria Simplex algorithms have been published. Their general structure follows the three phase scheme presented above. For pivoting among efficient bases it is necessary to identify efficient nonbasic variables. Other than those of Theorems 7.8 and 7.27 tests for nonbasic variable efficiency have been proposed by Ecker and Kouada (1978) and Zionts and Wallenius (1980). An alternative method to find an initial efficient extreme point is given in Benson (1981). Several proofs of the connectedness result of Theorem 7.10 are known, see e.g. Zeleny (1974), Yu and Zeleny (1975), and Isermann (1977). More on connectedness of efficient basic feasible solutions for degenerate MOLPs can be found in Schechter and Steuer (2005).

Algorithms based on the Simplex method are proposed by Armand (1993); Armand and Malivert (1991), Evans and Steuer (1973), Ecker *et al.* (1980); Ecker and Kouada (1978), Isermann (1977), Gal (1977), Philip (1972, 1977), Schönfeld (1964), Strijbosch *et al.* (1991), Yu and Zeleny (1975, 1976), Zeleny (1974). The algorithm by Steuer (1985) is implemented in the ADBASE Steuer (2000) code.

While all these algorithms identify efficient bases and extreme points, an algorithm by Sayin (1996) a top-down approach instead, that starts by finding the highest dimensional efficient faces first and then proceeds down to extreme points (zero dimensional faces).

In Proposition 7.25 we have shown how to decompose the weight space Λ to identify those weighting vectors that have an efficient face as optimal solutions of LP(λ). Such a partition can be attempted with respect to efficient bases of the MOLP or with respect to extreme points of $\mathcal{X}_E$ or $\mathcal{Y}_N$. Benson and Sun (2000) investigates the decomposition of the weight space according to the extreme points of $\mathcal{Y}_N$.

Interior point methods have revolutionized linear programming since the 1980's. However, they are not easily adaptable to multiobjective linear programming. Most methods proposed in the literature find one efficient solution, and involve the the elicitation of the decision makers preferences in an interactive fashion, see the work of Arbel (1997) and references therein. The only interior point method that is not interactive is Abhyankar *et al.* (1990).

The observation that the feasible set in objective space $\mathcal{Y}$ is usually of much smaller dimension than $\mathcal{X}$ has lead to a stream of research work on solving MOLPs in objective space. Publications on this topic include Dauer and Liu (1990); Dauer and Saleh (1990); Dauer (1993); Dauer and Gallagher (1990) and Benson (1998c,a,b).

Exercises

7.1 (Isermann (1977)). Let $\mathcal{J} \subset \mathcal{N}$ be an index set of nonbasic variables at efficient basis $\mathcal{B}$. Show that each variable $x_j, j \in \mathcal{J}$ is efficient if and only if the linear program

$$\begin{array}{ll} \max & e^T v \\ \text{subject to} & Rz - R^{\mathcal{J}}\delta + Iv = e \\ & z, \delta, v \geqq 0 \end{array}$$

has an optimal solution. Here $R^{\mathcal{J}}$ is the part of R pertaining to variables $x_j, j \in \mathcal{J}$. Hint: Use the definition of efficient nonbasic variable and look at the dual of the above LP.

7.2. A basis $\mathcal{B}$ is called weakly efficient, if $\mathcal{B}$ is an optimal basis of LP(λ) for some $\lambda \in \mathbb{R}^p_{\geq}$. A feasible pivot with nonbasic variable x_j entering the basis is called weakly efficient if the basis obtained is weakly efficient. Prove the following theorem.

Let x_j be nonbasic at weakly efficient basis $\mathcal{B}$. Then all feasible pivots with x_j as entering variable are weakly efficient if and only if the linear program

$$\begin{array}{ll} \max & v \\ \text{subject to} & Rz - r^j\delta + ev \geqq 0 \\ & z, \delta, v \geqq 0 \end{array}$$

has an optimal objective value of zero.

7.3. Solve the MOLP

$$\begin{array}{llll} \min & -3x_1 - x_2 & & \\ \min & x_1 - 2x_2 & & \\ \text{subject to} & 3x_1 + 2x_2 & \geq 6 & \\ & x_1 & \leq 10 & \\ & x_2 & \leq 3 & \\ & x_1, \quad x_2 & \geq 0 & \end{array}$$

using the multicriteria Simplex algorithm 7.1.

7.4. Determine, for each efficient extreme point x^i of the MOLP in Exercise 7.3, the set of all λ for which x^i is an optimal solution of LP(λ) and determine all maximal efficient faces.

7.5. Assume that the MOLP is not degenerate. Let x^1 and x^2 be efficient extreme points of $\mathcal{X}$ and assume that the corresponding bases are adjacent (i.e. one can be obtained from the other by an efficient pivot). Show that $\text{conv}(x^1, x^2)$ is an efficient edge.

7.6. Prove that a face $\mathcal{F}$ of $\mathcal{X}$ is efficient if and only if there is a $\lambda > 0$ such that all extreme points of $\mathcal{F}$ are optimal solutions of LP(λ) and $\lambda^T d = 0$ for all extreme rays of $\mathcal{F}$.

7.7. Let $\mathcal{X} = \{x \in \mathbb{R}^n : Ax = b, \ x \geqq 0\}$ and consider the MOLP $\min_{x \in \mathcal{X}} Cx$. An *improving direction* d direction at $x^0 \in \mathcal{X}$ is a vector $d \in \mathbb{R}^n$ such that $Cd \leq 0$ and there is some $t > 0$ such that $x^0 + \tau d \in \mathcal{X}$ for all $\tau \in [0, t]$.

Let $D := \{d \in \mathbb{R}^n \ : \ Cd \leq 0\}$ and $x^0 \in \mathcal{X}$. Prove that $x^0 \in \mathcal{X}_E$ if and only if $(x^0 + D) \cap \mathcal{X} = \{x_0\}$, i.e. if there is no improving direction at x^0. Illustrate the result for the problem of Exercise 7.3.

8

Multiobjective Combinatorial Optimization

The multicriteria optimization problems we have discussed up to now had a continuous feasible set described by constraints. In the remaining chapters of the book we will be concerned with discrete problems, in which the feasible set is a finite set. These are known as multiobjective discrete and combinatorial optimization problems. They arise naturally in many applications as we shall see in Example 8.1 below when variables are used to model yes/no decisions or objects that are not divisible. In many cases such problems can be understood as optimization problems on some combinatorial structure.

Example 8.1. An important problem in airline operations is the so-called "pairings" or "tours of duty" problem. Let $\{1, \ldots, m\}$ be a set of flights an airline intends to operate. In order to assign the necessary crew to all of these flights this set of flights has to be partitioned into sets of flights that can be operated in sequence by a crew member. Such a sequence of flight has to satisfy quite complex contractual and legal rules. Any sequence of flights that meets all rules is called a pairing or tour of duty. Operating it causes a cost, which consists of pay for the crew member and other costs such as hotel overnights, crew flying as passengers on certain flights, etc. The goal of the airline is to operate all scheduled flights at minimum cost, and it will therefore try to choose the set of pairings that allows to operate the scheduled flights at minimum cost, making sure that each flight is contained in one pairing. In general the number of legal pairings is huge.

However, disruptions caused by bad weather, mechanical problems etc. may occur during operation. These disruptions may result in missed connections and inevitably lead to additional costs caused by delays. An airline will therefore also be interested in selecting pairings that are robust in the sense that they are less susceptible to disruptions. This objective is in conflict with minimization of cost, because it favours longer breaks between flights to make it possible to compensate earlier delays. Longer breaks, on the other hand, increase cost due to unproductive time.

We can formulate this problem as follows. Let

$$x_i = \begin{cases} 1 & \text{if ToD } i \text{ is selected} \\ 0 & \text{otherwise.} \end{cases}$$

Furthermore, let c_i be a measure for the cost of ToD i and r_i be a measure of the possible delay caused by operating ToD i. We need to solve the bicriterion optimization problem

$$\begin{aligned} \min \sum_{i=1}^{n} c_i x_i & \\ \min \sum_{i=1}^{n} r_i x_i & \\ \text{subject to} \sum_{i=1}^{n} a_{ji} x_i, &= 1 \quad j = 1, \ldots, m \\ x &\in \{0,1\}^n. \end{aligned}$$

This is a bicriterion integer programming problem. Note that there are finitely many feasible solutions. More details about this problem can be found in Ehrgott and Ryan (2002). □

In this chapter we will give an introduction to multiobjective discrete optimization, including computational complexity, a general scalarization model for multiobjective integer programs, and solution algorithms for the case that the feasible set is explicitly given.

8.1 Basics

Combinatorial optimization problems have a finite set of feasible solutions. This has significant impact on the way we deal with these problems, both in theory and solution techniques. We shall first introduce combinatorial optimization problems and the multicriteria optimization classes we consider in the following chapters. Some basic observations show that in a multicriteria context combinatorial optimization is quite different from the general or linear optimization framework we have considered in earlier chapters of this text. In particular, we give a brief introduction to the concepts of computational complexity, such as $\mathcal{NP}$-completeness and $\#\mathcal{P}$-completeness. In the subsequent chapters we prove results on computational complexity. These chapters feature some selected combinatorial problems, which are chosen to illustrate one or more solution strategies. Thus, the reader can acquire an overview of available techniques for multicriteria combinatorial optimization. For a broader survey of the field we refer to a recent bibliographies Ehrgott and Gandibleux (2000) and Ehrgott and Gandibleux (2002a).

Let $\mathcal{E}$ be a finite set $\mathcal{E} := \{e_1, \ldots, e_n\}$ and let $c : \mathcal{E} \to \mathbb{Z}$ be a cost function on the elements of $\mathcal{E}$, which assumes integer values. A combinatorial optimization problem is given by a feasible set

$$\mathcal{X} \subset 2^{\mathcal{E}}$$

defined as a subset of the power set of $\mathcal{E}$ and an objective function $f : \mathcal{X} \to \mathbb{Z}$ to be minimized. It can therefore be written in the common notation

$$\min_{x \in \mathcal{X}} f(x).$$

Note that $x \in \mathcal{X}$ denotes a subset of $\mathcal{E}$, $x \subset \mathcal{E}$. We consider two types of objectives:

$$f(x) = \sum_{e \in x} c(e)$$
$$\text{and } f(x) = \max_{e \in x} c(e).$$

The problem

$$\min_{x \in \mathcal{X}} \sum_{e \in x} c(e) \tag{8.1}$$

is called *sum problem* and the problem

$$\min_{x \in \mathcal{X}} \max_{e \in x} c(e) \tag{8.2}$$

is called *bottleneck problem.*

Combinatorial problems are integer (in particular binary) linear programming problems. Defining $x \in \{0, 1\}^n$ by

$$x_i := \begin{cases} 1 \text{ if } e_i \in x \\ 0 \text{ otherwise} \end{cases} \tag{8.3}$$

for $i = 1, \ldots, n$ we can identify subsets $x \subset \mathcal{E}$ and binary vectors. Defining $c_i := c(e_i)$ problems (8.1) and (8.2) then read

$$\min_{x \in \mathcal{X}} \sum_{i=1}^{n} c_i x_i$$

and

$$\min_{x \in \mathcal{X}} \max_{i=1}^{n} c_i x_i,$$

respectively.

On the other hand, binary linear programs

$$\begin{aligned} \min c^T x \\ \text{subject to } Ax = b \\ x \in \{0,1\}^n \end{aligned}$$

are combinatorial problems. We will also consider general integer programs, where $x \in \{0,1\}^n$ is replaced by $x \in \mathbb{Z}^n, x \geqq 0$. We shall usually assume that the feasible set is bounded, and therefore finite.

We use the example of the spanning tree problem in this introductory section to illustrate our definitions.

Definition 8.2. *Let $\mathcal{G} = (\mathcal{V}, \mathcal{E})$ be a graph with vertex set $\mathcal{V} = \{v_1, \ldots, v_m\}$ and edge set $\mathcal{E} = \{e_1, \ldots, e_n\}$, where $e_j = [v_{i_1}, v_{i_2}]$ is an unordered pair of vertices. A* spanning tree *of $\mathcal{G}$ is a subset T of $m-1$ edges, such that each vertex $v \in \mathcal{V}$ is the endpoint of an edge of T and T is connected.*

Example 8.3. Let $\mathcal{G} = (\mathcal{V}, \mathcal{E})$ be a graph and $c : \mathcal{E} \to \mathbb{Z}$ be a cost function on the edge set $\mathcal{E}$. With

$$\mathcal{T} := \{T \subset \mathcal{E} : T \text{ defines a spanning tree of } \mathcal{G}\}$$

the problem of finding a spanning tree of minimal total cost is

$$\min_{T \in \mathcal{T}} \sum_{e \in T} c(e).$$

□

Since a (spanning) tree T is also a graph, we will write $\mathcal{E}(T)$ and $\mathcal{V}(T)$ for the sets of edges and vertices of T, respectively. Then $T = (\mathcal{E}(T), \mathcal{V}(T))$. We adopt this notation for other subgraphs later.

An *instance* of a combinatorial optimization problem is given by a specific set $\mathcal{E}$ and a specific cost function. The *size* of an instance is the length of the input needed to code the data of the instance, usually measured by $|\mathcal{E}| = n$.

Of course, in multicriteria optimization we use a vector-valued rather than a scalar-valued cost function, i.e.

$$c : \mathcal{E} \to \mathbb{Z}^p$$

consists of the scalar-valued component functions

$$c^k : \mathcal{E} \to \mathbb{Z}, \quad k = 1, \ldots, p.$$

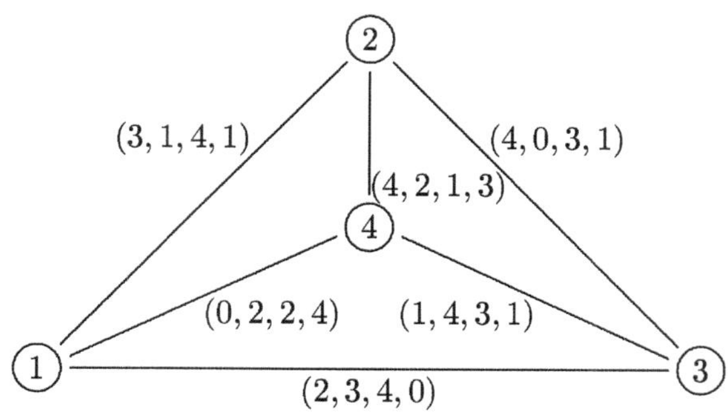

Fig. 8.1. A graph $(\mathcal{V}, \mathcal{E})$ with edge costs $c(e) \in \mathbb{Z}^4$.

Example 8.4. Continuing with spanning trees, the graph of Figure 8.1 defines an instance of a spanning tree problem with four objectives $f_k(T) = \sum_{e \in T} c^k(e)$, or $f_k(T) = \max_{e \in T} c^k(e)$.

□

At this stage, we have to mention the classes of multicriteria optimization problems (see Section 1.5) we consider in the following chapters. As before, most space will be devoted to efficient solutions, but occasionally we will also consider max-ordering and lexicographic max-ordering problems as well as lexicographic optimization problems.

Recall that a multiobjective optimization problem is identified by data, model map, and ordered set and written as

$$(\mathcal{X}, f, \mathbb{R}^p)/\theta/(\mathbb{R}^P, \preceq), \tag{8.4}$$

where $\mathcal{X}, f, \mathbb{R}^p$ are feasible set, objectives and objective space. For specific problems, $\mathcal{X}$ will be replaced by other letters to emphasize the combinatorial structure of the problem (such as $\mathcal{T}$ for the spanning tree problem). The vector valued function $f = (f_1, \ldots, f_p)$ will be composed of sum or bottleneck functions. We will explicitly note the number of sum and bottleneck functions by writing, e.g. 3-$\sum$ 2-max instead of f for a problem with three sum and two bottleneck objectives. $\theta : f(\mathcal{X}) \to \mathbb{R}^P$ is the model map, and $\mathbb{R}^P$ is ordered by means of the order relation $\preceq$. For combinatorial optimization problems $\mathbb{R}^p$ and $\mathbb{R}^P$ will usually be replaced by $\mathbb{Z}^p$ and $\mathbb{Z}^P$, respectively.

Recall that a feasible solution $\hat{x}$ is called an optimal solution of (8.4) if there is no $x \in \mathcal{X}, x \neq \hat{x}$ such that

$$\theta(f(x)) \preceq \theta(f(\hat{x})). \tag{8.5}$$

With the general definition of optimal solutions and optimal values of multicriteria optimization problems (see Definition 1.23) solving a multicriteria combinatorial optimization problem is to find, for each $y \in \text{Val}$, one element x of Opt. This corresponds to finding one optimal solution in the single objective case. For efficient solutions we will more formally define this later, see Definition 8.7.

Example 8.5. $(\mathcal{T}, 4\text{-}\sum, \mathbb{Z}^4)/\,\text{id}\,/(\mathbb{Z}^4, \leqq)$ denotes a spanning tree problem of the Pareto class with four sum objectives. The graph of Figure 8.1 has 16 spanning trees shown with their objective function vectors in Figure 8.2.

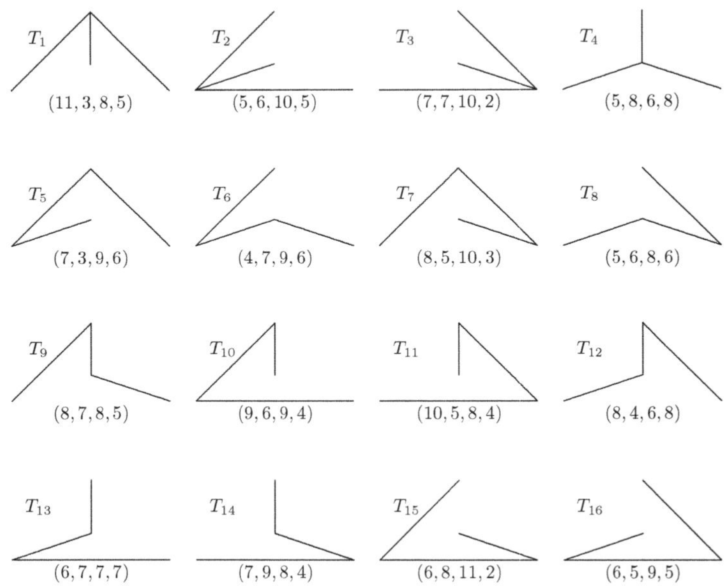

Fig. 8.2. All spanning trees of the graph of Figure 8.1.

It turns out that all spanning trees are efficient, i.e. $\mathcal{T}_E = \mathcal{T}$. What about optimal solutions according to other MCO classes?

For the max-ordering problem $(\mathcal{T}, 4\text{-}\sum, \mathbb{Z}^4)/\max/(\mathbb{Z}, <)$ the unique optimal solution is T_{13} with $f(T_{13}) = (6,7,7,7)$, so that $\mathcal{T}_{MO} = \{T_{13}\}$.

Next we consider lexicographic optimization problems $(\mathcal{T}, 4\text{-}\sum, \mathbb{Z}^4)/\pi/(\mathbb{Z}^4, <_{\text{lex}})$, where $\pi \in \Pi$ is a permutation of $\{1, \ldots, 4\}$ (see Section 5.1 for notational details). For all permutations that put f_1 first, we obtain T_6 with $f(T_6) = (4,7,9,6)$. For permutations with f_2 first we obtain T_1 with $f(T_1) = (11,3,8,5)$ or T_5 with $f(T_5) = (7,3,9,6)$, depending on the permutation. For permutations with f_3 first we obtain T_4 with $f(T_4) = (5, 8, 6, 8)$ or T_{12} with $f(T_{12}) = (8, 4, 6, 8)$, again depending on the permutation. For all permutations with f_4 first we obtain T_3 with $f(T_3) = (7, 7, 10, 2)$ or T_{15} with $f(T_{15} = (6, 8, 11, 2)$. Thus $\mathcal{T}_{\Pi} = \{T_1, T_3, T_4, T_5, T_6, T_{12}, T_{15}\}$.

For the lex-MO problem $(\mathcal{T}, 4\text{-}\sum, \mathbb{Z}^4)/\operatorname{sort}/(\mathbb{Z}^4, <_{\text{lex}})$, we obviously get $\mathcal{T}_{lexMO} = \{T_{13}\}$ again. In this example, the unique lex-MO solution is not lexicographically optimal for any permutation of the objectives. □

Example 8.5 illustrates (once again) that the optimal sets for a multicriteria optimization with the same data, but solved according to different problem classes intersect, see Chapter 5 for more details.

When we discuss the solution of multicriteria combinatorial optimization problems it is trivial, yet important, to note that $\mathcal{X}$, thus also $\mathcal{Y} = f(\mathcal{X})$ are finite. Consequently, whatever MCO class we consider, there always exist optimal solutions: $\operatorname{Opt}((\mathcal{X}, f, \mathbb{Z}^p)/\theta/\mathbb{Z}^P, \preceq)$ is never empty.

This seems to make these problems easier. However, we will need completely new methods. We illustrate this with an example that shows that the scalarization with weighted sums does not identify all efficient solutions of multiobjective discrete optimization problems. The reason for this is that combinatorial are problems nonconvex and we cannot expect to find all efficient solutions by weighted sum scalarization.

Example 8.6. Consider the multiobjective integer programming problem given by $\mathcal{X} = \{x \in \mathbb{Z}^2 : 2x_1 + 3x_2 \geq 11,\ x_1 \leq 4,\ x_2 \leq 4\}$ and $f_1(x) = x_1, f_2(x) = x_2$. Figure 8.3 shows the feasible set $\mathcal{X}$ as the integer grid points in the shaded feasible set of the LP relaxation of the problem, where $x \in \mathbb{Z}^2$ is replaced by $x \in \mathbb{R}^2$.

The efficient set is $\mathcal{X}_E = \{(0, 4), (1, 3), (3, 2), (4, 1)\}$ but there is no $\lambda > 0$ such that $(3, 2)$ is the optimal solution of a weighted sum problem. Note that this solution is not efficient for the LP relaxation of this problem. It is also worth mentioning that $x = (0, 4)$ is an efficient solution of the integer problem, but only a weakly efficient solution of the MOLP. □

Another particularity of combinatorial problems is that the distinction between efficient and properly efficient solutions vanishes. Recall that $\hat{x} \in \mathcal{X}$ is properly efficient (in the sense of Geoffrion, see Definition 2.39) if $\hat{x}$ is efficient and if there is a positive number $M > 0$ such that for all $i \in \{1, \ldots, p\}$ and $x \in \mathcal{X}$ with $f_i(x) < f_i(\hat{x})$ there is some $j \in \{1, \ldots, p\}$ such that $f_j(x) > f_j(\hat{x})$ and

$$\frac{f_i(\hat{x}) - f_i(x)}{f_j(x) - f_j(\hat{x})} \leq M.$$

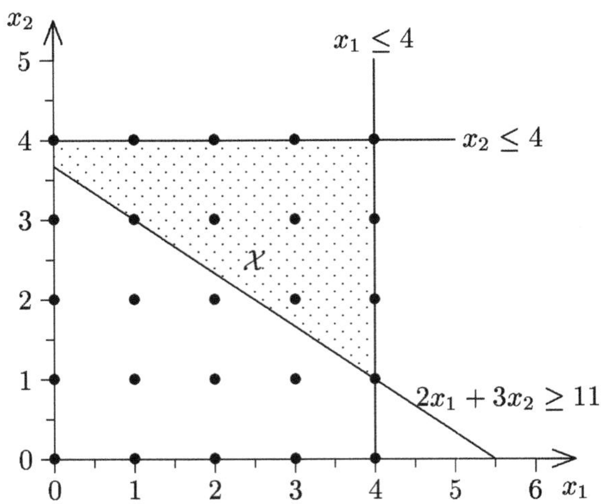

Fig. 8.3. Illustration of Example 8.6.

Because $\mathcal{Y}$ is a finite set and all objective function values are integers, there are only finitely many integer values for each objective. Therefore the denominator is never less than one and the trade off ratio is always bounded, so that the required M exists.

The above thoughts suggest to introduce some new definitions of (subsets of) efficient solutions and nondominated points, some of which are due to Hansen (1979).

Definition 8.7. *Let* $(\mathcal{X}, f, \mathbb{Z}^p)/\,\mathrm{id}\,/(\mathbb{Z}^p, \leq)$ *be a multiobjective optimization problem of the Pareto class and* $\mathcal{X}_E$ *be the efficient set and* $\mathcal{Y}_N$ *be the nondominated set.*

1. *Let* $x \in \mathcal{X}_E$. *If there is some* $\lambda \in \mathbb{R}^p_>$ *such that* $x \in \mathcal{X}_E$ *is an optimal solution of* $\min_{x \in \mathcal{X}} \lambda^T f(x)$ *then* x *is called a* supported efficient solution *and* $y = f(x)$ *is called* supported nondominated point. *The sets of all supported efficient solutions and supported nondominated points are denoted* $\mathcal{X}_{sE}$ *and* $\mathcal{Y}_{sN}$, *respectively. Otherwise* x *and* y *are called nonsupported, the notations are* $\mathcal{X}_{nE}$ *and* $\mathcal{Y}_{nN}$.
2. x^1 *and* x^2 *are called* equivalent *if* $f(x^1) = f(x^2)$. *A* complete set *of efficient solutions is any subset* $\mathcal{X}' \subset \mathcal{X}$ *such that* $f(\mathcal{X}) = \mathcal{Y}_N$. *A* minimal complete set *is a complete set without any equivalent solutions.* $\mathcal{X}_E$ *is also called the* maximal complete set.

3. *If x is a supported efficient solution and $y = f(x)$ is an extreme point of* $\operatorname{conv}\mathcal{Y}$ *then x is called an* extreme supported efficient solution. *y is an* extreme nondominated point.

In Figure 8.3 all three supported efficient solutions $(0,4),(1,3)$, and $(4,1)$ are extreme supported solutions.

Let us now introduce the main definitions to deal with computational complexity of multiobjective combinatorial optimization problems. This description is a very brief and informal summary. For an in depth introduction to computational complexity we refer to Garey and Johnson (1979). Computational complexity is a theory of "how difficult" it is to answer a decision problem DP, where a decision problem is a question that has either a yes or no answer. This difficulty is measured by the number of operations an algorithm needs to find the correct answer to the decision problem in the worst case.

To do this we use the "big O" notation. In this notation, the running time of an algorithm is $O(g(n))$ if there is a constant c, such that the number of operations performed by the algorithm is less than or equal to $cg(n)$ for all instances of the decision problem, where g is some function and n is the size of the instance.

Optimization and decision problems are closely related. Let $(\mathcal{X}, f, \mathbb{Z})/\operatorname{id}/(\mathbb{Z}, <)$ be a combinatorial optimization problem. The decision version of the optimization problem is the following question.

Given a constant $b \in \mathbb{Z}$, does there exist $x \in \mathcal{X}$ such that $f(x) \leq b$?

A decision problem belongs to the *class* $\mathcal{P}$ of problems, if there exists a deterministic algorithm that answers the decision problem and needs $O(p(n))$ operations, where p is a polynomial in n.

Example 8.8. Let $\mathcal{G} = (\mathcal{V}, \mathcal{E})$ be a graph and $c : \mathcal{E} \to \mathbb{Z}$. The decision version of the minimum spanning tree problem is: Given $b \in \mathbb{Z}$, does there exist a spanning tree T of $\mathcal{G}$ such that $\sum_{e \in T} c(e) \leq b$?

Since a spanning tree with minimal total edge weights can be found by Kruskal's algorithm (Kruskal, 1956) in $O(m^2)$ operations the problem is in class $\mathcal{P}$. Here m is the number of vertices of $\mathcal{G}$. □

An algorithm is called *polynomial time algorithm* if there is a a polynomial p such that the running time of the algorithm is $O(p(n))$. If there is no such polynomial the algorithm is said to be exponential. We will see that algorithms to solve multicriteria combinatorial optimization problems are often exponential.

Obviously all (decision versions of) combinatorial optimization problems for which an optimal solution can be found with a polynomial time algorithm belong to the class $\mathcal{P}$ of problems.

A decision problem belongs to the *class* $\mathcal{NP}$ if there is a *nondeterministic polynomial* time algorithm that solves the decision problem. Essentially this means that is possible to "guess" a solution, or, for the decision version of an optimization problem, given x it is possible to check whether $x \in \mathcal{X}$ and $f(x) \leq b$ in polynomial time. Clearly, $\mathcal{P} \subset \mathcal{NP}$.

Now let DP_1 and DP_2 be two decision problems. A *polynomial time transformation* of DP_1 to DP_2 is a polynomial time algorithm A that constructs an instance I_2 of DP_2 from an instance I_1 of DP_1 with the property that x_1 yields a "yes" answer for the instance I_1 if and only if $A(x_1)$ yields a "yes" answer for the instance of DP_2. We write $DP_1 \propto DP_2$. DP_1 and DP_2 are *equivalent* if $DP_1 \propto DP_2$ and $DP_2 \propto DP_1$.

A decision problem DP is $\mathcal{NP}$*-complete* if $DP \in \mathcal{NP}$ and $DP' \propto DP$ for all $DP' \in \mathcal{NP}$. Note that $\propto$ is transitive. Transitivity means that $\mathcal{NP}$-completeness of DP follows if $DP' \propto DP$ for one $DP' \in \mathcal{NP}$. Losely speaking, showing $DP' \propto DP$ actually means that DP' is a special case of DP, so DP is at least as difficult as DP'. A problem is called $\mathcal{NP}$*-hard* if $DP' \propto DP$ for all $DP' \in \mathcal{NP}$ but it is not known if $DP \in \mathcal{NP}$ (e.g. optimization problems).

Example 8.9. The travelling salesperson problem is an example of a $\mathcal{NP}$-hard problem. Given n cities and distances c_{ij} between all pairs (i,j) of cities, is there a tour that visits each city exactly once, returning to the starting point, with total length at most b? The decision version of the problem is $\mathcal{NP}$-complete. □

With each decision problem DP we we can associate the *counting problem* CP: How many "yes" answers does the decision problems have. Specifically for the decision version of an optimization problem:

Given $b \in \mathbb{Z}$, how many $x \in \mathcal{X}$ satisfy $f(x) \leq b$?

A counting problem belongs to the *class* $\#\mathcal{P}$ if there exists a nondeterministic algorithm that correctly "guesses" the answer to the counting problem and such that the longest computation that confirms a "yes" answer is bounded by a polynomial in the size of the instance.

A counting problem is $\#\mathcal{P}$*-complete*, if it is in $\#\mathcal{P}$ and for all $CP' \in \#\mathcal{P}$ there is a *parsimonious transformation* such that $CP' \propto_p CP$. $\propto_p$ denotes a parsimonious transformation , which is a polynomial time transformation that preserves the number of "yes" answers to decision problems.

Example 8.10.
- Counting the spanning trees of a graph is "easy", i.e. this counting problem belongs to $\#\mathcal{P}$, but it is not $\#\mathcal{P}$-complete. The number

of spanning trees of $\mathcal{G}$ is the determinant of AA^T, where A is the node-edge incidence matrix of $\mathcal{G}$, with a direction arbitrarily assigned to the edges, see e.g. Thulasiraman and Swamy (1992).
- Counting the perfect matchings of a bipartite graph is $\#\mathcal{P}$-complete, see Valiant (1979a). □

Note that even if a combinatorial optimization problem can be solved in polynomial time its counting version may be $\#\mathcal{P}$-complete. This is the case for the matching problem on bipartite graphs. Sometimes such problems are called "decision easy, counting hard" problems. In Welsh (1993) some examples are given.

To prove $\mathcal{NP}$- and $\#\mathcal{P}$-completeness of multiobjective combinatorial problems we need the following problems later on.

- KNAPSACK: Given $a \in \mathbb{Z}^n$ and $b \in \mathbb{Z}$, does there exist $x \in \{0,1\}^n$ such that $a^T x = b$?
- PARTITION: Given $c \in \mathbb{Z}^n$ with $\sum_{i=1}^n c_i = 2C$, does there exist $\mathcal{S} \subseteq \{1,\ldots,n\}$ such that $\sum_{i\in\mathcal{S}} c_i = \sum_{i\notin\mathcal{S}} c_i$?

Both problems are $\mathcal{NP}$-complete (Karp, 1972), the transformation used in the proofs are parsimonious, so they are also $\#\mathcal{P}$-complete (Welsh, 1993).

Lemma 8.11. *The problem* 0-1 KNAPSACK*: Does there exist $x \in \{0,1\}^n$ such that $(a^1)^T x \leq b_1$ and $(a^2)^T x \geq b_2$, where $a^1, a^2 \in \mathbb{Z}^n$ and $b_1, b_2 \in \mathbb{Z}$ are given? is $\mathcal{NP}$-complete and $\#\mathcal{P}$-complete.*

Proof. With $b_i^1 := b_i^2 := a_i$ for $i = 1,\ldots,n$ and $b_1 := b_2 := b$ we have a parsimonious transformation KNAPSACK $\propto$ 0-1 KNAPSACK. Therefore the result follows. □

Now consider a multiobjective combinatorial optimization problem $(\mathcal{X}, f, \mathbb{Z}^p)/\theta/(\mathbb{Z}^P, \preceq)$. The related decision problem is then defined as

Given constants $b_1,\ldots,b_P \in \mathbb{Z}$, does there exist a feasible solution $x \in \mathcal{X}$ such that $\theta(f(x)) \preceq (b_1,\ldots,b_P)$ or $\theta(f(x)) = (b_1,\ldots,b_P)$?

The related counting problem is

Given $b_1,\ldots,b_P \in \mathbb{Z}$, how many $x \in \mathcal{X}$ satisfy $\theta(f(x)) \preceq (b_1,\ldots,b_P)$ or $\theta(f(x)) = (b_1,\ldots,b_P)$?

With these definitions we can deal with $\mathcal{NP}$- and $\#\mathcal{P}$-completeness of MOCO problems. As a first example we consider the problem

$$
\begin{aligned}
\min \sum_{i=1}^{n} c_i^k x_i \quad & k = 1,\ldots,p \\
\text{subject to } x_i \in \{0,1\} \quad & i = 1,\ldots,n.
\end{aligned}
\tag{8.6}
$$

Note that for a single objective ($p = 1$) the solution is the empty set if all c_i are nonnegative or the set of all elements with negative cost. We call (8.6) the unconstrained multiobjective combinatorial optimization problem UMOCO.

Proposition 8.12. *The unconstrained multiobjective combinatorial optimization problem* (8.6) *is $\mathcal{NP}$- and #$\mathcal{P}$-complete even for $p = 2$.*

Proof. The decision version of (8.6) is: Given $c^1, c^2 \in \mathbb{Z}^n$ and $d_1, d_2 \in \mathbb{Z}$, does there exist $x \in \{0,1\}^n$ such that $\sum_{i=1}^n c_i^1 x_i \leq d_1$ and $\sum_{i=1}^n c_i^2 x_i \leq d_2$?

We show 0-1 KNAPSACK $\propto_p$ UMOCO. Let $c^1 := a^1, d_1 := b_1, c^2 := -a^2, d_2 := -b_2$. It is obvious that

$$\sum_{i=1}^n c_i^1 x_i \leq d_1 \iff (a^1)^T x \leq b_1,$$
$$\sum_{i=1}^n c_i^2 x_i \leq d_2 \iff (a^2)^T x \geq b_2.$$

The transformation is parsimonious. □

The final definition concerns problems for which exponential algorithms may be necessary.

Definition 8.13. *A MOCO problem $(\mathcal{X}, f, \mathbb{Z}^p)/\theta/(\mathbb{Z}^P, \preceq)$ is called* intractable, *if the size of* Val $(\mathcal{X}, f, \mathbb{Z}^p)/\theta/(\mathbb{Z}^P, \preceq)$ *(the set of optimal values) can be exponential in the size of an instance.*

For an intractable MOCO problem there is no polynomial p with the cardinality of the set of optimal values of order $O(p(n))$ and therefore no chance to find all optimal solutions in an efficient manner, i.e. by a polynomial time algorithm. Unfortunately, MOCO problems of the Pareto class are usually intractable. We illustrate this with the unconstrained MOCO problem.

Proposition 8.14. *The unconstrained MOCO problem is intractable.*

Proof. We define an instance in which all feasible solutions are efficient and have different objective function vectors. Since there are 2^n feasible solutions, we get $|\mathcal{Y}_N| = 2^n$, demonstrating the claim.

Let $c_i^k := (-1)^k 2^{i-1}$. Then

$$\mathcal{Y} = \mathcal{Y}_N = \left\{ \begin{pmatrix} 0 \\ 0 \end{pmatrix}, \begin{pmatrix} -1 \\ 1 \end{pmatrix}, \ldots, \begin{pmatrix} -2^n + 1 \\ 2^n - 1 \end{pmatrix} \right\}.$$

□

In the next section we will discuss very special multiobjective combinatorial optimization problem where $\mathcal{X} = \mathcal{E}$.

8.2 Problems with Explicitely Given Feasible Set

In combinatorial optimization problems the feasible set is usually given by constraint functions such as in $\mathcal{X} = \{x \in \mathbb{Z}^n : Ax = b, x \geqq 0\}$ or by a certain combinatorial structure such as $\mathcal{X} = \{T \subset \mathcal{E} : T$ defines a spanning tree of graph $\mathcal{G}\}$. We may, however, consider the most basic case that $\mathcal{X} = \mathcal{E}$. Solving a multicriteria optimization problem $(\mathcal{E}, f, \mathbb{Z}^p)/\theta/(\mathbb{Z}^P, \preceq)$ means finding the minimal elements of $\mathcal{Y} = \{f(e_1), \ldots, f(e_n)\}$ according to order $\preceq$. Since we consider f to be a sum or maximum as defined in (8.1) and (8.2), f is identical to c and the problem reduces to finding the minimal columns of the matrix $Y := (c(e_1), \ldots, c(e_n))$, the columns of which are the cost vectors of the elements e_i of $\mathcal{E}$, according to order $\preceq$. Below we present algorithms for the different MCO classes mentioned in Chapter 5 for problems with explicitly given finite set. Pairwise comparison of the columns will always give an answer, but not necessarily the most efficient. Column i of Y is denoted y_i.

The algorithm to find efficient solutions by pairwise comparison is shown as Algorithm 8.1.

Algorithm 8.1 (Efficient solutions for explicit feasible sets.)
Input: Y
Initialization: $\mathcal{X}_E := \mathcal{E}$
For $i := 1$ *to* $n - 1$ *do*
For $j := i + 1$ *to* n *do*
if $y_i \leq y_j$ *then* $\mathcal{X}_E := \mathcal{X}_E \setminus \{e_j\}$ *if* $y_j \leq y_i$ *then* $\mathcal{X}_E := \mathcal{X}_E \setminus \{e_i\}$
End for
End for
Output: $\mathcal{X}_E$

Algorithm 8.1 needs $O(n^2p)$ operations. Kung *et al.* (1975) prove the following result using a divide and conquer strategy.

Theorem 8.15 (Kung *et al.* (1975)). $(\mathcal{E}, c, \mathbb{Z}^p)/\operatorname{id}/(\mathbb{Z}^p, \leq)$ *can be solved in* $O(n \log n)$ *time if* $p \in \{1, 2, 3\}$. *If* $p \geq 4$ *it can be solved in* $O(n(\log n)^{p-2})$ *time.*

Due to p in the exponent in Theorem 8.15 algorithm of Kung *et al.* (1975) is faster than the straightforward Algorithm 8.1 only for a small number of objectives, $p \leq 5$. For arbitrary p, however, its worst case running time is exponential in p.

Max-ordering problems $(\mathcal{E}, c, \mathbb{Z}^p)/\max/(\mathbb{Z}, <)$ are even easier to solve. We have to find the maximal entries in the columns of Y, and look for the smallest of these maxima, which can be done in $O(np)$ time.

We present two algorithms for the solution of lexicographic max-ordering problems $(\mathcal{E}, c, \mathbb{Z}^p)/\operatorname{sort}/(\mathbb{Z}^p, <_{\text{lex}})$. These are taken from Ehrgott (1998). The first is based on sorting and lexicographic comparison of the columns of Y.

Algorithm 8.2 (Lex-MO solutions for explicit feasible sets I.)

Input: Y

For $i := 1$ *to* n *do* $y_i := \operatorname{sort}(y_j)$

$\alpha := 1,\ i := 1$

While $i \le n$ *do*

if $y_i <_{\text{lex}} y_\alpha$ *then* $\alpha := i$ $i := i + 1$

End While $\mathcal{X}_{lex-MO} := \{e_i : y_i = y_\alpha\}$

Output: $\mathcal{X}_{\text{lex}-\text{MO}}$

Algorithm 8.2 runs in $O(np \log p)$ time, i.e. it's worst case performance is bounded by the sorting of the n vectors y_i. Another approach is to first find the maxima of the columns, delete all columns where the maximum is greater than the smallest maximum and continue with a smaller matrix.

Algorithm 8.3 (Lex-MO solutions for explicit feasible sets II.)

Input: Y

$s := 1$

For $i := 1, \ldots, n$ *do* $\mathcal{Q}_i := \{1, \ldots, p\}$

$\mathcal{X}_{lexMO} := \mathcal{E}$

While $|\mathcal{X}_{lexMO}| > 1$ *and* $s \le p$

For all $e_i \in \mathcal{X}_{lexMO}$ *do*

$M_i := \max\limits_{k \in \mathcal{Q}_i} y_{ki}$

$\mathcal{I}_i := \operatorname{argmax}\{y_{ki} : k \in \mathcal{Q}_i\}$

$\mathcal{Q}_i := \mathcal{Q}_i \setminus \mathcal{I}_i$

End for

$M^* := \min\limits_{i=1,\ldots,n} M_i$

$\mathcal{X}_{lexMO} := \mathcal{X}_{lexMO} \setminus \{e_j : M_j > M^*\}$

For all $e_j \in \mathcal{X}_{lexMO}$ *do* $y_j := (y_{kj} : k \in \mathcal{Q}_i)$

$Y := (y_j : e_j \in \mathcal{X}_{lexMO})$

$s := s + 1$

End while
Output: $\mathcal{X}_{\text{lexMO}}$

Algorithm 8.3 needs $O(np^2)$ operations, more than Algorithm 8.2 in the worst case, but is often faster, in case the matrix Y becomes smaller quickly.

Example 8.16. Consider $\mathcal{X} = \mathcal{E} = \{e_1, \ldots, e_5\}$ with the following matrix

$$Y = \begin{pmatrix} 5 & 1 & 7 & 4 & 4 \\ 4 & 4 & 3 & 5 & 2 \\ 7 & 3 & 1 & 6 & 1 \\ 3 & 5 & 2 & 1 & 1 \\ 2 & 1 & 2 & 3 & 5 \end{pmatrix}.$$

Applying Algorithm 8.2 we first sort the columns of Y to get

$$Y = \begin{pmatrix} 7 & 5 & 7 & 6 & 5 \\ 5 & 4 & 3 & 5 & 4 \\ 4 & 3 & 2 & 4 & 2 \\ 3 & 1 & 2 & 3 & 1 \\ 2 & 1 & 1 & 1 & 1 \end{pmatrix}$$

and the lexicographically minimal column yields $\mathcal{X}_{lexMO} = \{e_5\}$.

With Algorithm 8.3 we proceed as follows: In the first iteration $M_1 = 7$, $M_2 = 5$, $M_3 = 7$, $M_4 = 6$, $M_5 = 5$. Therefore $\mathcal{Q}_1 = \{1,2,4,5\}, \mathcal{Q}_2 = \{1,2,3,5\}, \mathcal{Q}_3 = \{2,3,4,5\}, \mathcal{Q}_4 = \{1,2,4,5\}, \mathcal{Q}_5 = \{1,2,3,4\}$. Since $M^* = 5$ we delete e_1, e_3, e_4 from $\mathcal{X}_{lexMO}$ and $\mathcal{X}_{lexMO} := \{e_2, e_5\}$. The updated matrix Y is

$$Y = \begin{pmatrix} 1 & 4 \\ 4 & 2 \\ 3 & 1 \\ 1 & 1 \end{pmatrix}.$$

In the second iteration $M_2 = M_5 = 4$ and $\mathcal{Q}_2 = \{1,3,5\}, \mathcal{Q}_5 = \{2,3,4\}$. Thus, $\mathcal{X}_{lexMO}$ remains unchanged. The new matrix Y is

$$Y = \begin{pmatrix} 1 & 2 \\ 3 & 1 \\ 1 & 1 \end{pmatrix}.$$

Finally $M_2 = 3$, $M_5 = 2$ with $\mathcal{Q}_2 = \{1,5\}, \mathcal{Q}_5 = \{3,4\}$ and $\mathcal{X}_{lexMO} = \{e_2, e_5\} \setminus \{e_2\} = \{e_5\}$. The algorithm stops with the correct solution. □

Finally, we give an algorithm for lexicographic optimization problems with explicitly given feasible set. The algorithm finds lexicographically optimal solutions for all permutations of objectives, i.e it solves $(\mathcal{E}, Y, \mathbb{Z}^p)/\pi/(\mathbb{Z}^p, <_{\text{lex}})$ for all $\pi \in \Pi$. To do this efficiently it is impossible to simply consider the $p!$ permutations in turn. The algorithm is based on the following result, which shows that some e_i can be excluded as candidates for lexicographically optimal solutions.

Proposition 8.17 (Hamacher *et al.* (1999)). *For $i \in \{1, \dots, n\}$ let $\mathcal{M}(i) := \{k \in \{1, \dots, p\} : y_{ki} = \min_{j=1,\dots,n} v_{kj}\}$ and $\mathcal{J}(i) := \{j : \mathcal{M}(i) \subseteq \mathcal{M}(j)\}$. The following assertions hold.*

1. *If $\mathcal{M}(i) = \emptyset$ then $e_i \notin \mathcal{X}_\Pi$.*
2. *If $\mathcal{J}(i) = \{i\}$ then $e_i \in \mathcal{X}_\Pi$.*
3. *If $\{i\} \subset \mathcal{J}(i)$ then $e_i \in \mathcal{X}_\Pi$ if and only if there is a permutation π of $\{1, \dots, p\} \setminus \mathcal{M}(i)$ such that $(y_{\pi(k)i} : k \in \{1, \dots, p\} \setminus \mathcal{M}(i)$ is lexicographically minimal.*

Note that the sets $\mathcal{J}(i)$ are defined to have $\{i\} \subset \mathcal{J}(i)$ always.

Proof. 1. Let $l \in \{1, \dots, p\}$ and let $\pi \in \Pi$ be such that $l = \pi(1)$. Because $\mathcal{M}(i) = \emptyset$ there is some $j \neq i$ with $y_{lj} < y_{li}$. Therefore $y_{\pi(1)j} < y_{\pi(1)i}$ and $(y_{\pi(k)j} : k = 1, \dots, p) <_{\text{lex}} (y_{\pi(k)i} : k = 1, \dots, p)$, i.e. $e_i \notin \mathcal{X}_\Pi$.

2. Let $\pi \in \Pi$ and $m_i \in \mathbb{N}$ be such that $\{\pi^{-1}(l) : l \in \mathcal{M}(i)\} = \{1, \dots, m_i\}$. Then $(y_{\pi(k)i} : k \in \{1, \dots, p\}) <_{\text{lex}} (y_{\pi(k)j} : k \in \{1, \dots, p\})$ for all $j \neq i$, because otherwise $\mathcal{M}(i) \subseteq \mathcal{M}(j)$ for some $j \neq i$. Therefore $e_i \in \mathcal{X}_\Pi$.
3. For all $j \in \mathcal{J}(i)$ and for all $k \in \mathcal{M}(i)$ we have $y_{ki} = y_{kj}$. Therefore $e_i \in \mathcal{X}_\Pi$ if and only if there is a permutation of objectives in $\{1, \dots, p\} \setminus \mathcal{M}(i)$ such that $(y_{\pi(k)i} : k \in \{1, \dots, p\} \setminus \mathcal{M}(i))$ is lexicographically minimal in the set $\{(y_{\pi(k)j} : k \in \{1, \dots, p\} \setminus \mathcal{M}(i), j \in \mathcal{J}(i)\}$. □

The algorithm to solve $(\mathcal{E}, Y, \mathbb{Z}^p/\Pi/(\mathbb{Z}^p, <_{\text{lex}})$ can now be stated.

Algorithm 8.4 (Lexicographic solutions for explicit feasible sets.)

Input: Y
  $\mathcal{L} := \{1, \dots, n\}$
  For all $i \in \mathcal{L}$ *do*
    $\mathcal{M}(i) := \{k \in \{1, \dots, p\} : y_{ki} = \min_{j=1,\dots,n} y_{kj}\}$
    If $\mathcal{M}(i) = \emptyset$ *then* $\mathcal{L} := \mathcal{L} \setminus \{i\}$
  End for.
  $\mathcal{X}_\Pi := \emptyset$
  1: For all $i \in \mathcal{L}$ *do*

$\mathcal{J} := \mathcal{L}$, $\mathcal{Q} := \{1, \ldots, p\}$
2: For all $j \in \mathcal{J}$ *do* $\mathcal{M}(j) := \{k \in \mathcal{Q} : y_{kj} = \min\{y_{kl} : l \in \mathcal{J}\}\}$
If $\mathcal{M}(i) = \emptyset$ *then*
$\mathcal{L} := \mathcal{L} \setminus \{i\}$
Goto 1 with next i
End if
$\mathcal{J}(i) := \{j \in \mathcal{J} : \mathcal{M}(i) \subseteq \mathcal{M}(j)\}$
If $\mathcal{J}(i) = \{i\}$ *then*
$\mathcal{X}_\Pi := \mathcal{X}_\Pi \cup \{e_i\}$
Goto 1 with next i
$\mathcal{J} := \mathcal{J}(i)$
$\mathcal{Q} := \mathcal{Q} \setminus \mathcal{M}(i)$
Goto 2
Output: $\mathcal{X}_{\Pi(Q)}$

Example 8.18. As an example we use $\mathcal{E} = \{e_1, \ldots, e_5\}$ and the matrix

$$Y = \begin{pmatrix} 5 & 1 & 7 & 4 & 4 \\ 4 & 4 & 3 & 5 & 2 \\ 7 & 3 & 1 & 6 & 1 \\ 3 & 5 & 2 & 1 & 1 \\ 2 & 1 & 2 & 3 & 5 \end{pmatrix}$$

once again. The steps of the algorithm are summarized below.

First, $\mathcal{L} = \{1, 2, 3, 4, 5\}$ and $\mathcal{M}(1) = \emptyset, \mathcal{M}(2) = \{1, 5\}, \mathcal{M}(3) = \{3\}$, $\mathcal{M}(4) = \{4\}, \mathcal{M}(5) = \{2, 3, 4\}$ are computed. e_1 is deleted, so that $\mathcal{L} = \{2, 3, 4, 5\}$ and $\mathcal{X}_\Pi$ is set to $\emptyset$.

The checking of $i \in \mathcal{L}$ is as follows.

$$
\begin{aligned}
i = 2 : \;& \mathcal{J} = \{2,3,4,5\}, \mathcal{Q} = \{1,\dots,5\} \\
& \mathcal{J}(2) = \{2\}, \mathcal{X}_\Pi = \{e_2\} \\
i = 3 : \;& \mathcal{J} = \{2,3,4,5\}, \mathcal{Q} = \{1,\dots,5\} \\
& \mathcal{J}(3) = \{3,5\} \\
& \mathcal{J} = \{3,5\},\ \mathcal{Q} = \{1,2,4,5\} \\
& \qquad\qquad \mathcal{M}(3) = \{5\}, \mathcal{M}(5) = \{1,2,4\} \\
& \qquad\qquad \mathcal{J}(3) = \{3\}, \mathcal{X}_\Pi = \{e_2, e_3\} \\
i = 4 : \;& \mathcal{J} = \{2,3,4,5\}, \mathcal{Q} = \{1,\dots,5\} \\
& \mathcal{J}(4) = \{4,5\} \\
& \mathcal{J} = \{4,5\},\ \mathcal{Q} = \{1,2,3,5\} \\
& \qquad\qquad \mathcal{M}(4) = \{1,5\}, \mathcal{M}(5) = \{1,2,3\} \\
& \qquad\qquad \mathcal{J}(4) = \{4\}, \mathcal{X}_\Pi = \{e_2, e_3, e_4\} \\
i = 5 : \;& \mathcal{J} = \{2,3,4,5\}, \mathcal{Q} = \{1,\dots,5\} \\
& \mathcal{J}(5) = \{5\}, \mathcal{X}_\Pi = \{e_2, e_3, e_4, e_5\}
\end{aligned}
$$

□

The running time of Algorithm 8.4 is $O(p^2 n \log n)$. The inner loop takes $O(pn \log n)$ operations (one sorting instead of searching minima several times). The loop is carried out at most p times, because at least one element of $\mathcal{Q}$ is removed in each iteration.

8.3 Scalarization of Multiobjective Integer Programs

In the previous section we have shown how to solve combinatorial optimization problems in the special case that $\mathcal{X} = \mathcal{E}$. We will now turn to find efficient solutions of multiobjective integer programming problems using scalarization techniques. The following is from Ehrgott (2005). We have introduced a number of such techniques in Chapters 3 and 4.

A multiobjective integer program is the following optimization problem

$$
\begin{aligned}
\min\ & Cx \\
\text{subject to } & Ax = b \\
& x \geqq 0 \\
& x \in \mathbb{Z}^n,
\end{aligned}
\tag{8.7}
$$

where C is a $p \times n$ matrix of integers, A is a $m \times n$ matrix of integers, and $b \in \mathbb{Z}^n$.

Ideally, the scalarized single-objective version of (8.7) should have the following properties.

1. The scalarized problem is not harder to solve than (8.7) with $p = 1$. In particular, if (8.7) with a single objective is solvable in polynomial time, the scalarization of MOIP is not $\mathcal{NP}$-hard.

2. An optimal solution of the scalarization is an efficient solution of (8.7).
3. Every efficient solution of (8.7) is an efficient solution of the scalarized problem with appropriately selected parameters.
4. Since the constraints and objectives of (8.7) are linear, the objective and constraints of the scalarized problem are linear.

We review some scalarization techniques and state which of these properties they have.

The weighted sum scalarization

$$\min \left\{ \sum_{k=1}^{p} \lambda_k (c^k)^T x : x \in \mathcal{X} \right\} \tag{8.8}$$

with $\lambda_k > 0$ for all $k = 1, \ldots, p$ has properties 1, 2, and 4. Unfortunately Example 8.6 shows that there are efficient solutions which cannot be found by solving weighted sum problems.

The ε-constraint method

$$\min \left\{ (c^j)^T x : x \in \mathcal{X}, (c^k)^T x \leq \varepsilon_k, k \neq j \right\} \tag{8.9}$$

has properties 2, 3, and 4 (see Theorem 4.5). Unfortunately the constraints on objective values usually make the problem $\mathcal{NP}$-hard, even if $\min_{x \in \mathcal{X}} c^T x$ is polynomially solvable, e.g. for the shortest path problem, see Garey and Johnson (1979).

Given $x^0 \in \mathcal{X}$ Benson's scalarization (Section 4.4) is

$$\max \left\{ \sum_{k=1}^{p} z_k : Cx^0 - z - Cx = 0, x \in \mathcal{X}, z \geqq 0 \right\} \tag{8.10}$$

which is actually equivalent to

$$\min \left\{ \sum_{k=1}^{p} (c^k)^T x : Cx \leqq Cx^0, x \in \mathcal{X} \right\}. \tag{8.11}$$

The same comments as for the ε-constraint method apply.

The achievement function approach citepWierzbickiMakowskiWessels00 has been briefly mentioned in Section 4.6. We consider the following form

$$\min \left\{ \max_{k=1}^{p} \nu_k \left((c^k)^T x - \rho_k \right) + \gamma \sum_{k=1}^{p} \lambda_k \left((c^k)^T x - \rho_k \right) : x \in \mathcal{X} \right\}, \tag{8.12}$$

where $\rho \in \mathbb{R}^p$ is a reference point, $\nu \in \mathbb{R}^p_>$ is a vector of positive weights, $\gamma > 0$. It is easily seen that if $\gamma > 0$ an optimal solution $\hat{x}$ is efficient. Also, all efficient solutions can be found, see Theorem 4.29. In order to solve the scalarization using integer programming techniques, it is necessary to reformulate the problem using an additional variable and constraints to linearize the max term of the objective function:

$$\min_{x \in calX} \left\{ z + \gamma \sum_{k=1}^{p} \lambda_k \left((c^k)^T x - \rho_k \right) : z \geq \nu_k \left((c^k)^T x - \rho_k \right), k = 1, \ldots, p \right\}. \tag{8.13}$$

The author is not aware of a direct proof of $\mathcal{NP}$-completeness, but the presence of the min max component and the constraints in the reformulation strongly suggests that (8.13) is $\mathcal{NP}$-hard.

Finally, we mention the (weighted) max-ordering problem, which we have described as a nonscalarizing method in Chapter 5.

$$\min \left\{ \max_{k=1}^{p} \nu_k (c^k)^T x : x \in \mathcal{X} \right\}. \tag{8.14}$$

An optimal solution of this problem is at least weakly efficient and all efficient solutions can be found (see Theorem 5.10), however, 8.14 is again $\mathcal{NP}$-hard, see Chung *et al.* (1993) for the max-ordering version of the unconstrained problem UMOCO (8.6).

This review indicates that the scalarization techniques we have mentioned in this book either do not allow us to find all efficient solutions of (8.7) or lead to scalarized problems that are $\mathcal{NP}$-hard. We can formalize that some more if we consider only scalarizations that only use the maximum and the sum of linear terms (like all of the above). The maximum can be linearized using a variable z as before. We assume that no other new variables are introduced.

Let $\rho \in \mathbb{R}^p$ be a reference point, let $\lambda, \nu \in \mathbb{R}^p_{\geqq}$ be vectors of weights, and let $\varepsilon \in \mathbb{R}^p$ be a vector of right hand sides. We formulate the following general scalarized problem.

$$\begin{aligned} \min_{x \in \mathcal{X}} \max_{k=1}^{p} \nu_k \left((c^k)^T x - \rho_k \right) &+ \sum_{k=1}^{p} \lambda_k \left((c_k)^T x - \rho_k \right) \\ \text{subject to } \; c_k x \;\leq\; \varepsilon_k, &\quad k = 1, \ldots, p. \end{aligned} \tag{8.15}$$

In Table 8.1 we show which settings of the parameters $\rho, \nu, \lambda, \varepsilon$ yield the scalarizations discussed above. These settings are either given as specific values or as a set from which a parameter is to be chosen.

Additional scalarizations included in (8.15) are the compromise programming method $\min_{x \in X} ||Cx - y^I||$ for the l_1 and l_∞ norms (Section 4.5) and the hybrid method of Guddat *et al.* (1985),

$$\min \left\{ \sum_{k=1}^{p} \lambda_k (c_k)^T x : x \in \mathcal{X}, (c_k)^T x \leq \varepsilon_k, \; k = 1, \ldots, p \right\},$$

Table 8.1. Parameters in (8.15) and resulting scalarizations.

Equation	ρ	ν	λ	ε
(8.8)	0	0	$\lambda \in \mathbb{R}^p_>$	$\varepsilon_k = \infty \; k = 1, \ldots, p$
(8.9)	0	0	$\lambda_j = 1; \; \lambda_k = 0 \; k \neq j$	$\varepsilon_j = \infty; \; \varepsilon_k \in \mathbb{R} \; k \neq j$
(8.10)	0	0	$\lambda_k = 1 \; k = 1, \ldots, p$	$\varepsilon_k = c_k x^0 \; k = 1, \ldots, p$
(8.12)	$\rho \in \mathbb{R}^p$	$\nu \in \mathbb{R}^p_>$	$\lambda_k = \gamma \; k = 1, \ldots, p$	$\varepsilon_k = \infty \; k = 1, \ldots, p$
(8.14)	0	$\nu \in \mathbb{R}^p_>$	0	$\varepsilon_k = \infty \; k = 1, \ldots, p$

which generalizes the reformulation (8.11) of Benson's method.

We have the following result.

Proposition 8.19. *The general linear scalarization* (8.15) *is $\mathcal{NP}$-hard.*

Proof. We show that (8.15) includes the binary knapsack problem $\max\{cx : ax \leq b, x \in \{0,1\}^n\}$ as special case. To do that, we choose $\mathcal{X} = \{0,1\}^n, p = 2, \nu = \rho = 0, \lambda_1 = 1, \lambda_2 = 0, c_1 = -c \leq 0, c_2 = a \geq 0$, and $\varepsilon_1 = \infty, \varepsilon_2 = b > 0$. The rest follows from considering the decision versions of (8.15) and the binary knapsack problem together with Lemma 8.11 □

In order to linearize the max term in the objective, we introduce a new variable z and reformulate (8.15) as

$$
\begin{array}{lr}
\min z + \sum_{k=1}^{p} \lambda_k \left((c^k)^T x - \rho_k\right) & \\
\text{subject to} \qquad (c^k)^T x \leq \varepsilon_k, \; k = 1, \ldots, p & \\
\nu_k \left((c^k)^T x - \rho_k\right) \leq z, \quad k = 1, \ldots, p & \qquad (8.16) \\
x \in \mathcal{X} & \\
z \in \mathbb{R}. &
\end{array}
$$

Note that variable z is an unrestricted real variable, i.e. it can attain negative or positive values or be zero.

Since we have observed before that for many parameter settings, in which the constraints on objective functions are present (8.16) is $\mathcal{NP}$-hard (and may be very hard to solve computationally) a good idea seems to be to apply Lagrangian relaxation of those complicating constraints on objective functions. Using multipliers π_k for constraints $\nu_k(c_k x - \rho_k) \leq z$ and π'_k for $c_k x \leq \varepsilon_k$ the resulting problem is

$$\min_{z\in\mathbb{R},x\in\mathcal{X}} z\left(1-\sum_{k=1}^{p}\pi_k\right) - \sum_{k=1}^{p}(\pi_k\nu_k\rho_k + \pi_k'\varepsilon_k + \lambda_k\rho_k) + \sum_{k=1}^{p}(\pi_k\nu_k + \pi_k' + \lambda_k)(c^k)^T x. \tag{8.17}$$

For fixed $\pi, \pi', \lambda, \nu, \rho$, this problem is of the form

$$\min_{z\in\mathbb{R},x\in\mathcal{X}} z(1-\alpha) - \beta + \sum_{k=1}^{p}\omega_k(c^k)^T x, \tag{8.18}$$

where α and β are constants. Thus the optimal solution of (8.18) is obviously unbounded if $\alpha > 1$, otherwise it is attained as optimal solution of a problem

$$\min_{x\in X}\sum_{k=1}^{p}\omega_k c_k x,$$

i.e. a weighted sum problem. In addition, if $\alpha < 1$, the optimal value of z is equal to zero. We have thus shown:

Theorem 8.20. *Solving a scalarization of the MOIP* (8.7) *of the general form* (8.15) *by Lagrangian duality yields a supported efficient solution.*

With Proposition 8.19 and Theorem 8.20 we have shown that a general scalarization that uses maximum and sum terms of the objectives of (8.7) only and is able to find all efficient solutions is $\mathcal{NP}$-hard. In addition, if we want to overcome that problem by using Lagrangian relaxation we end up with scalarized problems that are equivalent to the weighted sum problem and thus cannot identify all efficient solutions. This is a strong argument for the use of methods other than pure scalarization to solve multiobjective combinatorial optimization problems. We will study such methods in the remaining chapters of the book.

8.4 Notes

It is worth mentioning that in quite a few publications on solution methods for multiobjective combinatorial optimization problems it is not mentioned whether algorithms find a complete set of efficient solutions, maximal, minimal, or otherwise. Most often, extreme supported efficient solutions are found as well as nonsupported solutions, excluding equivalent ones. Finding nonextreme supported solutions and equivalent solutions is inevitably an enumeration problem, because either the objective vectors or their weighted sum are the same than some other solution.

There has also been a major research effort in heuristic methods for multi-objective combinatorial optimization. Apart from the very last section this is beyond the scope of the book. A survey can be found in Ehrgott and Gandibleux (2004).

Exercises

8.1. Prove that for $\nu > 0$ and $\alpha > 0$ an optimal solution of the problem

$$\min_{x \in \mathcal{X}} \left(\max_{k=1,\ldots,p} \nu_k((c^k)^T x - y_k^I) + \alpha \sum_{k=1}^{p} ((c_k)^T x - y_k^I) \right) \tag{8.19}$$

is efficient. Why is $\alpha > 0$ necessary to obtain that result? Can you prove the converse, i.e. for every $\hat{x} \in \mathcal{X}_E$, there are $\nu \in \mathbb{R}^p, \mu \geq 0$ and $\alpha \in \mathbb{R}, \alpha \geq 0$ such that $\hat{x}$ is an optimal solution of (8.19)?

8.2. For the following problem types, give examples with two objective functions that show that unsupported efficient solutions can exist.

- The bicriterion shortest path problem.
- The bicriterion spanning tree problem.
- The bicriterion assignment problem.

Make these examples as small as possible.

9

Multiobjective Versions of Polynomially Solvable Problems

9.1 Algorithms for the Shortest Path Problem

This section is about the shortest path problem with multiple objectives. Let $\mathcal{G} = (\mathcal{V}, \mathcal{A})$ be a directed graph (or digraph) with $|\mathcal{V}| = m$ nodes (or vertices) and $|E| = n$ arcs. Let $c : \mathcal{A} \to \mathbb{Z}^p$ be a cost function on the arcs. In this section we consider the problems of finding efficient paths from a specified node s to another specified node t, or from node s to all other nodes of $\mathcal{G}$ with sum objectives. We will not discuss the problem of finding efficient paths between all pairs of nodes. This problem can always be solved by solving m shortest path problems with fixed starting node s and has not been addressed in the multicriteria literature as a distinct problem.

We show that the problem with fixed s and t is difficult in terms of finding and counting efficient solutions and that it is intractable. The algorithms we present are generalizations of the well known label-setting and label-correcting algorithms for single objective shortest path problems. At the end of this section we present a ranking algorithm for the biobjective shortest path problem. We present a generalization of this algorithm that can be used as a prototype to solve any MOCO problem.

Definition 9.1. *Let $\mathcal{G} = (\mathcal{V}, \mathcal{A})$ be a directed graph. Let $\mathcal{V} = \{v_1, \ldots, v_m\}$ and $\mathcal{A} = \{a_1, \ldots, a_n\}$ where $a_i = (v_{j_1}, v_{j_2})$ is an ordered pair of vertices.*

1. *A* path P *is a sequence of nodes and arcs $(v_{i_1}, a_{i_1}, v_{i_2}, \ldots, v_{i_{r-1}}, a_{i_{r-1}}, v_{i_r})$ such that either $a_{i_l} = (v_{i_l}, v_{i_{l+1}})$ or $a_{i_l} = (v_{i_{l+1}}, v_{i_l})$. In a* directed path *only the former is possible.*
2. *A* simple path *is a path without repetition of vertices. A* simple directed path *is a directed path without repetition of vertices.*
3. *A* cycle C *is a simple path together with the arc (v_{i_r}, v_{i_1}) or (v_{i_1}, v_{i_r}). A* directed cycle *is a directed simple path together with the arc (v_{i_r}, v_{i_1})*

We will often identify a (directed) path by its sequence of vertices $(v_{i_1}, \ldots, v_{i_r})$. Let $s, t \in \mathcal{V}$ be two vertices and let $\mathcal{P}$ denote the set of all directed paths with $v_1 = s$ and $v_r = t$. Let $c : \mathcal{A} \to \mathbb{Z}^p$ be a cost function. The multiobjective shortest path problem is to find all efficient directed paths from s to t, i.e.

$$\min_{P \in \mathcal{P}} \left\{ \sum_{a \in P} c(a) \right\}. \tag{9.1}$$

Theorem 9.2 (Serafini (1986)). *The bicriterion shortest path problem* (9.1) *is $\mathcal{NP}$-complete and #$\mathcal{P}$-complete in acyclic digraphs.*

Proof. The decision version of (9.1) is: Given $b \in \mathbb{Z}^2$, $\mathcal{G} = (\mathcal{V}, \mathcal{A})$, and $s, t \in \mathcal{V}$ does there exist a path P from s to t in $\mathcal{G}$ such that $\sum_{a \in P} c(a) \leqq k$. This problem is clearly in $\mathcal{NP}$. We give a parsimonious transformation 0-1 KNAPSACK $\propto_p (\mathcal{P}, 2\text{-}\sum, \mathbb{Z}^2)/\,\mathrm{id}\,/(\mathbb{Z}^2, \leq)$. Given an instance a^1, a^2, b_1, b_2 of 0-1 KNAPSACK we construct an instance of the shortest path problem as follows. Let

$$\begin{aligned}
\mathcal{V} &:= \{v_0, \ldots, v_n\}, \\
s &:= v_0, \\
t &:= v_n, \\
\mathcal{A} &:= \{(v_{i-1}, v_i) : i = 1, \ldots, n\} \cup \{(v_{i-1}, v_i)' : i = 1, \ldots, n\}, \\
c^1(a) &:= \begin{cases} a_i^1 & \text{if } a = (v_{i-1}, v_i) \\ 0 & \text{if } a = (v_{i-1}, v_i)', \end{cases} \\
c^2(a) &:= \begin{cases} 0 & \text{if } a = (v_{i-1}, v_i) \\ a_i^2 & \text{if } a = (v_{i-1}, v_i)'. \end{cases}
\end{aligned}$$

The graph defined here is shown in Figure 9.1.

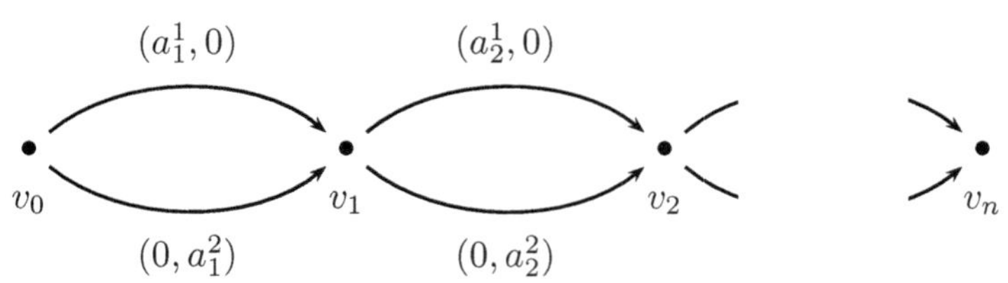

Fig. 9.1. Graph in the proof of Theorem 9.2.

Let $P \in \mathcal{P}$ be a path from s to t. Then

$$f_1(P) = \sum_{a \in P} c^1(a) \leq b_1 \tag{9.2}$$

$$\text{and } f_2(p) = \sum_{a \in P} c^2(a) \leq \sum_{i=1}^{n} a_i^2 - b_2 \tag{9.3}$$

if and only if there exists $x \in \{0,1\}^n$ such that $(a^1)^T x \leq b_1$ and $(a^2)^T(e-x) \leq \sum_{i=1}^n a_i^2 - b_2$, where $e = (1,\ldots,1) \in \mathbb{R}^n$. This is true if and only if there is $x \in \{0,1\}^n$ such that $(a^1)^T x \leq b_1$ and $(a^2)^T x \geq b_2$. Indeed, x is defined by $x_i = 1$ if and only if $(v_{i-1}, v_i) \in P$. The number of paths satisfying (9.2) and (9.3) is the same as the number of knapsack solutions. □

To demonstrate intractability, we need an example with an exponential number of efficient paths, which have incomparable objective function vectors.

Theorem 9.3 (Hansen (1979)). *The multicriteria shortest path problem* (9.1) *is intractable, even for* $p = 2$.

Proof. We construct an instance where $|\mathcal{Y}_N|$ is exponential in n. Let $\mathcal{V} = \{v_1, \ldots, v_n\}$, where n is odd, and define three sets of arcs as shown in Table 9.1. The graph with arc-weights alongside the arcs is shown in Figure 9.2.

Table 9.1. Arcs and their costs in Theorem 9.3.

Arcs	Costs
$a = (v_i, v_{i+2}), i = 1, 3, \ldots, n-2$	$c(a) = \left(2^{\frac{i-1}{2}}, 0\right)$
$a = (v_i, v_{i+1}), i = 1, 3, \ldots, n-2$	$c(a) = \left(0, 2^{\frac{i-1}{2}}\right)$
$a = (v_{i+1}, v_{i+2}), i = 1, 3, \ldots, n-2$	$c(a) = (0, 0)$

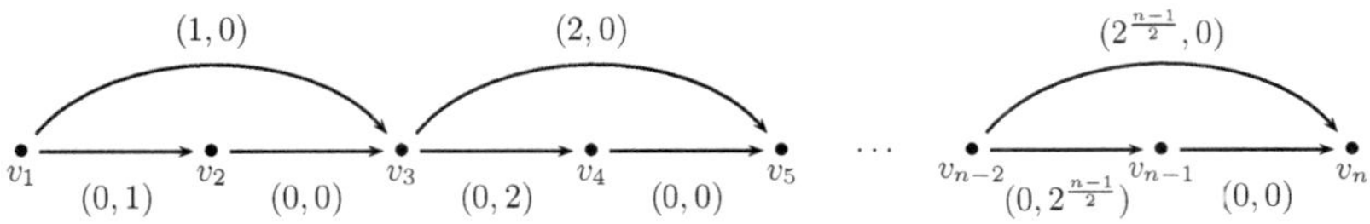

Fig. 9.2. A graph with exponentially many efficient paths.

Consider any path P from v_1 to v_n and observe that

$$\sum_{a \in P} (c^1(a) + c^2(a)) = \sum_{i=1,3,5,\ldots}^{n-2} 2^{\frac{i-1}{2}} = \sum_{i=0}^{\frac{n-3}{2}} 2^i = 2^{\frac{n-1}{2}} - 1$$

and that for each $z \in \{0, \ldots, 2^{\frac{n-1}{2}} - 1\}$ there is a path P from v_1 to v_n with $\sum_{a \in P} c^1(a) = z$. Therefore all $2^{\frac{n-1}{2}}$ paths are efficient and $|\mathcal{Y}_N| = 2^{\frac{n-1}{2}}$. □

To find paths a straightforward approach is to generalize single objective shortest path algorithms. Let us first assume that no negative cycles exist, i.e. for all cycles C in $\mathcal{G}$, and for all $k = 1, \ldots, p$

$$\sum_{a \in C} c^k(a) \geq 0. \tag{9.4}$$

and that this inequality is strict for at least one k.

Under this assumption, an efficient path will be simple, i.e. never visit any node more than once, because including a cycle increase the total cost of the path for at least one objective, whereas none of the other objectives is improved.

Proposition 9.4. *Let $\mathcal{G} = (\mathcal{V}, \mathcal{A})$ be a graph that satisfies the assumption above. Let P be an efficient path from s to t. Then P contains only efficient paths from s to intermediate nodes on the path.*

Proof. Let P_{st} be an efficient path from s to t. Assume P_{si} is a sub-path from s to node v_i and this path is not efficient. Then there is a path $P'_{si} \neq P_{si}$ with $f(P'_{si}) \leq f(P_{si})$.

Therefore P'_{si} together with P_{it}, the sub-path of P from i to t is a path from s to t with $f(P'_{si} \cup P_{it}) = f(P'_{si}) + f(P_{it}) < f(P_{si}) + f(P_{it}) = f(P_{st})$, contradicting efficiency of P_{st}. □

Actually, the proof of Proposition 9.4 is still valid if the path P_{si} is replaced by any subpath P_{uv} between nodes of the efficient path P_{st}. This is stated as a corollary.

Corollary 9.5. *Under the assumption of Proposition 9.4 let P_{st} be an efficient path from s to t. Then any subpath P_{uv} from u to v, where u and v are vertices on P_{st} is an efficient path from u to v.*

It is important to note that although an efficient path is always composed of efficient sub-paths between vertices along the path it is in general not true that compositions of efficient paths yield efficient paths again.

Example 9.6. In the graph of Figure 9.3 paths $P_{13} = (1,3)$ and $P_{34} = (3,4)$ are efficient, but their composition is not, because it is dominated by the path $(1,4)$ from node 1 to node 4.

□

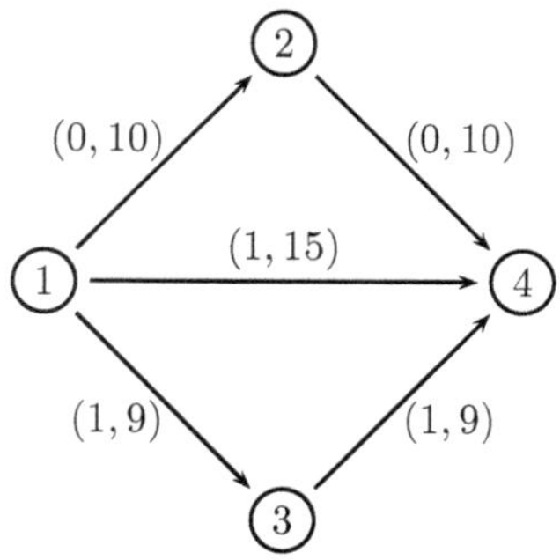

Fig. 9.3. Combining efficient paths.

Depending on whether negative weights may or may not occur in the underlying graph, label setting or label correcting algorithms can be designed to solve multicriteria shortest path problems. These algorithms will use sets of labels for each node rather than a single label, as the single objective labeling algorithms do. We present a label setting algorithm proposed by Martins (1984), which works under the stronger assumption that $c^k(a_i) \geq 0$ for all k and all arcs a_i.

Let v_i be a node of $\mathcal{G}$. A label of v_i is a $p+3$ tuple $(c^1, \ldots, c^p, v_j, l, k)$ where $v_j \neq v_i$ is a node of $\mathcal{G}$, l is the number of a label of node v_j, and k is the number of the label at node v_i. Thus, a label is a vector made up of a p-dimensional cost component, a node predecessor label, identifying the node from which the label was obtained, a further label indicating from which of the several labels of the predecessor it was computed, and a label number at the current node. We denote by $\mathcal{TL}$ a list of temporary labels, which is kept in lexicographic order, and a list $\mathcal{PL}$ of permanent labels, which will identify efficient paths.

Algorithm 9.1 (Multiobjective label setting algorithm.)

Input: *A digraph* $\mathcal{G} = (\mathcal{V}, \mathcal{A})$ *with* p *arc costs.*

Initialization: *Create label* $L = (0, \ldots, 0, 0, 0, 1)$ *at node* s *and let* $\mathcal{TL} := \{L\}$.

While $\mathcal{TL} \neq \emptyset$ *do*

Let label $L = (c^1, \ldots, c^p, v_h, l, k)$ *of node* v_i *be the lexicographically smallest label in* $\mathcal{TL}$.

Remove L *from* $\mathcal{TL}$ *and add it to* $\mathcal{PL}$.

For all $v_j \in \mathcal{V}$ *such that* $(v_i, v_j) \in \mathcal{A}$ *do*

Create label $L' = (c^1 + c^1(v_i, v_j), \ldots, c^p + c^p(v_i, v_j), v_i, k, t)$ *as the next label at node* v_j *and add it to* $\mathcal{TL}$.

Delete all temporary labels of node v_j dominated by L', delete L' if it is dominated by another label of node v_j.

End for.

End while.

Use the predecessor labels in the permanent labels to recover all efficient paths from s to other nodes of $\mathcal{G}$.

Output: *All efficient paths from node s to all other nodes of $\mathcal{G}$.*

To prove the correctness of Algorithm 9.1 we have to show that all efficient paths from s to any other node will be found, and that no permanent label defines a dominated path. The following lemma is useful.

Lemma 9.7. *If P_1 and P_2 are two paths between nodes s and t and $f(P_1) \leq f(P_2)$ then $f(P_1) <_{\text{lex}} f(P_2)$.*

The proof of this Lemma is an easy exercise, see Exercise 9.1.

Theorem 9.8. *At termination of Algorithm 9.1, the permanent labels at each node v_i define the set of efficient paths from node s to node v_i.*

Proof. 1. First we show that if a label is made permanent, it defines an efficient path. So let $L = (c^1, \ldots, c^p, v_h, l, k)$ be the label of node v_i which is lexicographically smallest in $\mathcal{TL}$. Let the corresponding path be $(s, \ldots, v_j, \ldots, v_h, v_i)$. Assume there is an efficient path P_{si} from s to v_i with $f(P_{si}) \leq c$. Then from Lemma 9.7 $f(P_{si}) <_{\text{lex}} (c^1, \ldots, c^p)$. Assume that v_j is the last node both paths have in common.
Because all arc costs are nonnegative, it follows that for all nodes v_l between v_j and v_i on path P_{si} it holds that $f(P_{sl}) \leqq f(P_{si})$ and therefore $f(P_{1l}) <_{\text{lex}} c$. Therefore, all nodes on path P_{ji} have labels that must have been found and made permanent before L has been made permanent. Finally, that label would either not have been created at all or been deleted from $\mathcal{TL}$ before being made permanent. This contradiction means that only efficient paths are found by the algorithm.

2. Now we need to show that all efficient paths are found, i.e. for each efficient path from s to a node v_i a permanent label is created. So suppose there is an efficient path P_{si} from s to some node v_i that has not been found. This is only possible if either the label L corresponding to this path has been found, but not made permanent, or if it has never been found.
The only way for a label in $\mathcal{TL}$ not to be made permanent is deletion due to domination. As the path P_{si} is efficient this is impossible. Note that if

a label is never deleted, it will eventually be lexicographically minimal in $\mathcal{TL}$.

So we may assume that L was never included in $\mathcal{TL}$. Let v_{i-1} be the node just before v_i along path P_{si}. According to Proposition 9.4 path $P_{s,i-1}$ from s to v_{i-1} is efficient. Since L has not been found, either the label L' corresponding to $P_{s,i-1}$ has not been found, or was not made permanent. Repeating this argument backwards along the path, we see that for the first node $v_1 \neq s$ on P_{si}, the label corresponding to the path from s to v_1, which is $(c^1(s, v_1), \ldots, c^p(s, v_1), s, 1, 1)$ has not been found, which is impossible, because it is created in the first iteration. This contradiction completes the proof. □

Example 9.9 (Martins (1984)). We apply algorithm 9.1 to the graph of Figure 9.4 with $s = 1$ and $t = 6$.

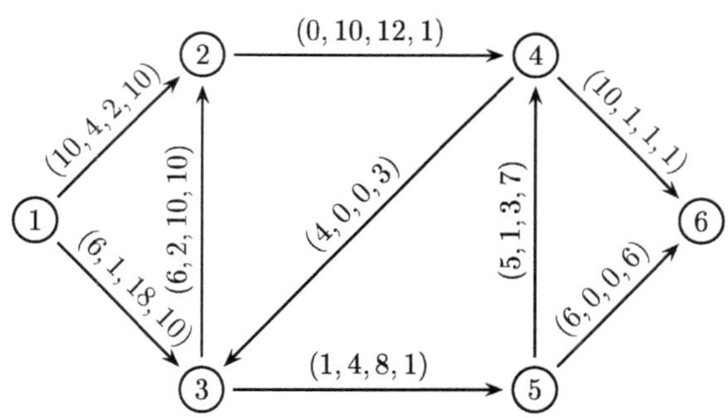

Fig. 9.4. Example graph for Algorithm 9.1.

The eleven iterations of Algorithm 9.1 are shown in Figure 9.5. Labels at nodes have a subscript, the iteration in which they are generated, and a superscript, the iteration in which they are chosen as permanent. Deleted dominated labels are crossed out. Arcs shown indicate directions of labeling with the respective iteration number alongside the arc. Initially $(0, \ldots, 0, 0, 0, 1)$ is set as label for node 1. In the first iteration it is selected and marked permanent. Note that in iterations 8 and 11 only labels at node 6 are identified as permanent, and no new labels are created.

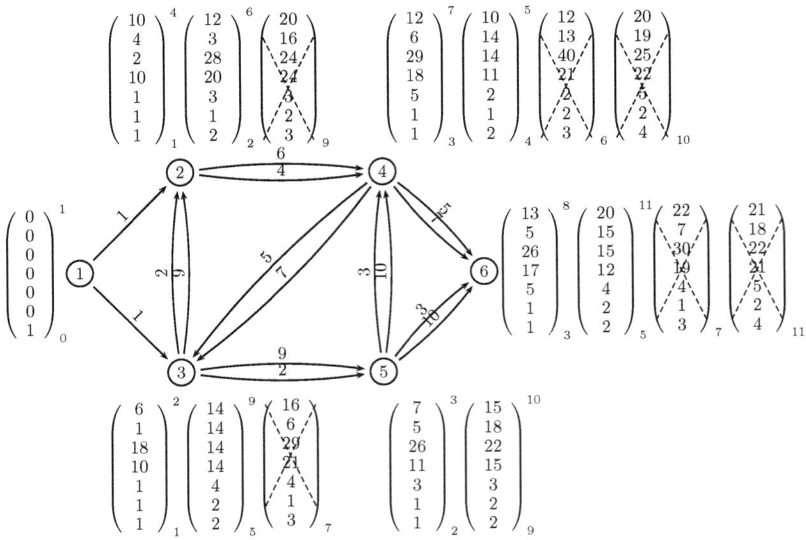

Fig. 9.5. The eleven iterations of 9.1 in Example 9.9.

Backtracking the predecessor labels, we obtain the efficient paths $(1, 3, 5, 6)$ and $(1, 2, 4, 6)$ from node 1 to node 6. Efficient paths to other nodes are obtained using permanent labels at these nodes. □

Let us remark on the difference between single and multiple objective label setting algorithms. When only interested in finding an optimal path from node s to some other node t, the single objective label correcting algorithm can be stopped once a label at node t is made permanent. Due to the possible existence of multiple efficient paths between a pair of nodes this is not possible for the multiobjective label setting algorithm. It will only stop once all efficient paths to all nodes are found.

To see that Algorithm 9.1 may not work when negative weights are allowed for the objectives the reader is asked to come up with an example in Exercise 9.2. If negative cycles are present, the whole problem becomes rather pathologic. Let C be a directed cycle, and assume there is a cost function c^k such that $\sum_{a \in C} c^k(a) < 0$. We may distinguish two cases. There might be another cost function c^j with $\sum_{a \in C} c^j(a) > 0$. Then for each pair of nodes for which a path containing any of the nodes of C exists, there is an infinite number of efficient paths: each move around the cycle decreases objective f_k while increasing f_j. If $\sum_{a \in C} c^j(a) \leq 0$ for all $j \neq k$, then there is no efficient path at all. In both cases Algorithm 9.1 would have an infinite loop.

This behaviour makes it clear that for digraphs with negative cost arcs it is important to have an algorithm which can detect negative cycles, and which either stops reporting the fact or produces the set of efficient paths.

We present a label correcting algorithm for this purpose. Algorithm 9.2 is from Corley and Moon (1985). Like 9.1 it finds efficient paths from s to all other nodes. Here, $\mathcal{L}(i,k)$ denotes the set of labels of node v_i in iteration k.

Algorithm 9.2 (Multiobjective label correcting algorithm.)
Input: *A digraph* $\mathcal{G} = (\mathcal{V}, \mathcal{A})$ *with* p *arc costs.*
Initialization: *Set* $d_{ii} := (0, \ldots, 0)$ *for* $i = 1, \ldots, n$.
  Set $d_{ij} := (\infty, \ldots, \infty)$ *if* $v_i \neq v_j$ *and* $(v_i, v_j) \notin \mathcal{A}$.
  Set $d_{ij} := (c^1(v_i, v_j), \ldots, c^p(v_i, v_j))$ *otherwise.*
  Set $\mathcal{L}(i, 1) := \{d_{1i}\}, i = 1, \ldots, n$.
  For $k := 1$ *to* $n - 1$ *do*
    For $i := 1$ *to* n *do*

$$\mathcal{L}(i, k+1) := \min \bigcup_{j=1}^{n} \{d_{ji} + l_j^k : l_j^k \in \mathcal{L}(j, k)\}$$

    End for.
    If $\mathcal{L}(i, k+1) = \mathcal{L}(i, k)$ *for all* $i = 1, \ldots, n$ *then*
      If $k = n - 1$ *then STOP, a negative cycle exists.*
    STOP
  End for.
Output: *All efficient paths from node* v_1 *to all other nodes.*

Algorithm 9.2 is in fact a generalization of Ford and Bellman's single objective label correcting method (Bellman, 1958).

Theorem 9.10. *At termination of Algorithm 9.2 for all* $v_i \in \mathcal{V}$ *either a negative cycle is detected, or* $\mathcal{L}(i,k)$ *is equal to the nondominated set of the multiobjective shortest path problem with* $s = v_1$ *and* $t = v_i$.

Proof. We show that $\mathcal{L}(i,k)$ is the set of objective vectors of efficient paths from node 1 to node i containing at most k arcs and that the termination criteria are correct.

The proof of the first claim is by induction on k. For $k = 1$ the claim is obviously true. Clearly, labels in $\mathcal{L}(i,k)$ pertain to paths with at most k arcs. Suppose the claim is true for $k = r$ and assume that there is a node v_t, such that $\mathcal{L}(t, r+1)$ is not the set of objective vectors of efficient paths from v_1 to v_t with no more than $r + 1$ arcs. This could occur for two reasons.

1. There is an efficient path P from v_1 to v_t with no more than $r+1$ arcs and objective vector $f(P) \notin \mathcal{L}(t, r+1)$. In this case there exists a node $v_s \in \{v_i : (v_i, v_t) \in \mathcal{A}\}$ such that $f(P) = d_{st} + f(P')$, where $f(P')$ is the cost of a path from v_1 to v_s with at most r arcs. Analogous to the proof of Proposition 9.4, we see that the path from v_1 to v_s is efficient (with at most r arcs), and from the induction hypothesis its objective vector $f(P') \in \mathcal{L}(s, r)$. Since P is an efficient path and $f(P') \in \mathcal{L}(s, r)$ Step 2 implies $f(P) \in \mathcal{L}(t, r+1)$, a contradiction.
2. $\mathcal{L}_(t, r+1)$ contains a vector u, which is not the cost of an efficient path from v_1 to v_t with at most $r+1$ arcs, but the weight of a (dominated) path with at most $r+1$ arcs. Then there is an efficient path P from v_1 to v_t with at most $r+1$ edges such that $f(P) \leq u$. But then again, as in Proposition 9.4, the sub-paths of P are efficient (with at most r arcs) and the induction hypothesis implies that $f(P) \in \mathcal{L}(t, r+1)$. But then $u \notin \mathcal{L}(t, r+1)$.

Consider termination in iteration $k = n-1$ with at least one label set changed, i.e. $\mathcal{L}(i, n) \neq \mathcal{L}(i, n-1)$ for some i. Since any simple path from v_1 to v_i can only contain $n-1$ arcs and because of the correctness of the labels, this implies that there is an efficient path from 1 to i with a negative cost component.

Termination with $\mathcal{L}(i, k) = \mathcal{L}(i, k+1)$ for all $i = 1, \ldots, n$ implies that no path with $k+1$ arcs dominates a path with at most k arcs, and paths with more arcs can only get worse. This implies also that no negative cycles exist. □

As far as complexity of the algorithm is concerned, we see that there are at most n iterations, where labels for n nodes are computed. The crucial part is the size of label sets. The minima can be computed by Algorithm 8.1 or the method of Kung *et al.* (1975), if p is small. However, in the worst case exponential running time can be encountered, because the problem is intractable. Below, we show an application of the algorithm in an example.

Example 9.11. We apply Algorithm 9.2 to the graph of Figure 9.6.

We initialize the distance matrix

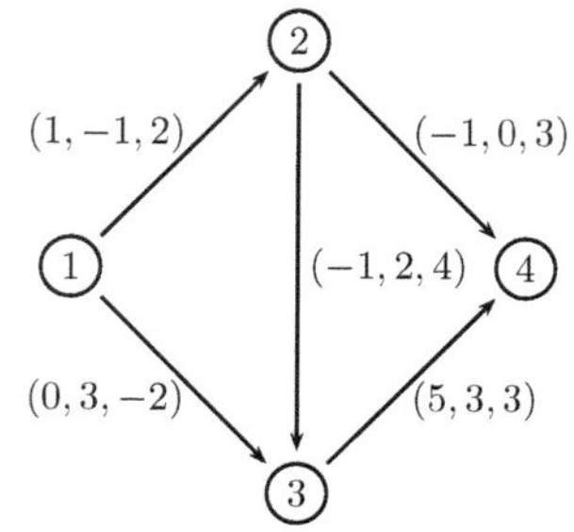

Fig. 9.6. Example graph for Algorithm 9.2.

$$d_{ij} = \begin{pmatrix} \begin{pmatrix} 0 \\ 0 \\ 0 \end{pmatrix} & \begin{pmatrix} 1 \\ -1 \\ 2 \end{pmatrix} & \begin{pmatrix} 0 \\ 3 \\ -2 \end{pmatrix} & \begin{pmatrix} \infty \\ \infty \\ \infty \end{pmatrix} \\ \begin{pmatrix} \infty \\ \infty \\ \infty \end{pmatrix} & \begin{pmatrix} 0 \\ 0 \\ 0 \end{pmatrix} & \begin{pmatrix} -1 \\ 2 \\ 4 \end{pmatrix} & \begin{pmatrix} -1 \\ 0 \\ 3 \end{pmatrix} \\ \begin{pmatrix} \infty \\ \infty \\ \infty \end{pmatrix} & \begin{pmatrix} \infty \\ \infty \\ \infty \end{pmatrix} & \begin{pmatrix} 0 \\ 0 \\ 0 \end{pmatrix} & \begin{pmatrix} 5 \\ 3 \\ 3 \end{pmatrix} \\ \begin{pmatrix} \infty \\ \infty \\ \infty \end{pmatrix} & \begin{pmatrix} \infty \\ \infty \\ \infty \end{pmatrix} & \begin{pmatrix} \infty \\ \infty \\ \infty \end{pmatrix} & \begin{pmatrix} 0 \\ 0 \\ 0 \end{pmatrix} \end{pmatrix}.$$

The main iterations of the algorithm are most easily visualized with illustrations (Figures 9.7 – 9.9) showing only arcs of paths with at most k arcs in iteration k.

At this stage $\mathcal{L}(i, 3) = \mathcal{L}(i, 2)$ for all nodes i and , the algorithm stops.

The actual paths are found by backtracking using additional labels as in Algorithm 9.1. There is one efficient path $(1, 2)$ to node 2, and two each to nodes 3 and 4, $(1, 3)$ and $(1, 2, 3)$, and $(1, 2, 4)$ and $(1, 3, 4)$, respectively. □

In the bicriterion case, another approach was proposed by Climaco and Martins (1982). It is for the problem of finding the efficient simple directed paths between two nodes s and t. Let $\mathcal{G} = (\mathcal{V}, \mathcal{A})$ be a directed graph and $c : \mathcal{A} \to \mathbb{Z}^2$ be a cost function.

The method starts by finding lexicographically shortest paths. For the two possible permutations of the two objectives a lexicographic optimization problem is solved, i.e.

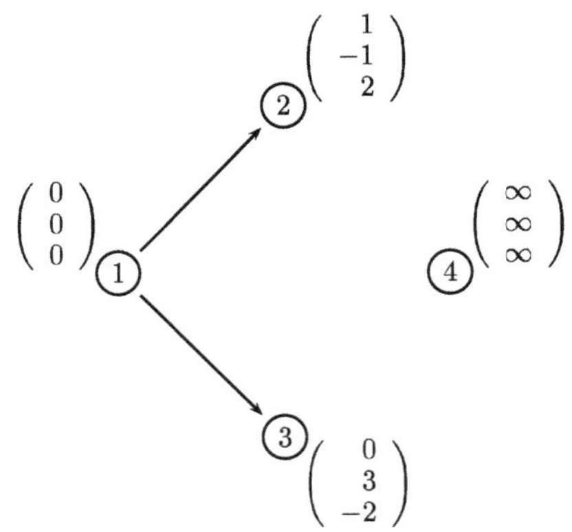

Fig. 9.7. Initialization, labels $\mathcal{L}(i,1)$.

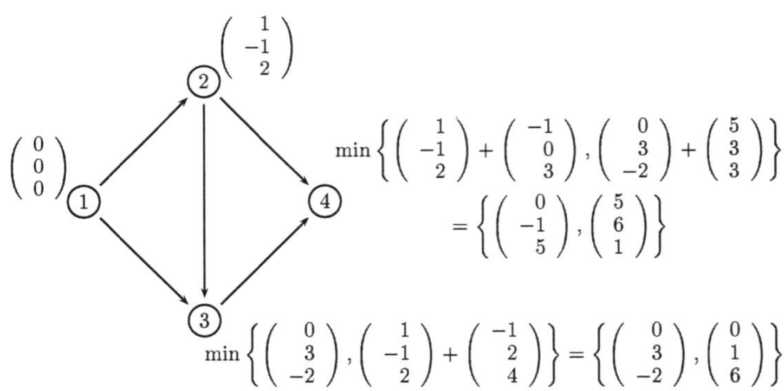

Fig. 9.8. First iteration, labels $\mathcal{L}(i,2)$.

$$\operatorname*{lexmin}_{P\in\mathcal{P}} \left(\sum_{a\in P} c^1(a), \sum_{a\in P}\right) c^2(a) \tag{9.5}$$

and

$$\operatorname*{lexmin}_{P\in\mathcal{P}} \left(\sum_{a\in P} c^2(a), \sum_{a\in P}\right) c^2(a). \tag{9.6}$$

We denote the values of the optimal solutions of (9.5) $(c_1^*, \hat{c}_2)$ and of (9.6) by $(c_2^*, \hat{c}_1)$, respectively. Lexicographically optimal paths are denoted $P^{1,2}$ and $P^{2,1}$. Problems (9.5) and (9.6) can be solved using variants of the label setting or label correcting algorithms (Algorithms 9.1 and 9.2). Exercise 9.4 asks the reader to describe these variants.

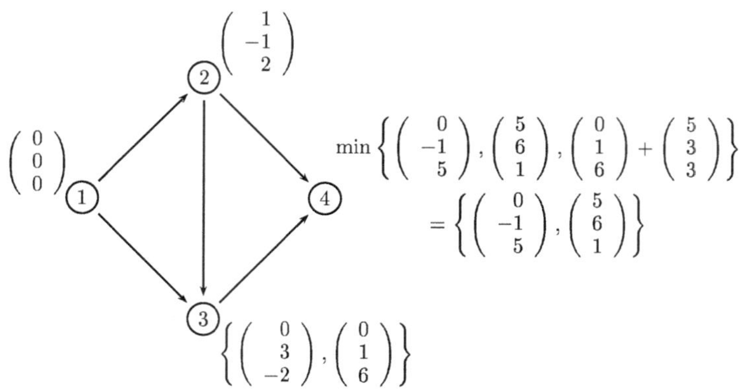

Fig. 9.9. Second iteration, labels $\mathcal{L}(i,3)$.

In fact, the vectors (c_1^*, c_2^*) and $(\hat{c}_1, \hat{c}_2)$ define the ideal and nadir points of the bicriterion shortest path problem, see Section 2.2. Also remember that $\mathcal{X}_\Pi \subset \mathcal{X}_E$, i.e. lexicographically optimal solutions are efficient (5.6).

Lemma 9.12. *If $P \in \mathcal{X}_E$ then $c_1^* \leq \sum_{a \in P} c^1(a) \leq \hat{c}_1$ and $c_2^* \leq \sum_{a \in P} c^2(a) \leq \hat{c}_2$.*

Proof. The lower bounds are trivial. Suppose there is an efficient path P such that $\sum_{a \in P} c^2(a) > \hat{c}_2$. From Proposition 5.6 we know that lexicographically optimal paths are efficient. Let $P^{1,2}$ be a lexicographically optimal path for permutation $(1, 2)$. Then

$$\left(\sum_{a \in P^{1,2}} c^1(a), \sum_{a \in P^{1,2}} c^2(a) \right) = (c_1^*, \hat{c}_2) \leq \left(\sum_{a \in P} c^1(a), \sum_{a \in P} c^2(a) \right),$$

which is clearly impossible if P is efficient.

The result for $P^{2,1}$ is proven analogously. □

The idea of the method is to start with a lexicographically optimal path, e.g. $P^{1,2}$, which is both an optimal path for the first objective and efficient. Then the second best path for the first objective is constructed, then the third best etc. until $P_{2,1}$ is reached. Throughout this process we need to ensure that all paths are efficient.

Let

$$\mathcal{P}^+ := \left\{ P \in \mathcal{P} : \sum_{a \in P} c^1(a) \geq c_1^* \right\}.$$

$\mathcal{P}^+ = \{P^1, \ldots, P^s\}$ is a finite set and can be ordered such that $c_1^* = c^1(P^1) \leq c^1(P^2) \leq \ldots \leq c^1(P^s)$. Let $\mathcal{P}_1^+, \ldots, \mathcal{P}_l^+$ be a partition of $\mathcal{P}^+$ into subsets of paths with equal first weight. Denote by P^{ri} the ith path of subset $\mathcal{P}_r^+$. Then $f_1(P^{ri}) = f_1(P^{rj})$ for all $P^{ri}, P^{rj} \in \mathcal{P}_r^+$ and $f_1(P^{ri}) < f_1(P^{tj}$ if and only if $r < j$.

Proposition 9.13. *Path P^{ri} is efficient if and only if*

1. $f_2(P^{ri}) = \min_{P \in \mathcal{P}_r^+} f_2(P)$ *and*
2. $\min_{P \in \mathcal{P}_r^+} f_2(P) < f_2(P^{tj})$ *for all* $P^{tj} \in \mathcal{P}_t^+$ *with* $t < r$.

Proof. "$\Longrightarrow$" Assume P^{ri} is a path with $f_2(P^{ri}) > \min_{P \in \mathcal{P}_r^+} f_2(P)$. Then $(f_1(P^{ri}), f_2(P^{ri})) \geq (f_1(P^{rj}), f_2(P^{rj}))$ for all $P^{rj} \in \mathcal{P}_r^+$ because $f_1(P^{rj}) = f_1(P^{ri})$. Therefore P^{ri} is not efficient.
Assume there is an index $t < r$ and a path $P^{tj} \in \mathcal{P}_t^+$ such that $f_2(P^{ri}) = \min_{P \in \mathcal{P}_r^+} f_2(P) \geq f_2(P^{tj})$. Since $t < r$ we know that $f_1(P^{tj}) < f_1(P^{ri})$ and again P^{ri} cannot be efficient.
"$\Longleftarrow$" Assume P^{ri} is not efficient. Then there is an efficient path P such that $(f_1(P^{ri}), f_2(P^{ri})) > (f_1(P), f_2(P))$. Since $P \in \mathcal{P}_t^+$ for some t we have two cases. Either $f_1(P^{ri}) = f^(P)$ and $t = r$, whence $f_2(P^{ri}) > f_2(P)$ and the first condition is violated, or $f_1(P^{ri}) > f_1(P)$, i.e. $r > t$ and the second condition is violated. □

As a consequence of Proposition 9.13, an algorithm to find "k-best" paths together with a routine checking Proposition 9.13 yield an algorithm for the bicriterion shortest path problem. Can it be efficient? Not in general, of course, because of the intractability we have shown in Theorem 9.3. For an example illustrating the approach see Exercise 9.5.

It should be noted that the "k-best shortest path problem" is $\mathcal{NP}$-hard (not known to be in $\mathcal{NP}$, however), see Garey and Johnson (1979). For fixed k it is polynomially solvable (see e.g. Eppstein (1998)), but unfortunately, in the case here, k might be very large and its value is not known in advance (we would have to choose k equal to the number of efficient solutions).

It is possible to generalize the idea of Martin's algorithm for any number of objectives, and for any multiobjective combinatorial optimization problem. Note, that we have not use any specifics of the shortest path problem in the above description. This is best done in the framework of the level sets introduced in Section 2.3.

Let $b \in \mathbb{R}^p$. Level sets and curves of objectives f_k at level b_k have been defined in Definition 2.28 as

$$\mathcal{L}_{\leq}(b_k) = \{x \in \mathcal{X} > f_k(x) \leq b_k\},$$
$$\mathcal{L}_{+}(b_k) = \{x \in \mathcal{X} > f_k(x) = b_k\}.$$

In Theorem 2.30 we have shown that $\hat{x} \in \mathcal{X}$ is efficient if and only if

$$\bigcap_{k=1}^{p} \mathcal{L}_{\leq}(f_k(\hat{x})) = \bigcap_{k=1}^{p} \mathcal{L}_{=}(f_k(\hat{x})).$$

We will use this result to derive an algorithm based on ranking the feasible solutions with respect to one objective function (without loss of generality we choose $k = 1$) and checking the intersection with other level sets to confirm efficiency.

Let $\mathcal{X}(b)_E := \{x \in \mathcal{X}_E : f_k(x) \leq b_k\}$. The goal of the algorithm is to find $\mathcal{X}(b)_E$. Choosing $b = y^N$, the nadir point, this includes finding the efficient set.

Let $\mathcal{L} = \{x^1, \dots, x^r\} \subset \mathcal{L}_{\leq}(b_1)$ be the r best solutions of $\min_{x\in\mathcal{X}} f_1(x)$, i.e. $f_1(x^1) \leq f_1(x^2) \leq \dots \leq f_1(x^r)$ and $f_1(x) \geq f_1(x^r)$ for all $x \in \mathcal{X} \setminus \mathcal{L}$. Furthermore denote $\mathcal{L}_E \subset \mathcal{L}$ the efficient set of $\mathcal{L}$. $\mathcal{L}_E = \{x^{i_1}, \dots, x^{i_t}\}$ is a set of potentially efficient solutions of the multiobjective optimization problem. However, there might exist solution $x \in \mathcal{X} \setminus \mathcal{L}$ that dominate some $x \in \mathcal{L}_E$.

Once we have found the $r+1$ best solution x of $\min x \in \mathcal{X} f_1(x)$ that also satisfies $f_k(x) \leq b_k$ for $k = 2, \dots, p$ we know that $f_1(x^{i_1}) \leq \dots \leq f_1(x^{i_t}) \leq f_1(x)$ and therefore $\{x^{i_1}, \dots, x^{i_t}, x\} \subset \mathcal{L}_{\leq}(f_1(x))$. Letting $i_{\max}$ be the largest index in $\{i_1, \dots, i_t\}$ such that $f_1(x^{i_{\max}}) < f_1(x)$ we have two possibilities.

If there is some $j \in \{i_1, \dots, i_{\max}\}$ such that $f_k(x^j) \leq f_k(x)$ for all $k = 2, \dots p$ then x^j dominates x and x is not efficient.

Otherwise $f(x)$ is incomparable with all $f(x^j), j = i_1, \dots, i_{\max}$. Then let $x^j \in \{x^{i_{\max}+1}, \dots, x^{i_t}\}$ and let $f' := (f_2, \dots, f_p)$. We compare $f'(x^j)$ with $f'(x)$ and distinguish the following cases.

- If $f'(x^j) \leq f'(x)$ then x is not efficient.
- If $f'(x) \leq f'(x^j)$ then x^j is not efficient.
- If $f'(x) = f'(x^j)$ then x and x^j are possibly efficient.
- Otherwise $f'(x)$ and $f'(x^j)$ are incomparable. If this is the case for all $x^j \in \{x^{i_{\max}+1}, \dots, x^{i_t}\}$ then x is possibly efficient.

Thus we need an algorithm to find r-best solutions of a combinatorial optimization problem $\min_{x\in\mathcal{X}} f_1(x)$. With that algorithm we can formulate the following level set algorithm for multiobjective combinatorial optimization (Ehrgott and Tenfelde-Podehl, 2002).

Algorithm 9.3 (Level set algorithm.)

Input: *An instance of a MOCO problem,* $b \in \mathbb{Z}^p$.

Find an optimal solution x^1 *of* $\min_{x \in \mathcal{X}} f_1(x)$.

If $f_1(x^1) > b_1$ *STOP,* $\mathcal{X}(b)_E = \emptyset$.

$r := 2, \mathcal{X}(b)_E := \{x^1\}$.

While $r \leq |\mathcal{X}|$

Find the rth *best solution* x^r *for* $\min_{x \in \mathcal{X}} f_1(x)$.

If $f_1(x^r) > b_1$*, STOP.*

If $f(x^r) > b_k$ *for some* $k \in \{2, \ldots, p\}$ *then* $r := r + 1$.

Else

For all $x \in \mathcal{X}(b)_E$ *do*

If x^r *dominates* x *then* $\mathcal{X}(b)_E := \mathcal{X}(b)_E \setminus \{x\}$.

Else if x *dominates* x^r *then* $r := r + 1$.

Else if $f(x^r) = f(x)$ *then* $\mathcal{X}(b)_E := \mathcal{X}(b)_E \cup \{x^r\}$ *and* $r := r + 1$.

$\mathcal{X}(b)_E := \mathcal{X}(b)_E \cup \{x^r\}$ *and* $r := r + 1$.

End while.

Output: $\mathcal{X}(b)_E$

In order to have the most efficient algorithm it is desirable to rank as few solutions as possible. A rule of thumb is to choose that objective for ranking which has the smallest range of objective values, i.e. the k for which $y_k^N - y_k^I$ is minimal.

9.2 The Minimum Spanning Tree Problem

In this chapter we consider first the problem of finding efficient spanning trees of a graph $\mathcal{G} = (\mathcal{V}, \mathcal{E})$, i.e. $\min_{T \in \mathcal{T}} f(T)$, where the components f_k of f are some functions. In other words $(\mathcal{T}, p\text{-}\sum, \mathbb{Z}^p)/\operatorname{id}/(\mathbb{Z}^p, \leq)$. $\mathcal{T}$ denotes the set of spanning trees of $\mathcal{G}$. First we prove some basic facts about edges of efficient spanning trees first stated in Corley (1985) and Hamacher and Ruhe (1994).

Theorem 9.14 (Hamacher and Ruhe (1994)). *Let T be an efficient spanning tree of $\mathcal{G}$.*

1. *Let $e \in \mathcal{E}(T)$ be an edge of T. Let $(\mathcal{V}(T_1), \mathcal{E}(T_1))$ and $(\mathcal{V}(T_2), \mathcal{E}(T_2))$ be the two connected components of $\mathcal{G} \setminus \{e\}$. Let $C(e) := \{f = (v_i, v_j) \in \mathcal{E} : v_i \in \mathcal{V}(T_1), v_j \in \mathcal{V}(T_2)\}$ be the* cut *defined by deleting e. Then $c(e) \in \min\{c(f) : f \in C(e)\}$.*

2. Let $f \in \mathcal{E} \setminus \mathcal{E}(T)$ and let $P(f)$ be the unique path in T connecting the end nodes of f. Then $c(f) \leq c(e)$ does not hold for any $e \in P(f)$.

Proof. In both cases we can define $\mathcal{E}(T') = (\mathcal{E}(T) \setminus \{e\}) \cup \{f\}$ for any $f \in C(e)$, respectively $e \in P(f)$, to obtain a new tree $T' = (\mathcal{V}, \mathcal{E}(T'))$. Since $f(T') = f(T) - c(e) + c(f)$ we have that $c(f) < c(e)$ for some $f \in C(e)$ or $e \in P(f)$ contradicts T being efficient. □

Figures 9.10 and 9.11 illustrate the construction of T'.

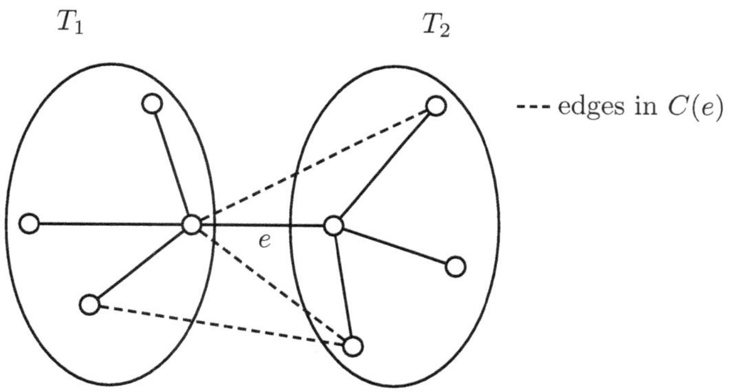

Fig. 9.10. Minimal edges in cuts ...

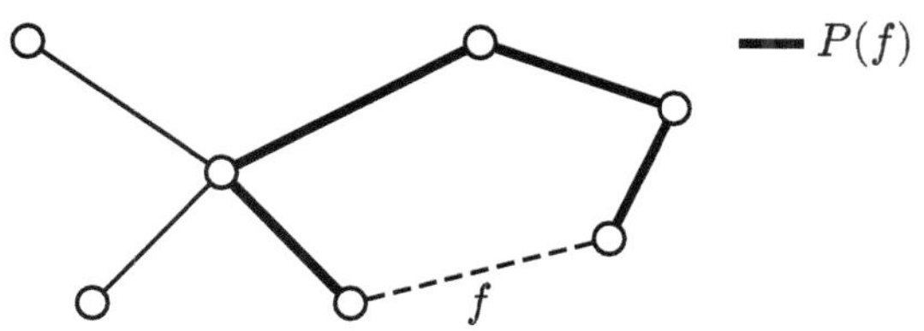

Fig. 9.11. ... and cycles.

The converse of Theorem 9.14 is not true, as shown in Example 9.15.

Example 9.15. The graph of Figure 9.12 has eight spanning trees, seven of which are efficient. We consider $T_0 = \{[1,2],[1,3],[3,4]\}$, which is dominated, e.g. by $T = \{[1,2],[2,3],[2,4]\}$).

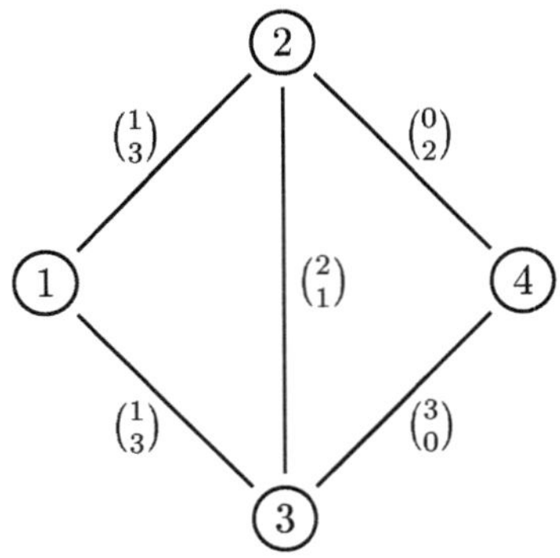

Fig. 9.12. The converse of Theorem 9.14 does not hold.

Consider first $e = [3,4]$. The cut obtained by deleting e contains edges $[2,4]$ and $[3,4]$ and e satisfies $c(e) \in \min\{c([2,4]), c([3,4])\}$.

With $f = [2,3]$ we have $P(f) = \{[1,2],[1,3]\}$ and there is no $e \in P(f)$ with $c([2,3]) < c(e)$. □

Nevertheless, Theorem 9.14 is a justification for a multiobjective version of Prim's algorithm Prim (1957). It maintains a list of potential subtrees of efficient trees starting with all edges the costs of which costs are nondominated. In each iteration it extends these subtrees by adding adjacent edges which have nondominated costs in the set of all edges connecting the current tree with the yet to be connected nodes. Because the condition of Theorem 9.14 is only necessary, but not sufficient, a dominance check among the resulting (sub)trees is necessary.

Algorithm 9.4 (Prim's spanning tree algorithm.)

Input: A graph $\mathcal{G} = (\mathcal{V}, \mathcal{E})$ with p edge costs.

$\mathcal{T}_1 := \text{argmin}\{c(e_1), \ldots, c(e_n)\}$

For $k := 2$ *to* $n-1$ *do*

$\mathcal{T}_k := \{E(T) \cup \{e_j\} : T \in \mathcal{T}_{k-1}, e_j \in \text{argmin}\{c(e) = c([v_i, v_j]) : v_i \in \mathcal{V}(T), v_j \in \mathcal{V} \setminus \mathcal{V}(T)\}\}$

$\mathcal{T}_k := \text{argmin}\{f(T) : T \in \mathcal{T}_k\}$

End for

Output: $\mathcal{T}_{n-1}$, all efficient spanning trees of $\mathcal{G}$.

Example 9.16. We apply Algorithm 9.4 to the graph of Figure 9.13, with the edge costs displayed alongside the edges.

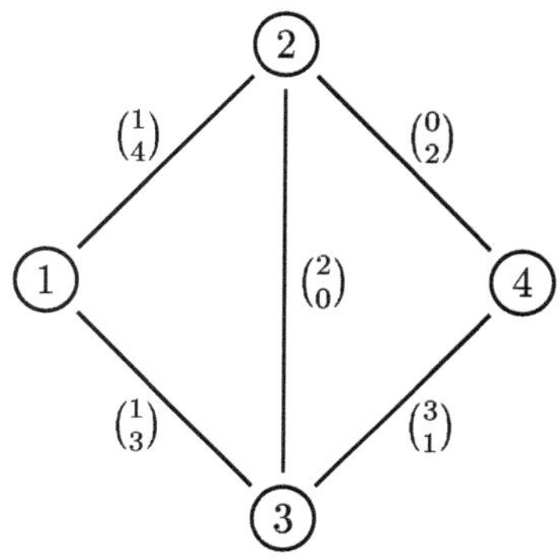

Fig. 9.13. Example graph for Algorithm 9.4.

The steps of the algorithm are summarized and illustrated below. $\mathcal{T}_1 = \{\{[2,3]\},\{[2,4]\}\}$. Edges connecting the one edge subtrees with the remaining nodes are shown in Figure 9.14. Broken lines show possible extensions of trees in $\mathcal{T}_1$.

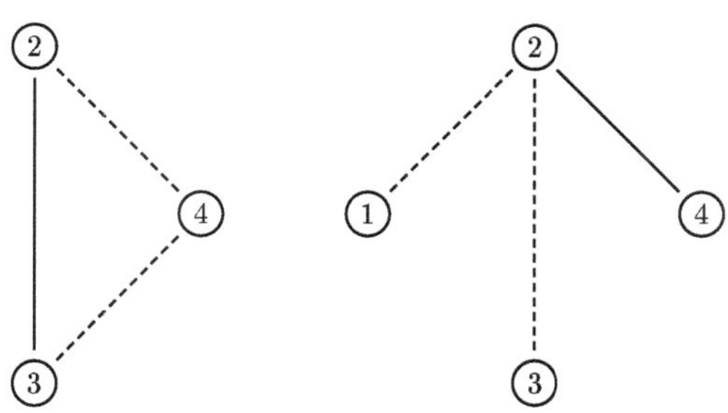

Fig. 9.14. Possible extensions of trees in $\mathcal{T}_1$.

Three of these are efficient subtrees, so

$$\mathcal{T}_2 = \{\{[2,3],[2,4]\},\{[2,3],[3,4]\},\{[2,4],[1,2]\}\}.$$

Their objective vectors are $\binom{2}{2}$, $\binom{5}{1}$ and $\binom{1}{6}$, respectively. Possible extensions of trees in $\mathcal{T}_2$ are shown in Figure 9.15.

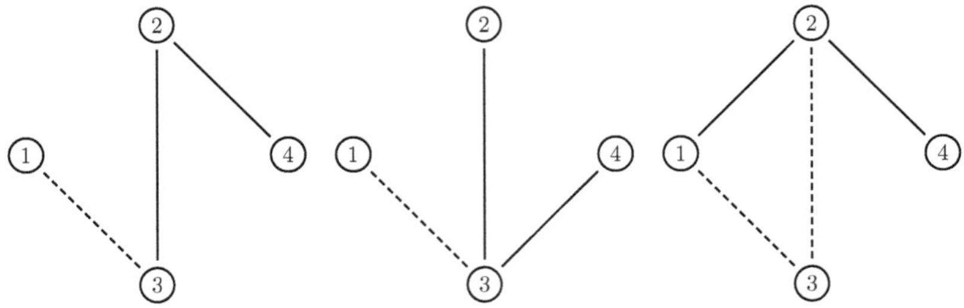

Fig. 9.15. Possible extensions of trees in $\mathcal{T}_2$.

For each tree in $\mathcal{T}_2$ only one of the possible extensions is efficient, so that

$$\mathcal{T}_3 = \{\{[2,3],[2,4],[1,3]\},\{[1,3],[2,3],[3,4]\},\{[2,4],[1,2],[1,3]\}\}$$

is the set of efficient trees $\mathcal{T}_\mathcal{E}$ of $\mathcal{G}$ and $\mathcal{Y}_N = \{\binom{3}{5},\binom{6}{4},\binom{2}{9}\}$. □

The worst case performance of Algorithm 9.4 is dominated by the domination test in Step 2 and depends on the size of the sets of objective vectors we might have to compare. We shall see later that the multicriteria spanning tree problem is intractable, and therefore an exponential worst case performance applies.

The greedy approach to the problem is very similar to the approach of Algorithm 9.4, which is in fact only a particular implementation of a greedy strategy for spanning trees. The greedy algorithm starts with a "cheapest" edge and adds "cheapest" available edges as long as the result does not contain a cycle. To clarify what "cheapest" means, recall that the set of edges $\{e_1,\ldots,e_n\}$ is partially ordered by the componentwise order on their cost vectors. We extend the partial order to a total order in the following way.

Definition 9.17. *A topological order of* $\{e_1,\ldots,e_n\}$*, partially ordered the componentwise order on the set of cost vectors* $\{c(e_1),\ldots,c(e_n)\}$ *is a total order* $\preceq$ *of* $\{e_1,\ldots,e_m\}$ *such that* $c(e_i) \leq c(e_j)$ *implies* $e_i \preceq e_j$.

The topological order explains what we termed a "cheapest" edge – a minimal edge according to a topological order. Note that for a given partial order several topological orders may exist. Using topological orders, we can apply the greedy algorithm to construct spanning trees. Unfortunately there are topological orders for which the result is not an efficient tree.

Example 9.18. Consider the graph of Figure 9.16.

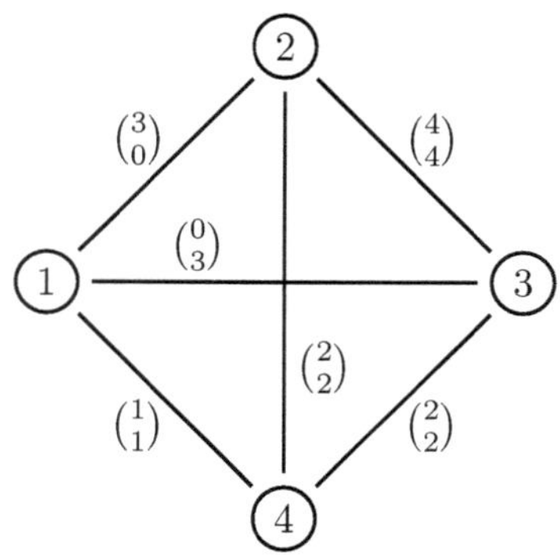

Fig. 9.16. Example for the greedy algorithm.

The partial order on the edge costs is given by $c([1,4]) \leq c([2,4]) = c([3,4]) \leq c([2,3])$, $c([1,2]) \leq c([2,3])$, and $c([1,3]) \leq c([2,3])$. Therefore

$$[1,4] \preceq [3,4] \preceq [2,4] \preceq [1,3] \preceq [1,2] \preceq [2,3]$$

is a topological order. The incomparabilities among weight vectors are resolved by the topological order and $\preceq$ preserves the componentwise order of weights.

The greedy algorithm applied to this example yields the tree $\{[1,4],[3,4],[2,4]\}$ with objective vector $\binom{5}{5}$ which is dominated by $\binom{4}{4}$, the objective vector of $T = \{[1,2],[1,3],[1,4]\}$. □

Despite Example 2.38 each efficient spanning tree can be identified using an appropriate topological order.

Theorem 9.19 (Serafini (1986)). *Let T be an efficient spanning tree. Then there exists a topological order of the edges of $\mathcal{G}$ such that the greedy algorithm applied to this order yields T.*

Proof. Define a digraph $\mathcal{D}$, where $\mathcal{V}(\mathcal{D}) = \mathcal{E}(\mathcal{G}) = \{e_1, \ldots, e_n\}$ and $(e_i, e_j) \in \mathcal{A}(\mathcal{D})$ if and only if $c(e_i) \leq c(e_j)$. $\mathcal{D}$ is acyclic, because "$\leq$" is a partial order. Let T be a spanning tree of $\mathcal{G}$ and define $\bar{\mathcal{E}} := \mathcal{E} \setminus \mathcal{E}(T)$. Let a new digraph $\mathcal{D}_T$ be defined by $\mathcal{V}(\mathcal{D}_T) = E(G)$ and $\mathcal{A}(\mathcal{D}_T) = \mathcal{A}(\mathcal{D}) \cup \{(e,f) : e \in C(f) \cap \mathcal{E}(T), f \in \bar{\mathcal{E}}\}$, where $C(f)$ is the unique cycle generated by adding f to T, i.e. $C(f) = P(f) \cup \{f\}$. If $\mathcal{D}_T$ is acyclic, it represents a partial order preserving $\leq$ (because $\mathcal{A}(\mathcal{D}) \subset \mathcal{A}(\mathcal{D}_T)$) and any topological order if its nodes has the required property. Note that due to the additional arcs only the edges of T can be chosen by the greedy algorithm.

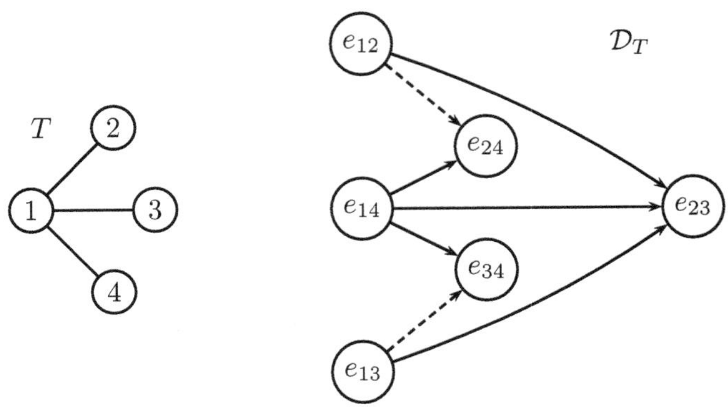

Fig. 9.17. Finding $\mathcal{D}_T$ for the graph of Figure 9.16 and T as shown.

So it remains to be shown that $\mathcal{D}_T$ is acyclic. Suppose $\mathcal{D}_T$ contains a cycle. Since $\mathcal{D}$ is acyclic the cycle contains one of the additional arcs. Because $\mathcal{D}$ is transitively closed by its definition (i.e. $(e,f) \in \mathcal{AD}$, $(f,g) \in \mathcal{A}(\mathcal{D}$ implies $(e,g) \in \mathcal{AD}$) we can assume that the cycle consists alternatingly of arcs (f_i, e_{i+1}) of $\mathcal{AD}$ and new arcs (e_i, f_i). Thus it can be written as a sequence

$$(e_1, f_1, e_2, f_2, \ldots, e_k, f_k, e_1)$$

of nodes of $\mathcal{D}_T$ and without loss of generality we may assume it is the smallest cycle containing e_1. Now construct a new spanning tree T' with $\mathcal{E}(T') = (\mathcal{E}(T) \cup \{f_i : i = 1' \ldots, k\}) \setminus \{e_i : i = 1' \ldots, k\}$. But since $c(f_i) < c(e_{i+1})$ and $c(f_k) < c(e_1)$ we get $f(T') < f(T)$, a contradiction. □

Corollary 9.20. *Let T be an efficient spanning tree of $\mathcal{G}$. Then T contains an edge $e \in$ argmin$\{c(e) : e \in \mathcal{E}\}$. We call such an edge a* minimal edge.

Proof. Consider a topological order such that the greedy algorithm finds T. Then T contains the (unique) minimal edge e according to this topological order. But e can only be chosen such that $c(e) \in \min\{c(e) : e \in E\}$, i.e. as a minimal edge. □

Due to Theorem 9.19 the greedy algorithm can be used in principle to find all efficient spanning trees of $\mathcal{G}$, but the problem of identifying the appropriate topological orders remains to be solved.

Theorem 9.14 may hint to another idea for solving the problem, namely neighbourhood search. The idea is to generate one efficient spanning tree T (e.g. a lexicographically optimal one, or an minimal spanning tree for the sum of costs $\bar{c}(e) = \sum_{k=1}^{p} c^k(e)$) and perform exchanges of edges to search for other efficient spanning trees in the "neighbourhood" of T.

Let T_1 and T_2 be spanning trees. We say T_1 and T_2 are *adjacent*, or *neighbours*, if they have $m - 2$ edges in common (m is the number of nodes of $\mathcal{G}$). Let $\mathcal{E}(T_1) \setminus \mathcal{E}(T_2) = \{e\}$ and $\mathcal{E}(T_2) \setminus \mathcal{E}(T_1) = \{f\}$. One would consider only exchanges of edges such that $c(e) - c(f)$ has positive as well as negative components, because efficient trees must have incomparable objective vectors. Does such an approach work? To answer this question, we represent adjacency among efficient spanning trees by a graph.

Definition 9.21. *The* Pareto graph $\mathcal{P}(\mathcal{G})$ *is a graph defined as follows.* $\mathcal{V}(\mathcal{P}(\mathcal{G}))$ *consists of the efficient spanning trees of* $\mathcal{G}$, *and* $(T_1, T_2) \in \mathcal{E}(\mathcal{P}(\mathcal{G}))$ *if* T_1 *and* T_2 *are adjacent.*

Then the neighbourhood search based on exchanges of edges can be used to find all efficient spanning trees if and only if $\mathcal{P}(\mathcal{G})$ is connected.

Example 9.22. For the first example in Section 8.1, $\mathcal{P}(\mathcal{G})$ is connected, because all spanning trees are efficient. For the graph $\mathcal{G}$ in Figure 9.13 the efficient spanning trees are $T_1 = \{[2,3],[3,4],[1,3]\}$, $T_2 = \{[1,3],[2,3],[2,4]\}$ and $T_3 = \{[2,4],[1,2],[1,3]\}$ and $\mathcal{P}(\mathcal{G})$, shown in Figure 9.18 is connected.

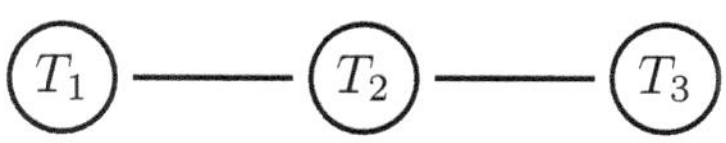

Fig. 9.18. Connected Pareto graph of the graph in Figure 9.13.

□

However, this connectedness result is not true in general.

Example 9.23 (Ehrgott and Klamroth (1997)). Consider graph $\mathcal{G}$ of Figure 9.19.

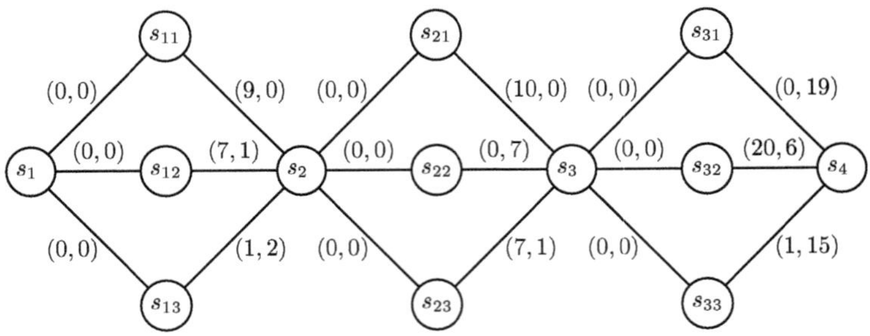

Fig. 9.19. Graph $\mathcal{G}$ has nonadjacent efficient spanning trees.

All efficient spanning trees of $\mathcal{G}$ contain all edges of zero cost and in addition exactly one of the three edges $[s_{ij}, s_{i+1}], j = 1, 2, 3$ connecting the nodes s_i, $i = 2, 3, 4$ to nodes on their left. The twelve efficient spanning trees are listed in Table 9.2, where the zero cost edges are omitted.

Table 9.2. Efficient spanning trees of graph $\mathcal{G}$ in Figure 9.19.

Efficient tree	Edges	Objective vector
T_1	$[s_{13}, s_2][s_{22}, s_3][s_{31}, s_4]$	(1,28)
T_2	$[s_{13}, s_2][s_{22}, s_3][s_{33}, s_4]$	(2,24)
T_3	$[s_{13}, s_2][s_{23}, s_3][s_{31}, s_4]$	(8,22)
T_4	$[s_{13}, s_2][s_{23}, s_3][s_{33}, s_4]$	(9,18)
T_5	$[s_{13}, s_2][s_{21}, s_3][s_{33}, s_4]$	(12,17)
T_6	$[s_{11}, s_2][s_{23}, s_3][s_{33}, s_4]$	(17,16)
T_7	$[s_{11}, s_2][s_{21}, s_3][s_{33}, s_4]$	(20,15)
T_8	$[s_{12}, s_2][s_{22}, s_3][s_{32}, s_4]$	(27,14)
T_9	$[s_{13}, s_2][s_{23}, s_3][s_{31}, s_4]$	(28,9)
T_{10}	$[s_{13}, s_2][s_{21}, s_3][s_{32}, s_4]$	(31,8)
T_{11}	$[s_{11}, s_2][s_{23}, s_3][s_{32}, s_4]$	(36,7)
T_{12}	$[s_{11}, s_2][s_{21}, s_3][s_{32}, s_4]$	(39,6)

Spanning tree T_8 is the only efficient tree containing edge $[s_{12}, s_2]$ and at least one of the other two nonzero cost edges is different from any other efficient spanning tree. $\mathcal{P}(\mathcal{G})$ is not connected, it has T_8 as an isolated node.

□

The bad news for neighbourhood search is not only that there are examples of graphs with nonconnected Pareto graphs. Moreover, every graph can be extended in such a way that the extended graph has a nonconnected Pareto graph. We only state the result here, for details and the (rather technical) proof the reader is referred to the original paper.

Theorem 9.24 (Ehrgott and Klamroth (1997)). *For any graph $\mathcal{G}$ there exists a graph $\mathcal{G}^*$ containing $\mathcal{G}$ as a subgraph such that $\mathcal{P}(\mathcal{G}^*)$ is not connected.*

Despite Theorem 9.24, some classes of graphs might have connected Pareto graphs, see Exercise 9.8 for a simple example. Another possibility to guarantee this is restriction to a subset of efficient solutions. We will see in the next section why $\mathcal{P}(\mathcal{G})$ is connected in Example 9.22. At the end of our discussion of the spanning tree problem we prove intractability and $\mathcal{NP}$-completeness of the problem.

Proposition 9.25 (Hamacher and Ruhe (1994)). *The bicriterion spanning tree problem $(\mathcal{T}, 2\text{-}\sum, \mathbb{Z}^2)/\operatorname{id}/(\mathbb{Z}^2, \leqq)$ is intractable.*

Proof. We consider $\mathcal{G} = K_n$, the complete graph on n nodes and edges $\{e_1, \ldots, e_m\}$, where $m = n(n-1)/2$. $\mathcal{G}$ contains n^{n-2} spanning trees. We show that it is possible that $|\mathcal{Y}_{\text{eff}}| = n^{n-2}$.

For e_i define $c^1(e_i) = 2^{i-1}, c^2(e_i) = 2^m - 2^{i-1}$ which implies $c^1(e_i) + c^2(e_i) = 2^m$ and thus $c^1(T) + c^2(T) = (n-1)2^m$ for all $T \in \mathcal{T}$. Furthermore, $c^1(T_1) \neq c^1(T_2)$ for all pairs of spanning trees $T_1, T_2 \in \mathcal{T}$ with $T_1 \neq T_2$.

Therefore all spanning trees have pairwise incomparable weights and are thus efficient. All weights being different, we know that $|\mathcal{Y}_N| = |\mathcal{T}| = n^{n-2}$.

□

The proof of Proposition 9.25 actually shows a more general statement. A multiobjective combinatorial optimization problem $(\mathcal{X}, 2\text{-}\sum, \mathbb{Z}^1)/\operatorname{id}/(\mathbb{Z}^2, <)$ with the additional property that $|x| = \gamma$ is the same for all feasible solutions $x \in \mathcal{X}$ is intractable if $|\mathcal{X}|$ is exponential in $|E|$.

The multicriteria spanning tree problem is also $\mathcal{NP}$-complete.

Proposition 9.26 (Camerini *et al.* (1984)). *$(\mathcal{T}, 2\text{-}\sum, \mathbb{Z}^2)/\operatorname{id}/(\mathbb{Z}^2, \leqq)$ is $\mathcal{NP}$-complete.*

Proof. The decision version of $(\mathcal{T}, 2\text{-}\sum, \mathbb{Z}^2)/\operatorname{id}/(\mathbb{Z}^2, \leq)$ is: Given $\mathcal{G}$ and $b_1, b_2 \in \mathbb{Z}$, does there exist a spanning tree T of $\mathcal{G}$ such that $\sum_{e \in T} c^1(e) \leq b_1$ and $\sum_{e \in T} c^2(e) \leq b_2$?

We prove a reduction KNAPSACK $\propto (\mathcal{T}, 2\text{-}\sum, \mathbb{Z}^2)/\operatorname{id}/(\mathbb{Z}^2, \leq)$. From an instance $(a_1, \ldots, a_n, b) \in \mathbb{Z}^{n+1}$ of KNAPSACK, we construct the graph of Figure 9.20.

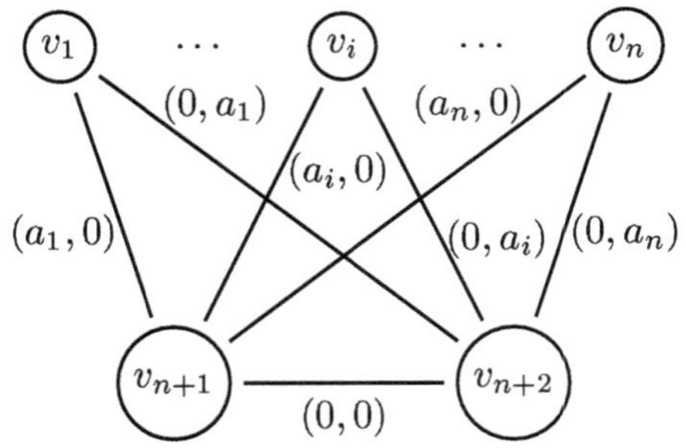

Fig. 9.20. Showing $\mathcal{NP}$-completeness of the biobjective spanning tree problem.

$\mathcal{G}$ is defined as follows:

$$\mathcal{V} := \{v_1, \ldots, v_{n+2}\}$$
$$\mathcal{E} := \{[v_i, v_{n+1}], [v_i, v_{n+2}] : i = 1, \ldots, n\} \cup \{[v_{n+1}, v_{n+2}]\}.$$

The costs are given by $c([v_i, v_{n+1}]) := \binom{a_i}{0}$, $c([v_i, v_{n+2}]) := \binom{0}{a_i}$, and $c([v_{n+1}, v_{n+2}]) = \binom{0}{0}$. Furthermore $b_1 := b$ and $b_2 := \sum_{i=1}^{n} a_i - b$.

Any $x \in \{0,1\}^n$ uniquely corresponds to a spanning tree T of $\mathcal{G}$, which contains $[v_{n+1}, v_{n+2}]$ and the other edges of which are defined by $x_i = 1$ if and only if $[v_i, v_{n+1}] \in T$ and $x_i = 0$ if and only if $[v_i, v_{n+2}] \in T$. Then

$$\sum_{e \in T} c^1(e) \leq b_1 \iff \sum_{i=1}^{n} a_i x_i \leq b$$
$$\sum_{e \in T} c^2(e) \leq b_2 \iff \sum_{i=1}^{n} a_i x_i \geq b.$$

Note that there may exist spanning trees T satisfying $\sum_{e \in T} c^1(e) \leq b_1$ and $\sum_{e \in T} c^2(e) \leq b_2$ not containing (v_{n+1}, v_{n+2}). In this case, one of the edges could be replaced by (v_{n+1}, v_{n+2}). However, it means that the transformation is not parsimonious. $\square$

$\#\mathcal{P}$-completeness of the multicriteria spanning tree problem is still open. It does not follow from intractability, because counting the spanning trees of a graph is easy, a formula was discovered by Kirchhoff (1847).

9.3 Matroids

Spanning trees of a graph are just one example of a general combinatorial structure called matroid. In the remaining part of this section, we study multicriteria problems on matroids. Those results on spanning trees, for which the graphical view was inessential remain valid for matroids. That concerns everything we presented about the greedy algorithm. And of course intractability and $\mathcal{NP}$-completeness are valid for the more general problem. It does, however, not include Algorithm 9.4 and Theorem 9.14.

Definition 9.27. *Let $\mathcal{E} = \{e_1, \ldots, e_n\}$ be a finite set and $\mathcal{X} \subset 2^{\mathcal{E}}$.*

1. *$\mathcal{J} = (\mathcal{E}, \mathcal{X})$ is called* independence system, *if $\mathcal{X} \neq \emptyset$ and if $x^1 \in \mathcal{X}, x^2 \subset x^1$ implies $x^2 \in \mathcal{X}$. $x \in \mathcal{X}$ is called an* independent set.
2. *Let $(\mathcal{E}, \mathcal{X})$ be an independence system. An independent set $x \in \mathcal{X}$ is called* maximal, *if $x \cup \{e\} \notin \mathcal{X}$ for all $e \in \mathcal{E} \setminus x$. A maximal independent set $\hat{x}$ is called* maximum, *if $|x| \leq |\hat{x}|$ for all independent subsets $x \in \mathcal{X}$. We let $m(\hat{x}) := \max\{|x| : x \subseteq \hat{x}, x \in \mathcal{X}\}$ for $\hat{x} \subset \mathcal{E}$ and call it the* rank *of $\hat{x}$.*
3. *$\mathcal{M} = (\mathcal{E}, \mathcal{X})$ is called* matroid, *if it is an independence system and if for any $T \subset \mathcal{E}$ and maximal independent set $x \subset T$ we have $|x| = m(T)$. (or if, equivalently, whenever $x, x' \in \mathcal{X}$, $|x'| = |x| + 1$ there is some $e \in x'$ such that $x \cup \{e\} \in \mathcal{X}$).*

The following example lists the matroids we will encounter in this section.

Example 9.28. 1. Let $\mathcal{E} = \{x^1, \ldots, x^n : x^i \in \mathbb{R}^n\}$. $\mathcal{X}$ is the set of all subsets of linear independent vectors of $\mathcal{E}$. $\mathcal{M} = (\mathcal{E}, \mathcal{X})$ is called *matric matroid.*

2. Let $\mathcal{E} = \{e_1, \ldots, e_n\}$. $\mathcal{X} = \{x \subset \mathcal{E} : |x| \leq k\}$. $\mathcal{M} = (\mathcal{E}, \mathcal{X})$ is called *uniform matroid.*
3. Let $\mathcal{E} = \mathcal{E}(\mathcal{G})$ be the edge set of a graph $\mathcal{G}$. $\mathcal{X}$ is the set of all subsets $x \subset \mathcal{E}$ that do not contain a cycle. $\mathcal{M} = (\mathcal{E}, \mathcal{X})$ is called *graphic matroid.*
4. $\mathcal{E} = \cup_{i=1}^n \mathcal{E}_i$, where $\mathcal{E}_i \cap \mathcal{E}_j = \emptyset$ for all i, j with $i \neq j$ and $x \in \mathcal{X}$ if and only if $|x \cap \mathcal{E}_i| \leq 1$ for $i = 1, \ldots, n$. $\mathcal{M} = (\mathcal{E}, \mathcal{X})$ is called *partition matroid.* □

A maximal independent subset of $\mathcal{E}$ is called *basis* of $\mathcal{M}$. Thus, bases in Example 9.28 are

1. bases of the linear space spanned by $x^1, \ldots, x^n$,

2. all subsets of $\mathcal{E}$ containing exactly k elements,
3. the spanning trees of $\mathcal{G}$,
4. and subsets consisting of exactly one member of each $\mathcal{E}_i$,

respectively. All bases of $\mathcal{M}$ have the same cardinality, which is one of many equivalent characterizations of matroids. For more about matroid theory we refer the reader to Oxley (1992). Most interesting for us is the following characterization in terms of optimal solutions of a combinatorial optimization problem associated with a matroid, see e.g. (Nemhauser and Wolsey, 1999, p. 667).

Theorem 9.29. *Let $(\mathcal{E}, \mathcal{X})$ be an independence system. Let $c : \mathcal{E} \to \mathbb{Z}_+$ be a weight function. Then $(\mathcal{E}, \mathcal{X})$ is a matroid if and only if the greedy algorithm finds a maximal independent set of minimal total weight for any choice of c.*

In particular, Theorem 9.29 says why the greedy algorithm works for the spanning tree problem.

Now, let $\mathcal{M} = (\mathcal{E}, \mathcal{X})$ be a matroid. We denote the set of bases of $\mathcal{M}$ by $\mathcal{B}$. We will discuss the problem $(\mathcal{B}, p\text{-}\sum, \mathbb{Z}^p)/\operatorname{id}/(\mathbb{Z}^p, \leqq)$ of finding efficient matroid bases and the max-ordering problem $(\mathcal{B}, p\text{-}\sum, \mathbb{Z}^p)/\max/(\mathbb{Z}, <)$. First we show that they are both $\mathcal{NP}$-complete, using a different example than that of spanning trees already given.

Theorem 9.30. *The problems of finding all efficient matroid bases and the max-ordering matroid basis problem are are $\mathcal{NP}$-hard.*

Proof. We will show the result for the special case where $\mathcal{M} = (\mathcal{E}, \mathcal{X})$ is a uniform matroid. Actually, for efficiency the result already follows from Proposition 9.26.

Consider the following decision problem

k-PARTITION: Given a set $\mathcal{E} = \{e_1, \ldots, e_n\}$ and integer numbers $c(e_i) \geq 0$, $\sum_{i=1}^{n} c(e_i) = 2C$. Does there exist a subset $\mathcal{S} \subset \mathcal{E}, |\mathcal{S}| = k$ such that $\sum_{e_i \in \mathcal{B}} c(e_i) = C$?

The problem is $\mathcal{NP}$-complete, because solving it for $k = 1, \ldots, n$ would yield a solution of PARTITION, a problem which is $\mathcal{NP}$-complete (see page 207 and (Garey and Johnson, 1979, p. 223)). Given an instance of k-Partition define costs on the elements e_i of $\mathcal{E}$ as follows.

$$
\begin{aligned}
c^1(e_i) &= \bar{c} + \frac{2C}{k} - c(e_i) \\
c^2(e_i) &= \bar{c} + c(e_i),
\end{aligned}
$$

where $\bar{c} > \max_{i=1}^{n} c(e_i)$.

Let $b_1 := b_2 := k\bar{c} + C$. Let $\mathcal{S} \subset \mathcal{E}$ with $|\mathcal{S}| = k$ be a basis of the uniform matroid $\mathcal{M} = (\mathcal{E}, \mathcal{X})$. Then $(c^1(\mathcal{S}), c^2(\mathcal{S})) \leq (b_1, b_2)$ if and only if $\max\{c^1(\mathcal{S}), c^2(\mathcal{S})\} \leq b_1 = b_2$, so that both claims will be proved at the same time.

Furthermore

$$\begin{aligned}
f_1(\mathcal{S}) \leq b_1 = C + k\bar{c} &\iff k\bar{c} + 2C - \sum_{e_i \in \mathcal{S}} c(e_i) \leq C + k\bar{c} \\
&\iff \sum_{e_i \in \mathcal{S}} c(e_i) \geq C, \\
f_2(\mathcal{S}) \leq b_2 = C + k\bar{c} &\iff k\bar{c} + \sum_{e_i \in \mathcal{S}} c(e_i) \leq k\bar{c} + 2C \\
&\iff \sum_{e_i \in \mathcal{S}} c(e_i) \leq C.
\end{aligned}$$

Therefore these conditions are satisfied if and only if $\mathcal{S}$ solves k-Partition. □

In Section 8.1 we have shown that solving scalarizations of combinatorial problems using weighted sums is not useful for finding efficient solutions. Here, we will investigate how weighted sum problems can be used to solve max-ordering matroid problems

$$\min_{B \in \mathcal{B}} \max_{k=1}^{p} \sum_{e \in B} c^k(e). \tag{9.7}$$

Later on we will also see that they can be used to solve lexicographic problems and investigate in greater detail their limited effectiveness for finding efficient solutions.

For $\lambda \in \mathbb{R}^p_>$ define $\bar{c}(e) = \sum_{k=1}^{p} \lambda_k c^k(e)$.

For the objective value of a basis B in the single objective matroid optimization problem

$$\min_{B \in \mathcal{B}} \sum_{e \in B} \bar{c}(e) \tag{9.8}$$

we have

$$\begin{aligned}
\sum_{e \in B} \bar{c}(e) &= \sum_{e \in B} \sum_{k=1}^{p} \lambda_k c^k(e) = \sum_{k=1}^{p} \lambda_k \sum_{e \in B} c^k(e) \\
&\leq \left(\sum_{k=1}^{p} \lambda_k \right) \max_{k=1}^{p} \left(\sum_{e \in B} c^k(e) \right) = \max_{k=1}^{p} \left(\sum_{e \in B} c^k(e) \right)
\end{aligned} \tag{9.9}$$

We apply (9.9) to an optimal solution $\hat{B}$ of the max-ordering matroid problem (9.7) to get

$$\min_{B\in\mathcal{B}} \sum_{e\in B} \bar{c}(e) \leq \sum_{e\in B^*} \bar{c}(e) \leq \max_{k=1}^{p}\left(\sum_{e\in B^*} c^k(e)\right).$$

If we now apply a ranking of the solutions of the weighted sum problem (9.8), we get a sequence $\{B_1,\ldots,B_r\}\subset\mathcal{B}$ of bases such that

$$\sum_{e\in B_1} \bar{c}(e) \leq \sum_{e\in B_2} \bar{c}(e) \leq \ldots \leq \sum_{e\in B_r} \bar{c}(e)$$

and obtain the following result, relating the weighted sum (9.8) and max-ordering (9.7) problems.

Proposition 9.31. *Let r be the smallest index such that*

$$c^* := \min_{l=1,\ldots,r-1}\left\{\max_{k=1,\ldots,p} \sum_{e\in B_l} c^k(e)\right\} \leq \sum_{e\in B_r} \bar{c}(e). \tag{9.10}$$

Then any $B^*\in\mathcal{B}$ with $\max_{k=1,\ldots,p}\sum_{e\in B^*} c^k(e) = c^*$ is an optimal solution of (9.7). In particular, $\{B_1,\ldots,B_{r-1}\}$ contains a max-ordering solution.

Proof. From the choice of $B_1,\ldots,B_r$, (9.10), and (9.9) it follows that

$$\max_{k=1}^{p} \sum_{e\in B^*} c^k(e) \leq \sum_{e\in B_r} \bar{c}(e) \leq \sum_{e\in B} \bar{c}(e) \leq \max_{k=1}^{p} \sum_{e\in B} c^k(e).$$

for all $B\in\mathcal{B}\setminus\{B_1,\ldots,B_r\}$. □

It is possible that the condition of Proposition 9.31 is not satisfied, even if $r=|\mathcal{B}|$. The choice of λ is crucial for the success of the ranking approach to max-ordering problems. No general rule for generating a scalarizing vector λ that guarantees the existence of r as in Proposition 9.31 is known. The reader is asked to check the result on an example in Exercise 9.9.

Now let us consider efficiency again. From the general discussion of the weighted sum method in Chapter 3, we know that efficient solutions are found.

Lemma 9.32. *Let B be an optimal solution of the weighted sum problem (9.8). Then B is efficient.*

In Definition 8.7 we introduced the term *supported efficient solutions* for optimal solutions of weighted sum problems. . Among supported efficient bases we distinguish those for which $f(B)$ is an extreme point of $\mathrm{conv}(\mathcal{Y})$ and those which are not. The former are called *extreme efficient bases.*

An important difference between extreme efficient bases and all efficient bases is the following. Let $\mathcal{P}(\mathcal{M})$ be the Pareto graph of matroid $\mathcal{M}$, where the nodes are the efficient bases and there exists an edge between two bases B_1 and B_2 if $|B_1\setminus B_2| = |B_2\setminus B_1| = 1$. This is just the obvious generalization of the Pareto graph $\mathcal{P}(\mathcal{G})$ we introduced for the spanning tree problem in Definition 9.21.

Theorem 9.33 (Ehrgott (1996)). *The subgraph of $\mathcal{P}(\mathcal{M})$ generated by the extreme efficient bases is connected.*

The proof is rather lengthy and omitted here, it can be found in Ehrgott (1997) and Ehrgott (1996).

Theorem 9.33 also explains, why $\mathcal{P}(G)$ is connected in Example 9.22. All three efficient spanning trees are extreme efficient spanning trees, see Exercise 9.6.

Example 9.34. We consider a partition matroid for $\mathcal{E} = \{1, \ldots, 6\} = \{1, 5, 6\} \cup \{3\} \cup \{2, 4\}$. A basis is a subset of $\mathcal{E}$ consisting of exactly one element of the three given sets. The costs are

$$c(1) = \begin{pmatrix} 1 \\ 5 \end{pmatrix}, c(2) = \begin{pmatrix} 1 \\ 4 \end{pmatrix}, c(3) = \begin{pmatrix} 1 \\ 1 \end{pmatrix}, c(4) = \begin{pmatrix} 3 \\ 1 \end{pmatrix}, c(5) = \begin{pmatrix} 4 \\ 1 \end{pmatrix}, c(6) = \begin{pmatrix} 2 \\ 2 \end{pmatrix}.$$

The matroid has six bases, four of which are (extreme) efficient ones. The graph of all bases (solid lines) and efficient bases is shown in Figure 9.21. Edges of $\mathcal{P}(\mathcal{M})$ are shown as broken lines.

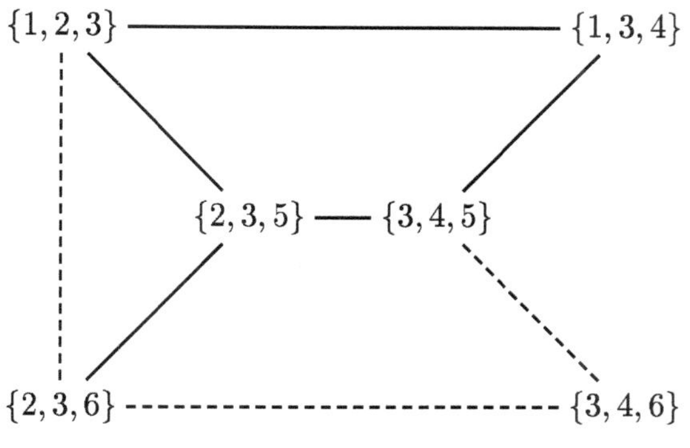

Fig. 9.21. Pareto graph in Example 9.34.

□

Another interesting fact is that lexicographically optimal bases are extreme efficient bases if $c^k(e) \geq 0$ for all $k = 1, \ldots, p$ and all $e \in \mathcal{E}$.

Proposition 9.35 (Hamacher and Ruhe (1994)). *Let B_π be a lexicographically optimal basis for* $\min_{B\in\mathcal{B}} \operatorname{lexmin} \pi(f(x))$. *Then there exists* $\lambda \in \mathbb{R}^p_>$ *such that B_π is an optimal solution of* $\min_{B\in\mathcal{B}} \sum_{e\in B} \sum_{k=1}^p \lambda_k c^k(e)$.

Proof. Without loss of generality we assume $\pi = \text{id}$. Let $\lambda_1 := 1-\varepsilon(1-\varepsilon^{p-1})$, $\lambda_k := \varepsilon^{k-1}(1-\varepsilon)$ where $\varepsilon = 1/(2M)$ with $M > \max_{k=1,\ldots,p} \max_{B\in\mathcal{B}} f_k(B)$ and $M > 1$. Then

$$\begin{aligned}\sum_{k=1}^p \lambda_k &= 1-\varepsilon+\varepsilon^p + \sum_{k=2}^p \varepsilon^{k-1}(1-\varepsilon)\\ &= 1-\varepsilon+\varepsilon^p + \sum_{k=1}^{p-1} \varepsilon^k - \sum_{k=2}^p \varepsilon^k\\ &= 1-\varepsilon+\varepsilon^p+\varepsilon-\varepsilon^p \;=\; 1.\end{aligned}$$

Now let B be any not lexicographically minimal basis and $i := \min\{k : f_k(B_\pi) < f_k(B)\}$. Assume $i \geq 2$. Then

$$\begin{aligned}&\sum_{k=1}^p \lambda_k f_k(B_\pi) - \sum_{k=1}^p \lambda_k f_k(B)\\ &= \lambda_i \underbrace{(f_i(B_\pi) - f_i(B))}_{\leq -1} + \sum_{k=i+1}^p \lambda_k \underbrace{(f_k(B_\pi) - f_k(B))}_{\leq M}\\ &\leq \varepsilon^{i-1}(1-\varepsilon)(-1) + \sum_{k=i+1}^p \varepsilon^{k-1}(1-\varepsilon)M\\ &\leq \varepsilon^{i-1}(1-\varepsilon)(-1) + (1-\varepsilon)\varepsilon^i \frac{1-\varepsilon^{p-1}}{1-\varepsilon} M\\ &= \varepsilon^{i-1}(\varepsilon-1) + \varepsilon^{i-1}\left(1-\varepsilon^{p-1}\right)\frac{M}{2M}\\ &= \varepsilon^{i-1}\left(\varepsilon - 1 + \frac{1}{2} - \frac{1}{2}\varepsilon^{p-1}\right) \;<\; 0\end{aligned}$$

since $\varepsilon < 1/2$. A similar calculation shows the result in case $i = 1$. □

As a final comment, we remark that both Proposition 9.31 and Proposition 9.35 are valid for any multiobjective combinatorial optimization problem, as we never used the fact that the feasible solutions are bases of a matroid. Therefore, for general MOCO problems, max-ordering optimal solutions and lexicographically optimal solutions can be found by the weighted sum method.

9.4 The Assignment Problem and the Two Phase Method

In this section, we show that the multicriteria assignment problem is $\mathcal{NP}$-complete, $\#\mathcal{P}$-complete, and intractable. We present an algorithm to find all efficient solutions in the bicriterion case, which might of course need exponential time.

The multicriteria assignment problem can be formulated as a 0-1 programming problem.

$$\min \sum_{i=1}^{n} \sum_{j=1}^{n} c_{ij}^k x_{ij} \qquad k = 1, \ldots, p \tag{9.11}$$

$$\text{subject to } \sum_{i=1}^{n} x_{ij} = 1; \quad j = 1, \ldots, n \tag{9.12}$$

$$\sum_{j=1}^{n} x_{ij} = 1; \quad i = 1, \ldots, n \tag{9.13}$$

$$x_{ij} \in \{0, 1\}; \quad i, j = 1, \ldots, n \tag{9.14}$$

In addition to the 0-1 programming formulation (9.11) – (9.14) we can also formulate it as an optimization problem on graphs, revealing the combinatorial structure. We assume that all $c_{ij}^k \in \mathbb{Z}_{\geqq}$.

Definition 9.36. *1. A* complete bipartite graph $\mathcal{G} = \mathcal{K}_{n,n}$ *on n nodes is defined by the node set $\mathcal{V} = \mathcal{L} \cup \mathcal{R}$, where $\mathcal{L} = \{v_1, \ldots v_n\}$ and $\mathcal{R} = \{v'_1, \ldots, v'_n\}$ and edge set $\mathcal{E} = \{[v_i, v'_j] : v_i \in \mathcal{L}, v'_j \in \mathcal{R}\}$.*
2. A matching *is subset $M \subset \mathcal{E}$ of the edges of a graph $\mathcal{G}$ such that no two edges of M have a node in common. A matching is called* perfect *if $|M| = n$.*

Let $c : \mathcal{E} \to \mathbb{Z}^p$ be a cost function on the edges of a complete bipartite graph $\mathcal{K}_{n,n}$ and $f(x) = \sum_{e \in x} c(e)$. The bicriterion assignment problem (9.11) – (9.14) is to find the efficient matchings of $\mathcal{K}_{n,n}$ according to objective function f. The cost coefficients c_{ij}^k of the p objective functions are represented by p matrices $C^1, \ldots, C^p$ and $f_k(x) = \sum_{i=1}^n \sum_{j=1}^n c_{ij}^k x_{ij}$ denotes the kth objective function. The set of all feasible assignments (perfect matchings) will be denoted by $\mathcal{M}$, for matchings.

Theorem 9.37. *The biobjective assignment problem* (9.11) – (9.14) *is $\mathcal{NP}$-complete, $\#\mathcal{P}$-complete, and intractable.*

Proof. 1. First we show $\mathcal{NP}$-completeness. This part of the proof is from Serafini (1986). We provide a reduction from EQUI-PARTITION, which is $\mathcal{NP}$-complete according to (Garey and Johnson, 1979, p. 223). The EQUI-PARTITION problem is defined as follows. Given $(c_1, \ldots, c_{2n}) \in \mathbb{Z}^{2n}$ with $\sum_{i=1}^{2n} c_i = 2C$, does there exist a subset $\mathcal{S} \subset \{1, \ldots, 2n\}, |\mathcal{S}| = n$, such that $\sum_{i \in \mathcal{S}} c_i = \sum_{i \notin \mathcal{S}} c_i = C$?
For the reduction we use $(c_1, \ldots, c_{2n})$ to construct an instance of the bicriterion assignment problem. Let $\hat{c} > \max_{i=1,\ldots,2n} c_i$. We define $\mathcal{V} = \mathcal{L} \cup \mathcal{R} := \{l_1, \ldots, l_{2n}\} \cup \{r_1, \ldots, r_{2n}\}$ as node set of a complete bipartite graph $\mathcal{K}_{2n,2n}$ and choose costs

$$c([l_i, r_j]) := \begin{cases} (\hat{c} + c_i, \hat{c} - c_i) & \text{if } j \text{ is odd} \\ (\hat{c}, \hat{c}) & \text{if } j \text{ is even.} \end{cases}$$

Then there exists a subset $\mathcal{S} \subset \{1, \ldots, 2n\}$ with $\sum_{i \in \mathcal{S}} c_i = C$ if and only if there is a perfect matching $\mathcal{M}$ with $\sum_{[l_i,r_j] \in \mathcal{M}} c^1([l_i, r_j]) \leq 2n\hat{c} + C$ and $\sum_{([l_i,r_j]) \in \mathcal{M}} c^2([l_i, r_j]) \leq 2n\hat{c} - C$.
Because this transformation is not parsimonious, it does not prove $\#\mathcal{P}$-completeness.

2. The proof of $\#\mathcal{P}$-completeness and intractability is based on counting perfect matchings. Valiant (1979b) showed that counting the number of perfect matchings of a bipartite graph is $\#\mathcal{P}$-complete. Therefore, we have to find an instance of the problem where all perfect matchings are efficient.

   Let $\{e_1, \ldots, e_{n^2}\}$ be the edges of $\mathcal{K}_{n,n}$. Choosing $c(e_i) = (2^{i-1}, 2^{n^2} - 2^{i-1})$ for all $i = 1, \ldots, n^2$ we get, as in Proposition 9.25 that all feasible solutions have incomparable objective vectors. Thus, a matching is perfect if and only if it is efficient. For intractability, note that there are $n!$ perfect matchings in a complete bipartite graph. □

The first part of the proof is from Serafini (1986). For an independent proof of the second, we refer to Neumayer (1994).

Let us now discuss the solution of multiobjective assignment problems. We restrict this discussion to the biobjective case. For more than two objectives, no algorithm is known. With a single objective it is well known that the problem can be solved as a linear program because total unimodularity of the constraint matrix guarantees an optimal solution of the LP to be integral. It can be solved very efficiently with the Hungarian method Papadimitriou and Steiglitz (1982). This appealing property cannot be fully exploited in the multicriteria case. It would only be possible to find optimal solutions of weighted sum scalarizations this way. However, the following example shows that there may exist efficient points in the interior of $\text{conv}(\mathcal{Y})$.

Example 9.38 (Ulungu and Teghem (1994)). The problem is defined by two cost matrices C^1 and C^2

$$C^1 := (c^1_{ij}) = \begin{pmatrix} 5 & 1 & 4 & 7 \\ 6 & 2 & 2 & 6 \\ 2 & 8 & 4 & 4 \\ 3 & 5 & 7 & 1 \end{pmatrix}, \qquad C^2 := (c^2_{ij}) = \begin{pmatrix} 3 & 6 & 4 & 2 \\ 1 & 3 & 8 & 3 \\ 5 & 2 & 2 & 3 \\ 4 & 2 & 3 & 5 \end{pmatrix}.$$

The 24 feasible assignments can be represented by their cost vectors in objective space as shown in Figure 9.22.

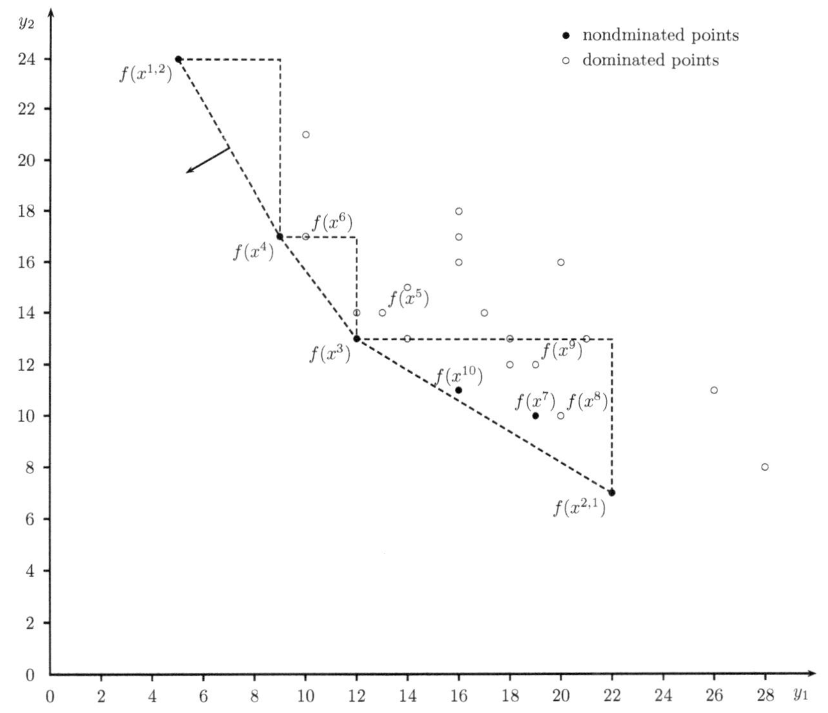

Fig. 9.22. Example 9.38: Feasible set in objective space.

The problem has six efficient solutions, four of which are supported (see Definition 8.7). x^6 and x^7 are nonsupported, their objective vectors $f(x^6)$ and $f(x^7)$ are in the interior of the convex hull of $\mathcal{Y}$. All of the supported solutions are extreme solutions, they define extreme points of $\operatorname{conv}\mathcal{Y}$. Note that supported efficient solutions can be found solving weighted sum problems with weight vector corresponding to a normal to the line connecting two "consecutive" extreme nondominated points.

$x^{1,2}$ and $x^{2,1}$ are the lexicographically optimal solutions. There are also two weakly efficient solutions, indicated by the points above and to the right of $f(x^3)$. (The numbering of solutions is according to the order in which they are generated by the algorithm described later, see Example 9.40). □

Definition 9.39. *The* nondominated frontier *of a multicriteria combinatorial optimization problem is the nondominated set of* $\text{conv}(\mathcal{Y})$.

With $p = 2$, the efficient frontier is a piecewise linear curve, as seen in Figure 9.22. For the bicriterion assignment the efficient frontier actually consists of the efficient set of the MOLP defined by the linear relaxation of the combinatorial problem. However, this is due to the fact that any weighted sum scalarization of (9.11) – (9.14) has an integer optimal solution, it does not hold for general MOIPs (8.7).

We will now present the two-phase method of Ulungu and Teghem (1994) to solve the biobjective assignment problem. It exploits the unimodularity property as far as possible. In Phase 1 it finds the supported efficient assignments. In Phase 2 it then proceeds to find the nonsupported efficient solutions in the interior of $\text{conv}(\mathcal{Y})$. To that end it searches regions of $\text{conv}(\mathcal{Y})$, where nondominated points can possibly be found. These are the triangles defined by "consecutive" points of $\mathcal{Y}_N$ on the efficient frontier, as shown in Figure 9.22.

Phase 1 starts with the computation of both lexicographically optimal solutions. This is done by solving two single objective assignment problems. These are added to a list of efficient solutions, maintained in order of increasing value of $f_1(x)$. The main part of the algorithm is formulated recursively. Given two consecutive solutions from the list, a weighted sum problem is formulated and solved. If new efficient solutions are found they are added to the list. The recursion stops if no further efficient solutions are found.

Algorithm 9.5 (Assignment problem Phase 1 algorithm.)

Input: Cost matrices C^1, C^2.

Solve the assignment problem with cost matrix C^1.

Let $\hat{x}^1$ *be an optimal solution.*

Find an optimal solution $x^{1,2}$ *of the assignment problem with cost matrix* $f_2(\hat{x}^1)C^1 + C^2$.

Solve the assignment problem with cost matrix C^2.

Let $\hat{x}^2$ *be an optimal solution.*

Find an optimal solution $x^{2,1}$ of the assignment problem with cost matrix $C^1 + f_1(\hat{x}^2)C^2$.
$\mathcal{X}_{sE} = \{x^{1,2}, x^{2,1}\}$.
`SolveRecursion`$(x^{1,2}, x^{2,1}, \mathcal{X}_{sE})$
Output: $\mathcal{X}_{sE}$

Algorithm 9.6 (Solve recursion.)

Input: $x^r, x^s \in \mathcal{X}_{sE}$ *with* $f_1(x^r) < f_1(x^s)$.
$\lambda_1 := (f_2(x^r) - f_2(x_s)), \lambda_2 := (f_1(x^s) - f_2(x^r))$.
Compute the set $\mathcal{R}$ *of all optimal solutions of the assignment problem with cost matrix* $\lambda_1 C^1 + \lambda_2 C^2$.
$\mathcal{X}_{sE} := \mathcal{X}_{sE} \cup \mathcal{R}$.
If $\lambda_1 f_1(x) + \lambda_2 f_2(x) < \lambda_1 f_1(x^r) + \lambda_2 f_2(x^r)$ *for some* $x \in \mathcal{R}$ *then*
  Let $x^{t_1} \in \operatorname{argmin}\{f_1(x) : x \in \mathcal{R}\}$
  Let $x^{t_2} \in \operatorname{argmin}\{f_2(x) : x \in \mathcal{R}\}$
  `SolveRecursion`$(x^r, x^{t_1}, \mathcal{X}_{sE})$
  `SolveRecursion`$(x^{t_2}, x^s, \mathcal{X}_{sE})$
End if

The lexicographic solutions are found by first solving an assignment problem with a single objective. In case an optimal solution, which is weakly efficient, is found, a second single objective problem is solved – with weights that guarantee that a lexicographic solution is then found.

It is easy to see that (λ_1, λ_2) as defined in Algorithm 9.6 defines a normal to the line segment connecting $f(x^r)$ and $f(x^s)$ (see Exercise 9.12). The single objective assignment problems can be solved very efficiently using the Hungarian method, see e.g. Papadimitriou and Steiglitz (1982). However, *all* optimal solutions have to be found, as they may have different objective vectors, i.e. define different nondominated points, with the same weighted sum objective value. An algorithm to do that is given in Fukuda and Matsui (1992). Note that the condition of the "if" statement in Algorithm 9.6 holds for either all or no $x \in \mathcal{R}$. At termination of Algorithm 9.5 we can write $\mathcal{X}_{sE} = \{x^1, \ldots, x^t\}$ and assume that $f_1(x^i) \leq f_1(x^{i+1})$ and therefore $f_2(x^i) \geq f_2(x^{i+1})$ for all $i = 1, \ldots, t-1$. There will always exist $\lambda \in \mathbb{R}^2_>$ such that both x^i and x^{i+1} are optimal solutions of the weighted sum problem with weight λ.

We now analyze the solution of the weighted sum assignment problems in some greater detail. The following cases may occur for the set $\mathcal{R}$.

1. $\mathcal{R} \cap \{x^r, x^s\} = \emptyset$. In this case there must be an assignment x with $f_1(x^r) < f_1(x) < f_1(x^s)$ and $f_2(x^s) < f_2(x) < f_2(x^r)$ as well as $\lambda_1 f_1(x) + \lambda_2 f_2(x) < \lambda_1 f_1(x^r) + \lambda_2 f_2(x^r)$. This situation is shown in Figure 9.23. New solution x^{t_1} and x^{t_2} coincide if the weighted sum problem has a unique optimal solution. It is then necessary to solve two new weighted sum problems with weights corresponding to the normal between x^r and x^{t_1} as well as x^{t_2} and x^s.

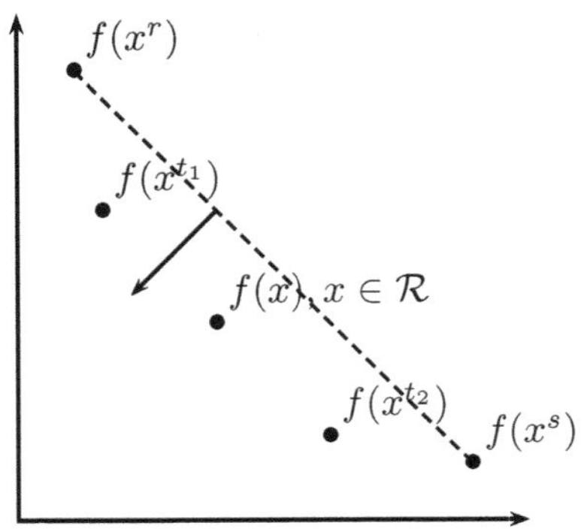

Fig. 9.23. New supported efficient solutions defining new extreme points.

2. $\{x^r, x^s\} \subset \mathcal{R}$. This includes the case of equality. Then all solutions of $\mathcal{R}$ except x^r and x^s are new supported efficient solutions. However, $\lambda_1 f_1(x) + \lambda_2 f_2(x) = \lambda_1 f_1(x^r) + \lambda_2 f_2(x)$ for all $x \in \mathcal{R}$, so that there is no need to solve any further weighted sum problem.

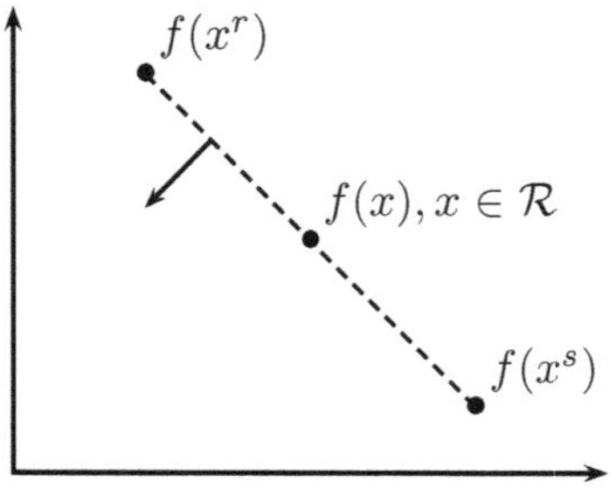

Fig. 9.24. A new supported efficient solution on $\text{conv}\{f(x^r), f(x^s)\}$.

Algorithm 9.5 determines all supported efficient solutions. Note that if only one optimal solution of the weighted sum problems is computed, there can be no guarantee that all efficient solutions for which $f(x)$ is on a line between two extreme supported efficient solutions are found.

Now let $\mathcal{X}_{sE} = \{x^1, \ldots, x^t\}$ be the result of Algorithm 9.5, ordered as before. Phase 2 determines efficient solutions with $f_1(x^i) < f_1(x) < f_1(x^{i+1})$ and $f_2(x^i) > f_2(x) > f_2(x^{i+1})$ and also $\lambda_1 f_1(x) + \lambda_2 f_2(x) > \lambda_1 f_1(x^j) + \lambda_2 f_2(x^j)$, $j \in \{i, i+1\}$, where λ defines a normal to the line between $f(x^i)$ and $f(x^{i+1})$.

In order to find these solutions we will fix some variables to 1. We are interested in finding lower bounds on the value of $f_1(x)$, $f_2(x)$, or $\lambda_1 f_1(x) + \lambda_2 f_2(x)$ after fixing a variable. So let C be a cost matrix and let x be a feasible solution of the assignment problem

$$\min \left\{ \sum_{i=1}^{n} \sum_{j=1}^{n} c_{ij} x_{ij} : \sum_{i=1}^{n} x_{ij} = 1, j = 1, \ldots, n; \sum_{j=1}^{n} x_{ij} = 1, i = 1, \ldots, n; x \geqq 0 \right\}. \tag{9.15}$$

The dual of the linear program (9.15) is

$$\max \left\{ \sum_{i=1}^{n} u_i + \sum_{j=1}^{n} v_j : u_i + v_j \leq c_{ij}, i, j = 1, \ldots, n \right\}. \tag{9.16}$$

C will be either C^1 or C^2 or $\lambda_1 C^1 + \lambda_2 C^2$ in the algorithm for Phase 2.

Let x be an assignment and let $\bar{C}$ be a matrix of reduced costs associated with x. This can be the reduced cost matrix of a basis for which x is a basic feasible solution in the LP (9.15) or defined by $\bar{c}_{ij} = c_{ij} - u_i - v_j$ with $u, v \in \mathbb{R}^n$ such that $\bar{c}_{ij} = 0$ for all i, j with $x_{ij} = 1$ (from the dual (9.16), this is used in the Hungarian method). Let us assume we want to impose $x_{i^*j^*} = 1$, where that variable is 0 in x. This can be done by continuing the Hungarian method or the simplex algorithm to restore feasibility. We can derive lower bounds on the value of $f(x)$ after fixing a variable, depending on whether x is an optimal solution of (9.15) or not.

1. If x is optimal, we can choose $\bar{C}$ so that $\bar{c}^i_{ij} \geq 0$ for all i, j. Now for solution x there must be i_k, j_k such that $x_{i_k j^*} = x_{i^* j_k} = 1$. Thus, after fixing $x_{i^*j^*} = 1$ and restoring feasibility, one variable in row i_k and one variable in column j_k other than $x_{i_k j^*}$ and $x_{i^* j_k}$ will be equal to 1.
   - If these two variables are one and the same we will have $x_{i_k j_k} = 1$ and the objective function will increase by at least $\bar{c}_{i_k j_k}$.
   - Otherwise there will be an increase of at least

$$\gamma = \min_{j \neq j^*} \bar{c}_{i_k j} + \min_{i \neq i^*} \bar{c}_{i j_k}.$$

Therefore a lower bound on the objective value after fixing a variable is

$$\alpha = f(x) + \bar{c}_{i^*j^*} + \min\{\bar{c}_{i_k j_k}, \gamma\}. \tag{9.17}$$

2. If x is not optimal then $\bar{C}$ has at least one negative entry. Hence, the lower bound is computed from the minimal entries in all rows and columns of $\bar{C}$ and we obtain

$$\alpha = f(x) + \bar{c}_{i^*j^*} + \max\left\{\sum_{i \neq i^*} \min_{j \neq j^*} \bar{c}_{ij}, \sum_{j \neq j^*} \min_{i \neq i^*} \bar{c}_{ij}\right\}. \tag{9.18}$$

Let $\lambda > 0$ be such that supported efficient solutions x^i and x^{i+1} are optimal solutions of the weighted sum assignment problem and let $\bar{C}$ be a reduced cost matrix for an optimal solution of that problem. Any unsupported efficient solution x with $f(x)$ in the triangle defined by $f(x^i)$, $f(x^{i+1})$, and $(f_1(x^{i+1}), f_2(x^i))$ is not an optimal solution of the weighted sum problem. Such a solution can be obtained by choosing $(i^*, j^*) \in \mathcal{L} := \{(i,j) : \bar{c}_{ij} > 0\}$ and forcing $x_{i^*,j^*} = 1$. After restoring feasibility we obtain an assignment x' with $\lambda_1 f_1(x') + \lambda_2 f_2(x') > \lambda_1 f_1(x^i) + \lambda_2 f_2(x^i)$. x' is a candidate for a new efficient solution. The best assignment with $x_{i^*,j^*} = 1$ can be found by re-optimizing, i.e. solving a smaller assignment problem. We can perform three tests to determine whether $f(x')$ is in the triangle. These tests are for the value of objectives $\lambda_1 f_1 + \lambda_2 f_2$, f_1, and f_2, respectively.

Test 1: Let α^i be the lower bound (9.17) obtained from a reduced cost matrix for the weighted sum LP with respect to x^i and let α^{i+1} be that obtained from a reduced cost matrix with respect to x^{i+1}. Define $\alpha^* = \max\{\alpha_1, \alpha_2\}$. Then $f(x')$ is not in the triangle if

$$\alpha^* \geq \lambda_1 f_1(x^{i+1}) + \lambda_2 f_2(x^i).$$

With Test 1 we check, if $f(x')$ is in the hatched area of Figure 9.25.

Test 2: This tests checks the objective value for f_1. If $x^i = x^{1,2}$ is a lexicographically optimal solution then let α be defined by (9.17) with reduced cost matrix of C^1 for x^i. Otherwise let it be defined by (9.18). Then $f(x')$ is not in the triangle if

$$\alpha \geq f_1(x^{i+1}). \tag{9.19}$$

The test determines if $f(x')$ is in the hatched area in Figure 9.26.

Test 3: Finally, this test checks objective f_2. It is analogous to Test 2, only that α is computed from a reduced cost matrix of C^2 for x^{i+1}. Then $f(x')$ is not in the triangle if

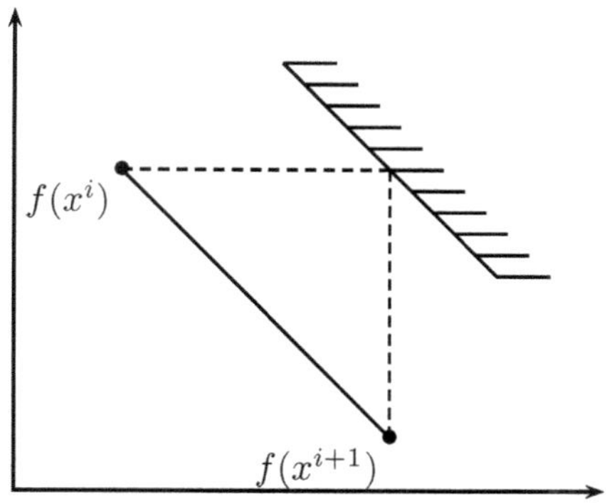

Fig. 9.25. Test 1 for the weighted sum of objectives.

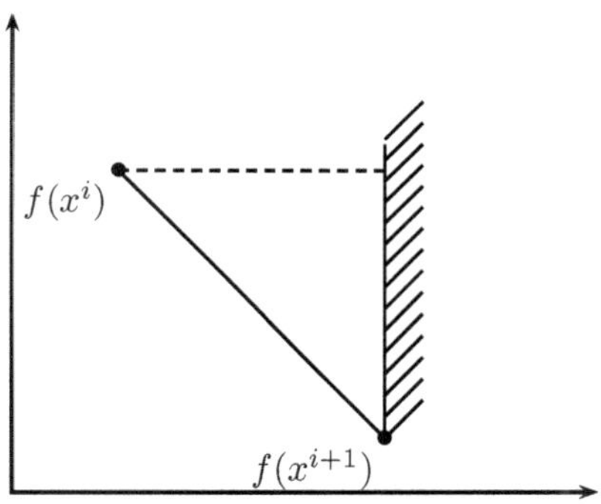

Fig. 9.26. Test 2 for objective function 1.

$$\alpha \geq f_2(x^i). \tag{9.20}$$

This test checks, if $f(x')$ in the hatched area of Figure 9.27.

Note that the bigger f_1 or f_2 get, the more negative elements appear in the reduced cost matrices $\bar{C}^1$ respectively $\bar{C}^2$ and the more inefficient Tests 2 and 3 are. Since the tests apply to any feasible solution x' obtained from fixing $x_{i^*j^*} = 1$ we can apply them to exclude variables from $\mathcal{L}$.

During the course of the algorithm it may be necessary to fix more than one variable to find all efficient solutions. This is done in a depth-first search to avoid redundancies.

Let again $x^i, x^{i+1} \in \mathcal{X}_{sE}$ as before. In Phase 2, a set $\mathcal{P}$ of potentially efficient assignments with $f(x)$ in the triangle defined by $f(x^i), f(x^{i+1})$, and

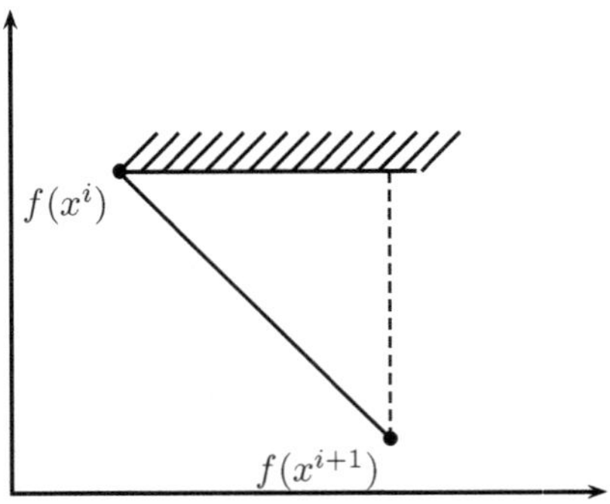

Fig. 9.27. Test 3 for objective function 2.

$(f_1(x^{i+1}), f_2(x^i))$ is maintained. Solutions in $\mathcal{P}$ are also numbered in increasing order of $f_1(x)$. Initially, $\mathcal{P} = \emptyset$. New solutions, which are optimal solutions of assignment problems, where one or more variables from the list $\mathcal{L}$ have been fixed to one are added to this list. They may be removed again if they are proved to be dominated. At termination, $\mathcal{P}$ is the set of nonsupported efficient solutions in the triangle. Let $\mathcal{P} = \{x^1, \ldots, x^s\}$ at some stage of the algorithm and define

$$\gamma := \max_{j=1}^{s-1}\{\lambda_1 f_1(x^{j+1}) + \lambda_2 f_2(x^j)\}.$$

Then any feasible assignment with $\lambda_1 f_1(x)+\lambda_2 f_2(x)$ bigger than the upper bound

$$\max\{\gamma, \lambda_1 f_1(x^1) + \lambda_2 f_2(x^i), \lambda_1 f_1(x^{i+1}) + \lambda_2 f_2(x^s)\} \tag{9.21}$$

is dominated. This upper bound has been proposed in Tuyttens *et al.* (2000).

Finally we need to describe the procedure for fixing variables. This is done using the pairs (i, j) in $\mathcal{L}$ in a depth first strategy. We assume that $\mathcal{L} = \{(i, j)_1, \ldots, (i, j)_l\}$. $(i, j) \in \mathcal{L}$ is selected, Tests 1 – 3 are performed, if any of them fails, (i, j) is removed from the list. Otherwise, the weighted sum assignment problem is solved with $x_{ij} = 1$ and the upper bound (9.21) is checked. If the resulting assignment x' is potentially efficient, it is added to $\mathcal{P}$ and the upper bound (9.21) is updated. The next pair from $\mathcal{L}$ is chosen and that variable fixed in addition to the ones already fixed, if these pairs are compatible (e.g. it is not possible to impose $x_{15} = 1$ and $x_{13} = 1$, so $(1, 5)$ and $(1, 3)$ are incompatible). This procedure stops and backtracks when there is no pair in $\mathcal{L}$ left or the bound (9.21) is violated. Note that fixing more variables can only increase the objective value. Whenever an assignment dominating any $x \in \mathcal{P}$ is found, $\mathcal{P}$ is updated.

Phase 2 is now summarized in Algorithm 9.7.

Algorithm 9.7 (Assignment problem Phase 2 algorithm.)

Input: $\mathcal{X}_{sE}$, C^1 *and* C^2.

  $\mathcal{X}_E := \mathcal{X}_{sE}$. *For all* $x^i, x^{i+1} \in \mathcal{X}_{sE}$ *do*

    Find an optimal solution $\tilde{x}$ *of the assignment problem with cost matrix* $C^\lambda = (f_2(x^i) - f_2(x^{i+1}))C^1 + (f_1(x^{i+1}) - f_1(x^i))C^2$.

    Let $\bar{C}^\lambda$ *be a reduced cost matrix of* C^λ *for* $\tilde{x}$.

    Let $\bar{C}^1$ *be a reduced cost matrix of* C^1 *for* x^i.

    Let $\bar{C}^2$ *be a reduced cost matrix of* C^2 *for* x^2.

    $\mathcal{L} := \{(i,j) : \bar{c}_{ij} > 0\}$.

    $\mathcal{P} = \emptyset$

    Compute upper bound u *according to* (9.21)

    If $u > \lambda_1 f_1(x^i) + \lambda_2 f_2(x^{i+1})$ *then*

      For all $(i,j) \in \mathcal{L}$ *do*

        If Tests 1 – 3 fail then $\mathcal{L} := \mathcal{L} \setminus \{i,j\}$.

      End for.

      If $\mathcal{L} \neq \emptyset$ *then*

        For all $(i,j) \in \mathcal{L}$ *do*

          `Impose`$((i,j), \emptyset)$

        End for.

      End if.

    End if. $\mathcal{X}_E = \mathcal{X}_E \cup \mathcal{P}$

  End for.

Output: $\mathcal{X}_E$.

Algorithm 9.8 (Impose.)

Input: $(k,l), \mathcal{L}'$

  ($\mathcal{L}'$ *is a list of variables already assigned 1.) Find all optimal solutions of the assignment problem with cost matrix* C^λ *and* $x_{ij} = 1$ *for all* $(i,j) \in \mathcal{L}' \cup \{i,j\}$.

  $\mathcal{L}' = \mathcal{L}' \cup \{(k,l\}$.

  If $\lambda_1 f_1(x) + \lambda_2 f_2(x) \leq u$ *then*

    If $f(x)$ *is in the triangle and* $f(x)$ *is not dominated then*

      $\mathcal{P} := \mathcal{P} \cup \{x\}$.

      Remove dominated solutions from $\mathcal{P}$.

      Compute u *according to* (9.21).

    End if.

    For all $(i,j) \in \mathcal{L} \setminus \mathcal{L}'$ *do*

If the pair (i,j) *is compatible with* $\mathcal{L}'$ *then*
Impose$((i,j),\mathcal{L}')$
$\mathcal{L}' := \mathcal{L}' \setminus \{k,l\}$
End if.
End for
End if

Example 9.40. We solve the problem already introduced in Example 9.38, where

$$C^1 := \begin{pmatrix} 5 & 1 & 4 & 7 \\ 6 & 2 & 2 & 6 \\ 2 & 8 & 4 & 4 \\ 3 & 5 & 7 & 1 \end{pmatrix}, \quad C^2 := \begin{pmatrix} 3 & 6 & 4 & 2 \\ 1 & 3 & 8 & 3 \\ 5 & 2 & 2 & 3 \\ 4 & 2 & 3 & 5 \end{pmatrix}.$$

Phase 1 (Algorithm 9.5) first finds lexicographically optimal assignments $x^{1,2}$ and $x^{2,1}$. The former is defined by

$$x_{12} = x_{23} = x_{31} = x_{44} = 1, f(x^{1,2}) = \begin{pmatrix} 6 \\ 24 \end{pmatrix},$$

the latter by

$$x_{14} = x_{21} = x_{33} = x_{42} = 1, f(x^{2,1}) = \begin{pmatrix} 22 \\ 7 \end{pmatrix}.$$

$\mathcal{X}_{sE}$ is initialized with $\{x^{1,2}, x^{2,1}\}$.

The recursion solves the following problems.

1. For $f(x)$ between $f(x^{1,2})$ and $f(x^{2,1})$. $\lambda_1 = 24 - 7 = 17, \lambda_2 = 22 - 6 = 16$. Solving the weighted sum assignment problem yields x^3 with
$$x_{11} = x_{22} = x_{33} = x_{44} = 1, f(x^3) = \begin{pmatrix} 12 \\ 13 \end{pmatrix}.$$
$\mathcal{X}_{sE} = \{x^{1,2}, x^3, x^{2,1}\}$
2. For $f(x)$ between $f(x^{1,2})$ and $f(x^3)$. $\lambda_1 = 24 - 13 = 11, \lambda_2 = 12 - 6 = 6$. Solving the weighted sum assignment problem yields x^4 with
$$x_{13} = x_{22} = x_{31} = x_{44} = 1, f(x^4) = \begin{pmatrix} 9 \\ 17 \end{pmatrix}.$$
$\mathcal{X}_{sE} = \{x^{1,2}, x^4, x^3, x^{2,1}\}$

3. For $f(x)$ between $f(x^{1,2})$ and $f(x^4)$. $\lambda_1 = 24 - 17 = 7, \lambda_2 = 9 - 6 = 3$. The weighted sum assignment problem has optimal solutions $x^{1,2}$ and x^4.
4. For $f(x)$ between $f(x^4)$ and $f(x^3)$. $\lambda_1 = 17 - 13 = 4, \lambda_2 = 12 - 9 = 3$. The weighted sum assignment problem has optimal solutions x^4 and x^3.
5. For $f(x)$ between $f(x^3)$ and $f(x^{2,1}$. $\lambda_1 = 13 - 7 = 6, \lambda_2 = 22 - 12 = 10$. The weighted sum assignment problem has optimal solutions x^3 and $x^{2,1}$.

In Phase 2 we investigate the three triangles defined by the four supported efficient solutions.

1. Checking $x^{1,2}, x^4$. For $\lambda_1 = 7, \lambda_2 = 3$

$$\bar{C^\lambda} = \begin{pmatrix} 9 & 0 & 0 & 20 \\ 12 & 0 & 0 & 18 \\ 0 & 43 & 0 & 8 \\ 11 & 29 & 31 & 0 \end{pmatrix}.$$

is a reduced cost matrix for both optimal solutions of the weighted sum assignment problem. Thus we have

$$\mathcal{L} = \{(3,4), (1,1), (4,1), (2,1), (2,4), (1,4), (4,2), (4,3), (3,2)\}.$$

We perform Test 1 and eliminate pairs $(2,4), (1,4), (4,2), (4,3)$, and $(3,2)$ are eliminated. Test 2 (with a reduced cost matrix of C^1 for $x^{1,2}$) eliminates $(3,4), (1,1), (4,1)$, and $(2,1)$ and now $\mathcal{L} = \emptyset$.
2. Checking x^4, x^3. For $\lambda_1 = 4, \ \lambda_2 = 3$

$$\bar{C} = \begin{pmatrix} 0 & 0 & 0 & 6 \\ 3 & 0 & 9 & 10 \\ 0 & 22 & 0 & 3 \\ 4 & 13 & 18 & 0 \end{pmatrix}.$$

is a reduced cost matrix for both optimal solutions of the weighted sum assignment problem. Thus we have

$$\mathcal{L} = \{(2,1), (3,4), (4,1), (1,4), (2,3), (2,4), (4,2), (4,3), (3,2)\}.$$

Test 1 eliminates $(2,4), (4,2), (4,3)$, and $(3,2)$. To perform Test 2 we can use

$$\bar{C} = \begin{pmatrix} 1 & -3 & 0 & 4 \\ 4 & 0 & 0 & 4 \\ 0 & 6 & 20 & 2 \\ 2 & 4 & 6 & 0 \end{pmatrix},$$

as reduced cost matrix of C^1 for x^4. $(2,1)$ and $(1,4)$ are eliminated. Test 3 does not lead to any further eliminations. We are left with $\mathcal{L} = \{(3,4),(4,1)(2,3)\}$. Imposing either $x_{34} = 1$ or $x_{41} = 1$ yields solution x^5 defined by

$$x_{13} = x_{22} = x_{34} = x_{41} = 1, f(x^5) = \begin{pmatrix} 13 \\ 14 \end{pmatrix},$$

which is dominated by x^3. The same is true for imposing both at the same time. Imposing $x_{23} = 1$ gives $x^{1,2}$. Imposing $x_{23} = 1$ together with one or two of the other pairs leads to x^6 with

$$x_{12} = x_{23} = x_{34} = x_{41} = 1, f(x^6) = \begin{pmatrix} 10 \\ 17 \end{pmatrix},$$

which is dominated by x^4. Thus, no potentially efficient solutions are found.

3. Checking $x^3, x^{2,1}$. For $\lambda_1 = 6, \lambda_2 = 10$

$$\bar{C} = \begin{pmatrix} 0 & 10 & 4 & 0 \\ 0 & 0 & 46 & 18 \\ 18 & 28 & 0 & 8 \\ 4 & 0 & 18 & 0 \end{pmatrix}.$$

is a reduced cost matrix for both optimal solutions of the weighted sum assignment problem. Thus we have

$$\mathcal{L} = \{(1,2),(1,3),(2,3),(2,4),(3,1),(3,2),(3,4),(4,1),(4,3)\}.$$

Test 1 does not eliminate any pairs, neither does Test 2. Test 3 with a reduced cost matrix of C^2 for $x^{2,1}$ does eliminate $(1,2)$, $(3,1)$, and $(2,3)$. So $\mathcal{L} = \{(1,3),(2,4),(3,2),(3,4),(4,1),(4,3)\}$.
We consider imposing all single pairs first. Forcing $x_{13} = 1$ or $x_{34} = 1$ yields solution x^7 with

$$x_{13} = x_{21} = x_{34} = x_{42} = 1, f(x^7) = \begin{pmatrix} 19 \\ 10 \end{pmatrix}.$$

Forcing $x_{24} = 1$ or $x_{43} = 1$ yields solution x^8 with

$$x_{11} = x_{24} = x_{33} = x_{42} = 1, f(x^8) = \begin{pmatrix} 20 \\ 10 \end{pmatrix}$$

or x^9 with

$$x_{11} = x_{22} = x_{34} = x_{43} = 1, f(x^9) = \begin{pmatrix} 18 \\ 12 \end{pmatrix}.$$

x^8 is and x^9 are both dominated by x^7.
Forcing $x_{32} = 1$ also yields a dominated solution.
Forcing $x_{41} = 1$ yields solution x^{10} with

$$x_{14} = x_{22} = x_{33} = x_{41} = 1, f(x^{10}) = \begin{pmatrix} 16 \\ 11 \end{pmatrix}.$$

We have two potentially efficient solutions and update $\mathcal{P} = \{x^{10}, x^7\}$ With these we can update the upper bound from $6 \times 22 + 10 \times 13 = 262$ (from $f_2(x^3)$ and $f_1(x^{2,1})$) to $\max\{6 \times 16 + 10 \times 13, 6 \times 19 + 10 \times 11, 6 \times 22 + 10 \times 10\} = 232$ according to (9.21). Imposing any two pairs at the same time yields either a solution already found before, a solution dominated by one of the existing solutions, or a solution that violates the new upper bound. For example, if $x_{24} = x_{32} = 1$ then $\bar{C}$ implies that the weighted sum objective will increase by at least $18 + 28 = 46$. Since the value at x^3 and $x^{2,1}$ is 202, the new solution will have a value greater than 232.

□

9.5 Notes

There is a rich literature on multicriteria shortest path problems. Here are some additional references to those cited in Section 9.1. $\mathcal{NP}$-completeness of the max-ordering problem is mentioned in Murthy and Her (1992), Warburton (1987) describes an approximation algorithm. A variety of algorithms based on dynamic programming (e.g. Henig (1985); Kostreva and Wiecek (1993); Sniedovich (1988)), label setting (Hansen, 1979; Martins, 1984) and label correcting methods (e.g. Brumbaugh-Smith and Shier (1989); Mote *et al.* (1991); Skriver and Andersen (2000)) are available with computational experiments (Brumbaugh-Smith and Shier, 1989; Huarng *et al.*, 1996; Skriver and Andersen, 2000) comparing different methods.

The level set algorithm 9.3 requires algorithms to rank feasible solutions, i.e. to find k-best solutions of combinatorial optimization problems are necessary. Such algorithms are known for a number of problems. The largest amount of research on ranking solutions is available for the shortest path problem. Algorithms by Azevedo *et al.* (1993) Martins *et al.* (1999a), and Eppstein (1998) are very efficient. Ranking algorithms for simple paths are e.g. proposed by Carraresi and Sodini (1983) and Martins *et al.* (1999b). Several methods are also known for the minimum spanning tree problem. We mention papers by Gabow (1977) and Katoh *et al.* (1981). Algorithms for ranking matroid bases are found in Hamacher and Queyranne (1985) and

Ehrgott (1996). Chegireddy and H.W. (1987) present an algorithm to find k-best perfect matchings, Brucker and Hamacher (1989) discuss k-best solutions for polynomially solvable scheduling problems, and an algorithm to rank (integer) network flows is presented in Hamacher (1995).

In Section 9.3 we have indicated that ranking algorithms can also be used to solve max-ordering problems. An algorithm that problems with two objectives is given in Ehrgott and Skriver (2003).

Algorithms other than those cited in Section 9.2 to find efficient trees range from minimizing weighted sums (Punnen and Nair, 1996; Schweigert, 1990) to approximation (Hamacher and Ruhe, 1994) and genetic algorithms (Zhou and Gen, 1999). The complexity status of a variety of multiobjective spanning tree problems, involving other than the typical sum and bottleneck objectives is studied in Camerini *et al.* (1984); Dell'Amico and Maffioli (1996, 2000).

The assignment problem is among the first MOCO problems studied (Dathe, 1978). We have explained the two-phase method for the assignment problem. This algorithmic template has also been applied to other problems. These include Lee and Pulat (1993); Sedeño-Noda and González-Martín (2001) for integer network flow, Ulungu (1993); Visée *et al.* (1998) for knapsack, and Ramos *et al.* (1998) for spanning tree problems. While Algorithm 9.5 for Phase 1 can be used essentially unchanged, problem specific modifications for the Phase 2 algorithm are always necessary. It is interesting to observe that the ranking algorithms can also be used in the two phase method. Finding unsupported efficient solutions is the same as finding non-optimal, or k-best, solutions of particular weighted sum problems.

Exercises

9.1. Prove Lemma 9.7.

9.2. Find an example of a digraph with negative weights, but no negative cycles (i.e. $\sum_{a \in C} w_q(a) \geq 0$ for all q and all cycles$C$) for which the label setting Algorithm 9.1 constructs a dominated path from node 1 to some other node i.

9.3. Apply the label correcting Algorithm 9.2 to the graph of Example 9.9.

9.4. 1. Modify the label setting and label correcting algorithms (Algorithms 9.1 and 9.2 so that they find a lexicographically minimal path from s to t, i.e. a path P such that

$$(\sum_{a \in P} c^1(a), \ldots, \sum_{a \in P} c^p(a)) \leq_{lex} (\sum_{a \in P'} c^1(a), \ldots, \sum_{a \in P'} c^p(a))$$

for all s-t paths P'.

2. Is it possible to use a modified label setting algorithm to solve the max-ordering shortest path problem

$$\min_{P \in \mathcal{P}} \max_{k=1,\ldots,p} \sum_{a \in P} c^k(a)$$

in directed graphs with nonnegative weights $c^k(a) \geq 0$ for all k and all a? Write down such a modified algorithm and either give an argument that it works (like the correctness proof for the label setting algorithm) or give an example to show that it does not work.

9.5. Illustrate the ranking approach to the bicriterion shortest path problem (with $s = 1$ and $t = 6$) for the following graph. The double arrow on arc $(3, 4)$ indicates that there is an arc in both directions, with the same cost both ways.

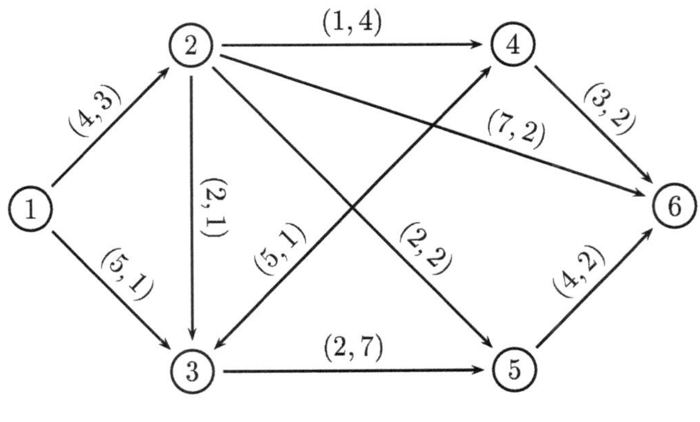

9.6. Show that all efficient spanning trees of the graph in Figure 9.13 are extreme efficient spanning trees (see also Example 9.22.

9.7. Kruskal's algorithm (Kruskal, 1956) for the (single objective) spanning tree problem works as follows:

- $\mathcal{E}(T) := \emptyset$
- For $i = 1$ to $m - 1$ do
  $\mathcal{E}(T) := \mathcal{E}(T) \cup \text{argmin}\{c(e) : \mathcal{E}(T) \cup e \text{ does not contain a cycle }\}$.

Formulate a multiobjective version of this algorithm and apply it to the graphs of Figure 9.12 and Figure 9.16.

9.8. Show that if $\mathcal{G}$ is a 1-tree (i.e. a tree plus an additional edge, or a graph with exactly one cycle), the Pareto graph $\mathcal{P}(\mathcal{G})$ defined for the spanning tree problem is always connected. Can you identify other classes of graphs for which this is true?

9.9. Use Proposition 9.31 to find max-ordering spanning trees in the graph of Figure 8.1. Try $\lambda = (1/4, 1/4, 1/4, 1/4)$ and λ with $\lambda_i = 1$ for some $i \in \{1, \ldots, 4\}, \lambda_j = 0, i \neq j$.

9.10. Solve the bicriterion assignment problem with cost matrices

$$C^1 = \begin{pmatrix} 2 & 1 & 0 \\ 0 & 2 & 1 \\ 2 & 0 & 3 \end{pmatrix} \quad C^2 = \begin{pmatrix} 0 & 1 & 2 \\ 2 & 0 & 2 \\ 2 & 2 & 0 \end{pmatrix}$$

by the two phase method with Algorithms 9.5 and 9.7. There are two supported and two unsupported efficient solutions.

9.11. Explain why the Phase 1 algorithm can be modified so that in Step 1 instead of lexicographically optimal solutions, solutions of $\min_{x \in \mathcal{X}} f_1(x)$ and $\min_{x \in X} f_2(x)$, which may be only weakly efficient, are computed. In which way would you have to modify the algorithm to guarantee that all efficient solutions are found?

9.12. Show that $\lambda_1 = f_2(x^i) - f_2(x^{i+1}), \lambda_2 = f_1(x^{i+1} - f_1(x^i)$ defines a normal to the line connecting $f(x^i)$ and $f(x^{i+1})$, where x^i and x^{i+1} are two optimal solutions of the weighted sum problem with objective $\lambda_1 f_1(x) + \lambda_2 f_2(x)$.

10

Multiobjective Versions of Some $\mathcal{NP}$-Hard Problems

10.1 The Knapsack Problem and Branch and Bound

As for the assignment problem in the previous section, we consider only finding efficient solutions. And we also restrict ourselves to the bicriterion case. The bicriterion knapsack problem is the binary integer program

$$\max f_1(x) = \sum_{i=1}^{n} c_i^1 x_i \tag{10.1}$$

$$\max f_2(x) = \sum_{i=1}^{n} c_i^2 x_i \tag{10.2}$$

$$\text{subject to } \sum w_i x_i \leq W \tag{10.3}$$

$$x_i \in \{0,1\}; \quad j = 1, \ldots, n. \tag{10.4}$$

The problem is obviously $\mathcal{NP}$-hard, as a counterpart of an $\mathcal{NP}$-hard single objective problem (see Lemma 8.11). Whether the problem is $\#\mathcal{P}$-complete or intractable is yet unknown.

We will present a branch and bound algorithm. To avoid trivial solutions and to have a meaningful problem we make some basic assumptions on the parameters of the knapsack problem. We assume that all values c_i^k, all weights w_i as well as the capacity W are nonnegative. Furthermore, no single weight exceeds capacity, i.e. $w_i \leq W$ for all $i = 1, \ldots, n$, but the total weight of all items is bigger than W, $\sum_{i=1}^{n} w_i > W$.

For the solution of knapsack problems the value to weight ratios c_i^k / w_i are of essential importance. In the single objective linear knapsack problem (where $x_i \in \{0,1\}$ is replaced by $0 \leq x_i \leq 1$),

$$\begin{aligned} \max \sum_{i=1}^{n} c_i x_i & \\ \text{subject to} \sum_{i=1}^{n} w_i x_i \leq W & \\ x_i \leq 1 \quad & i = 1, \ldots, n \\ x_i \geq 0 \quad & i = 1, \ldots, n \end{aligned}$$

they are used to easily find an optimal solution.

Assume that items $1, \ldots, n$ are ordered such that

$$\frac{c_1}{w_1} \geq \frac{c_2}{w_2} \geq \ldots \geq \frac{c_n}{w_n}. \tag{10.5}$$

Let $i^* := \min\{i : \sum_{j=1}^{i} w_j > W\}$ be the smallest index such that the weight of items 1 to i exceeds the total capacity. Item i^* is called the *critical item.* The solution of the continuous knapsack problem is simply given by taking all items 1 to $i^* - 1$ and a fraction of the critical item, that is $x_i = 1$ for $i = 1, \ldots, i^* - 1$ and

$$x_{i^*} = \frac{\left(W - \sum_{i=1}^{i^*-1} w_i \right)}{w_{j^*}}.$$

Good algorithms for the single objective problem use this fact and focus on optimization of items around i^*, see e.g. Martello and Toth (1990); Pisinger (1997); Kellerer *et al.* (2004). Ideas of such algorithms have been adapted to the bicriterion case by Ulungu and Teghem (1997).

The two criteria induce two different sequences of value to weight ratios. Let $\mathcal{O}_k$ be the ordering (10.5) according to c_i^k / w_i, $k = 1, 2$. Let r_i^k be the rank or position of item i in order $\mathcal{O}_k$ and let $\mathcal{O}$ be the order according to increasing values of $(r_i^1 + r_i^2)/2$, the average rank of an item.

The branch and bound method will create partial solutions by assigning zeros and ones to subsets of variables denoted $\mathcal{B}_0$ and $\mathcal{B}_1$, respectively. These *partial solutions* constitute nodes of the search tree. Variables not assigned either zero or one are called *free variables* for a partial solution and define a set $\mathcal{F} \subseteq \{1, \ldots, n\}$ such that $\{1, \ldots, n\} = \mathcal{B}_1 \cup \mathcal{B}_0 \cup \mathcal{F}$. A solution formed by assigning all free variables a value is called *completion* of a partial solution. Variables of a partial solution will be assigned a value according to the order $\mathcal{O}$. It is convenient to number the items in that order so that we will have

$$\mathcal{B}_1 \cup \mathcal{B}_0 = \{1, \ldots, l - 1\}, \mathcal{F} = \{l, \ldots, n\}$$

for some l. Furthermore we shall denote r_k the index of the first variable in $\mathcal{F}$ according to order $\mathcal{O}_k$, for $k = 1, 2$.

Vector valued bounds will be used to fathom a node of the tree when no completion of the partial solution $(\mathcal{B}_1, \mathcal{B}_0)$ can possibly yield an efficient solution. A lower bound $(\underline{z}_1, \underline{z}_2)$ at a partial solution is simply given by the value of the variables which have already been assigned value 1,

$$(\underline{z}_1, \underline{z}_2) = \left(\sum_{i \in \mathcal{B}_1} c_i^1, \sum_{j \in \mathcal{B}_1} c_i^2 \right). \tag{10.6}$$

For the computation of upper bounds we define

$$\overline{W} := W - \sum_{i \in \mathcal{B}_1} w_i \geq 0,$$

the remaining free capacity of the knapsack after fixing variables in $\mathcal{B}_1$. Furthermore, we denote

$$s_k := \max \left\{ l_k \in \mathcal{F} : \sum_{j_k = r_k}^{l_k} w_{j_k} < \overline{W} \right\}$$

to be the last item that can be chosen to be added to a partial solution according to $\mathcal{O}_k$. Thus, $s_k + 1$ is in fact the critical item in order $\mathcal{O}_k$, taking the already fixed variables in $\mathcal{B}_0$ and $\mathcal{B}_1$ into account.

The upper bound for each objective value at a partial solution is computed according to the rule of Martello and Toth (1990).

$$\overline{z}_k = \underline{z}_k + \sum_{j_k = r_k}^{s_k} c_{j_k}^k + \max \left\{ \left\lfloor \overline{W}_k \frac{c_{s_k+2}^k}{w_{s_k+2}} \right\rfloor, \left\lfloor c_{s_k+1}^k - (w_{s_k+1} - \overline{W}_k) \frac{c_{s_k}^k}{w_{s_k}} \right\rfloor \right\}, \tag{10.7}$$

where $\overline{W}_k = \overline{W} - \sum_{j_k = r_k}^{s_k} w_{j_k}$. The bound can be justified from the observation that $x_{s_{k+1}}$ cannot assume fractional values. Therefore it must be either zero or one. It is computed from the value of the already assigned variables ($\underline{z}_k$), plus the value of those items that fit entirely, in order $\mathcal{O}_k$, plus the maximum term. The first term in the maximum in (10.7) comes from setting $x_{s_{k+1}} = 0$ and filling remaining capacity with $x_{s_{k+2}}$, while the second means setting $x_{s_{k+1}} = 1$ and removing part of x_{s_k} to satisfy the capacity of the knapsack. Another way of computing an upper bound using (10.7) is indicated in Exercise 10.1.

Given a partial solution, an assignment of zeros and ones to the free variables is sought, to find potentially efficient solutions of the whole problem. This problem related to the partial solution $(\mathcal{B}_1, \mathcal{B}_0)$ is again a bicriterion knapsack problem:

$$\max \sum_{i \in \mathcal{F}} c_i^1 x_i + \sum_{i \in \mathcal{B}_1} c_i^1$$
$$\max \sum_{i \in \mathcal{F}} c_i^2 x_i + \sum_{i \in \mathcal{B}_1} c_i^2$$
$$\text{subject to } \sum_{i \in \mathcal{F}} w_i x_i \leq \overline{W}$$
$$x_i \in \{0, 1\}.$$

We now have all prerequisites to formulate the algorithm. The algorithm pursues a depth first strategy. That is, movement down a branch of the search tree, and therefore having more variables fixed in $\mathcal{B}_1$ or $\mathcal{B}_0$, is preferred to investigating partial solutions with fewer fixed variables. Successors of a node in a tree (branching) are distinguished by different sizes of $\mathcal{B}_1$ and $\mathcal{B}_0$. Actually, the tree can be drawn so that the sets $\mathcal{B}_1$ of successors of a node will be become smaller, as variables are moved from $\mathcal{B}_1$ to $\mathcal{B}_0$, see Figure 10.1. The idea is that by fixing many variables first according to their value to weight ratios, a good feasible solution is obtained fast, so that many branches of the tree can be fathomed early.

Throughout, we keep a list $\mathcal{L}$ of potentially nondominated points identified so far (note that due to maximization y^1 dominating y^2 means $y^1 > y^2$ here), and a list of nodes $\mathcal{N}$ still to be processed.

The list $\mathcal{N}$ is maintained as a last-in-first-out queue. Nodes are fathomed if the bounds show that they can only lead to dominated solutions, if they have been completely investigated (they represent a complete solution), or if no further succeeding node can be constructed.

When a node is fathomed, the algorithm backtracks and creates a new node by moving the last item of $\mathcal{B}_1$ to $\mathcal{B}_0$, removing all items after this new item from $\mathcal{B}_0$. If, however, the last item in $\mathcal{B}_1$ was n then the algorithm chooses the smallest v such that all items $\{v, \ldots, n\}$ were in $\mathcal{B}_1$, removes them all, and defines $\mathcal{B}_0$ to be all previous elements of $\mathcal{B}_0$ up to $v - 1$ and to include v.

When a node is not fathomed, the algorithm proceeds deeper down the tree in that it creates a new successor node. This may again be done in two ways. If $\mathcal{B}_1$ allows addition of the first item l of $\mathcal{F}$, as many items as possible are included in $\mathcal{B}_1$, according to order $\mathcal{O}$, i.e. as they appear in $\mathcal{F}$. If, on the other hand, the remaining capacity $\overline{W}$ does not allow item l to be added to $\mathcal{B}_1$, the first possible item r of $\mathcal{F}$, which can be added to $\mathcal{B}_1$ is sought and item r is added to $\mathcal{B}_1$. Of course all items $i, \ldots, r - 1$ must be added to $\mathcal{B}_0$. Since the current node has not been fathomed such an r must exist.

Algorithm 10.1 (Branch and bound for the knapsack problem.)

Input: Values c_i^1 and c_i^2, weights w_i for $i = 1, \ldots, n$ and capacity W.
Initialization: Create root node N_0 as follows.
  Set $\mathcal{B}_1 := \emptyset$, $\mathcal{B}_0 := \emptyset$, $\mathcal{F} := \{1, \ldots, n\}$,
  Set $\underline{z} := \binom{0}{0}, \overline{z} := \binom{\infty}{\infty}$, $\mathcal{L} := \emptyset$, $\mathcal{N} := \{N_0\}$.
  While $\mathcal{N} \neq \emptyset$
    Choose the last node $N \in \mathcal{N}$.
    Compute $\overline{W}$ and $\underline{z}$.
    Add $\underline{z}$ to $\mathcal{L}$ if it is not dominated.
    Compute $\overline{z}$.
    If $\{i \in \mathcal{F} : w_i \leq \overline{W}\} = \emptyset$ or $\overline{z}$ is dominated by some $y \in \mathcal{L}$
      Fathom node N. $\mathcal{N} := \mathcal{N} \setminus \{N\}$.
      Create a new node N' as follows.
      Let $t := \max\{i : i \in \mathcal{B}_1\}$.
      If $t < n$ do
        $\mathcal{B}_1 := \mathcal{B}_1 \setminus \{t\}$, $\mathcal{B}_0 := (\mathcal{B}_0 \cap \{1, \ldots, t-1\}) \cup \{t\}$, $\mathcal{F} := \{t+1, \ldots, n\}$
      End if.
      If $t = n$ do
        Let u be $\min\{j : \{j, j+1, \ldots, t-1, t\} \subset \mathcal{B}_1\}$.
        Let v be $\max\{j : | \in \mathcal{B}_1 \setminus \{u, \ldots, t\}\}$.
        $\mathcal{B}_1 := \mathcal{B}_1 \setminus \{v, u, u+1, \ldots, t-1, t\}$, $\mathcal{B}_0 := (\mathcal{B}_0 \cap \{1, \ldots, v-1\}) \cup \{v\}$,
        $\mathcal{F} := \{v + 1, \ldots, n\}$
      End if. $\mathcal{N} := \mathcal{N} \cup \{N\}$
      If set $\mathcal{B}_1$ of N' is smaller than $\mathcal{B}_1$ of predecessor nodes of N, which are not predecessors of N' then fathom these nodes.
      If no new node can be created ($\mathcal{B}_1 = \emptyset$), STOP.
    Otherwise
      Create a new node N' as follows.
      Let $s := \max\left\{ t \in \mathcal{F} : \sum_{j=1}^{t} w_j < \overline{W} \right\}$ according to order $\mathcal{O}$.
      If $w_l > \overline{W}$ let $s := l - 1$.
      If $s \geq l$ do
        $\mathcal{B}_1 := \mathcal{B}_1 \cup \{i, \ldots, s\}$, $\mathcal{B}_0 := \mathcal{B}_0$, $\mathcal{F} := \mathcal{F} \setminus \{i, \ldots, s\}$.
      End if
      If $s = i - 1$ do
        Let $r := \min\{j : j \in \mathcal{F}, w_j < \overline{W}\}$ according to order $\mathcal{O}$.
        $B_1 := B_1 \cup \{r\}$, $\mathcal{B}_0 := \mathcal{B}_0 \cup \{i, \ldots, r-1\}$, $\mathcal{F} := \mathcal{F} \setminus \{i, \ldots, r\}$

End if

$\mathcal{N} := \mathcal{N} \cup \{N\}$

End otherwise.

End while.

Output: All efficient solutions.

We illustrate Algorithm 10.1 with an example also used in Ulungu and Teghem (1997). Following the iterations along the branch and bound tree of Figure 10.1 will make clear how the algorithm works.

Example 10.1 (Ulungu and Teghem (1997)). We consider the problem

$$\begin{array}{lrl} \max & 11x_1 + 5x_2 + \ \ 7x_3 + 13x_4 + 3x_5 & \\ \max & 9x_1 + 2x_2 + 16x_3 + \ \ 5x_4 + 4x_5 & \\ \text{subject to} & 4x_1 + 2x_2 + \ \ 8x_3 + \ \ 7x_4 + 5x_5 \leq 16 & \\ & x_j \in \{0,1\}, j = 1, \ldots, 5 & \end{array}$$

The orders are

$$\begin{aligned} \mathcal{O}_1 &= \{x_1, x_2, x_4, x_3, x_5\} \\ \mathcal{O}_2 &= \{x_1, x_3, x_2, x_5, x_4\} \\ \mathcal{O} &= \{x_1, x_2, x_3, x_4, x_5\}. \end{aligned}$$

First, node N_0 is created with $\mathcal{B}_1 = \emptyset$, $\mathcal{B}_0 = \emptyset$, $\mathcal{F} = \{1, 2, 3, 4, 5\}$ and the bounds are initialized as $\underline{z} = \binom{0}{0}$, $\overline{z} = \binom{\infty}{\infty}$ and $\mathcal{L} := \emptyset$, $\mathcal{N} = \{N_0\}$

1. Node N_0 is selected and a new node N_1 is created with $\mathcal{B}_1 = \{1, 2, 3\}$, $\mathcal{B}_0 = \emptyset$, $\mathcal{F} = \{4, 5\}$. Thus $\mathcal{N} = \{N_0, N_1\}$.
2. Node N_1 is selected. We compute $\overline{W} = 2$, $\underline{z} = \overline{z} = \binom{23}{27}$ and add this to $\mathcal{L}$. $\mathcal{L} = \{\binom{23}{27}\}$.
   Since $\{j \in \mathcal{F} : w_j < \overline{W}\} = \emptyset$, node N_1 is fathomed and we create node N_2. Since$t = 3$ we get $\mathcal{B}_1 = \{1, 2\}$, $\mathcal{B}_0 = \{3\}$, $\mathcal{F} = \{4, 5\}$.
   $N = \{N_0, N_2\}$.
3. Node N_2 is selected. We compute $\overline{W} = 10$, $\underline{z} = \binom{16}{11}$, check that $\{j \in F : w_j < \overline{W}\} = \{4, 5\} \neq \emptyset$. The upper bound is $\overline{z} = \binom{29}{18}$.
   Node N_3 is created with $s = 4$ and $\mathcal{B}_1 = \{1, 2, 4\}$, $\mathcal{B}_0 = \{3\}$, $\mathcal{F} = \{5\}$.
   $\mathcal{N} = \{N_0, N_2, N_3\}$
4. N_3 is selected. $\overline{W} = 3$ and $\underline{z} = \overline{z} = \binom{29}{16}$ is added to $\mathcal{L}$ so that $\mathcal{L} = \{\binom{23}{27}, \binom{29}{16}\}$. Since $\{j \in F : w_j < \overline{W}\} = \emptyset$ node N_3 is fathomed.
   Node N_4 is created with $t = 4$, $\mathcal{B}_1 = \{1, 2\}$, $\mathcal{B}_0 = \{3, 4\}$, and $\mathcal{F} = \{5\}$.
   $\mathcal{N} = \{N_0, N_2, N_4\}$.

5. We select N_4 and compute $\overline{W} = 10$, $\underline{z} = \binom{16}{11}$ and $\overline{z} = \binom{16+3}{11+4} = \binom{19}{15}$. $\overline{z}$ is dominated, so node N_4 fathomed.
   Node N_5 is created with $t = 2$, $\mathcal{B}_1 = \{1\}$, $\mathcal{B}_0 = \{2\}$, and $\mathcal{F} = \{3, 4, 5\}$.
   $\mathcal{B}_1$ at N_5 is smaller than $\mathcal{B}_1$ at node N_2 and N_2 is fathomed.
   $\mathcal{N} = \{N_0, N_5\}$.
6. Node N_5 is selected. At this node $\overline{W} = 12$, $\underline{z} = 119$. Since $\{j \in F : w_j < \overline{W}\} \neq \emptyset$, $\overline{z} = \binom{27}{27}$ is not dominated.
   Node N_6 is created with $s = 3$, $\mathcal{B}_1 = \{1, 3\}$, $\mathcal{B}_0 = \{2\}$, and $\mathcal{F} = \{4, 5\}$.
   $\mathcal{N} = \{N_0, N_5, N_6\}$.
7. Select N_6. $\overline{W} = 4$ and $\underline{z} = \overline{z} = \binom{18}{25}$ Because $\{j \in \mathcal{F} : w_j < \overline{W}\} = \emptyset$ node N_6 is fathomed.
   We create N_7 with $t = 3$, $\mathcal{B}_1 = \{1\}$, $\mathcal{B}_0 = \{2, 3\}$, $\mathcal{F} = \{4, 5\}$.
   $\mathcal{N} = \{N_0, N_5, N_7\}$.
8. Select N_7. $\overline{W} = 12$, $\underline{z} = \binom{11}{9}$. $\{j \in \mathcal{F} : w_j < \overline{W}\} \neq \emptyset$. Upper bound $\overline{z} = \binom{27}{18}$ is not dominated.
   We create N_8 with $s = 5$, $\mathcal{B}_1 = \{1, 4, 5\}$, $\mathcal{B}_0 = \{2, 3\}$, and $\mathcal{F} = \emptyset$.
   $\mathcal{N} = \{N_0, N_5, N_7, N_8\}$
9. Node N_8 is selected. $\overline{W} = 0$ and $\underline{z} = \overline{z} = \binom{27}{18}$ is added to $\mathcal{L}$ to give $\mathcal{L} = \{\binom{23}{27}, \binom{29}{16}, \binom{27}{18}\}$. Since obviously $\{j \in F : w_j < \overline{W}\} = \emptyset$ node N_8 fathomed.
   Node N_9 is created with $t = 5$, $u = 4$, $v = 1$, and $\mathcal{B}_1 = \emptyset$, $\mathcal{B}_0 = \{1\}$, $\mathcal{F} = \{2, 3, 4, 5\}$.
   $\mathcal{B}_1$ at N_9 is smaller than $\mathcal{B}_1$ at N_7 and N_5 so that N_5, N_7 are fathomed.
   $\mathcal{N} = \{N_0, N_9\}$.
10. Select N_9. $\overline{W} = 16$, $\underline{z} = \binom{0}{0}$. To compute $\overline{z}$ use $s_1 = 4$, $s_2 = 5$ so that

$$\overline{z} = \begin{pmatrix} 0 + 5 + 13 \qquad + \max\{[7\frac{3}{5}], [7 - (8 - 7)\frac{13}{7}]\} \\ 0 + 16 + 2 + 4 + \max\{[0], [5 - (7 - 1)\frac{4}{5}]\} \end{pmatrix} = \begin{pmatrix} 23 \\ 22 \end{pmatrix}.$$

   The upper bound $\overline{z}$ is dominated, N_9 is fathomed. Because $\mathcal{B}_1 = \emptyset$ at N_9, N_0 can be fathomed. Thus $\mathcal{N} = \emptyset$ and the algorithm stops.

Graphically, the solution process can be depicted as in Figure 10.1. Each node shows the node number and the sets $\mathcal{B}_1$ and $\mathcal{B}_0$. $\mathcal{L}$ is shown for a node whenever it is updated at that node.

There are three efficient solutions, x^1 with $x_1 = x_2 = x_3 = 1$, x^2 with $x_1 = x_2 = x_4 = 1$, and x^3 with $x_1 = x_4 = x_5 = 1$. □

Algorithm 10.1 finds a complete set of efficient solutions. If no duplicates of nondominated points are kept in $\mathcal{L}$ it will be a minimal complete set, otherwise the maximal complete set.

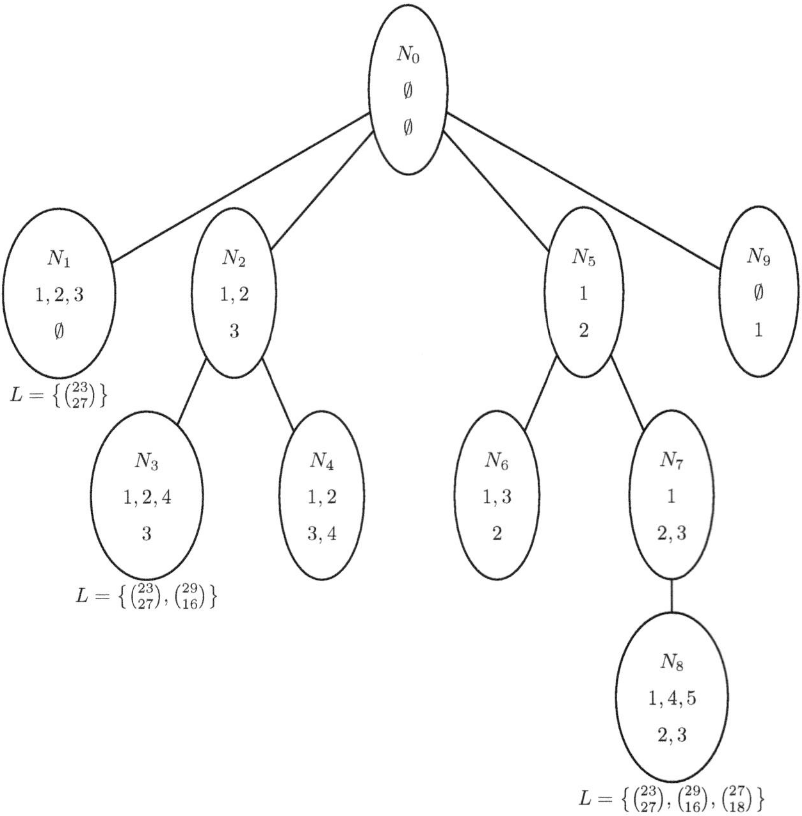

Fig. 10.1. Branch and bound tree in Example 10.1.

10.2 The Travelling Salesperson Problem and Heuristics

The travelling salesperson problem (TSP) consists in finding a shortest tour through n cities. Given a distance matrix $C = (c_{ij})$, find a cyclic permutation π of $\{1, \ldots, n\}$ such that $\pi(n) = 1$ and $\sum_{i=1}^{n} c_{i\pi(i)}$ is minimal. It can also be formulated as an optimization problem on a complete graph. Let $\mathcal{K}_n = (\mathcal{V}, \mathcal{E})$ be a graph and $C : \mathcal{E} \to \mathbb{R}_{\geqq}$ be a cost (or distance) function on the edges. The TSP is to find a simple cycle that visits every vertex exactly once and has the smallest possible total cost (distance). Such cycles are called Hamiltonian cycles, and we shall use the notation HC.

In the multicriteria case we have p distance matrices $C^k, k = 1, \ldots, p$ and the problem is to find cyclic permutations of $\{1, \ldots, n\}$ that minimize $(f_1(\pi), \ldots, f_p(\pi))$, where $f_k(\pi) = \sum_{i=1}^{n} c^k_{i\pi(i)}$. The problem is $\mathcal{NP}$-hard for one objective, so also in the multicriteria case. $\#\mathcal{P}$-completeness is open, but we can prove intractability.

Proposition 10.2 (Emelichev and Perepelitsa (1992)). *The multicriteria TSP is intractable, even if $p = 2$.*

Proof. We consider the TSP on the graph $\mathcal{G} = \mathcal{K}_n$ with edge set $\{e_1, \ldots, e_{n(n-1)/2}\}$ and assign the costs $c(e_i) = (2^i, 2^{n^2} - 2^i)$. As shown earlier (Theorem 9.37) all feasible solutions have incomparable weights. Because there are $(n-1)!$ feasible solutions, the claim follows. □

Methods for finding efficient solutions usually imply solving many single objective TSPs, i.e. $\mathcal{NP}$-hard problems. The TSP is therefore well suited to establish and exemplify results on approximation algorithms for multiobjective combinatorial optimization problems. The results discussed in this section have been obtained in Ehrgott (2000).

To illustrate the idea of approximation algorithms, we first review briefly approximation algorithms for single objective combinatorial optimization problems $\min_{x \in \mathcal{X}} f(x)$.

Let $x \in \mathcal{X}$ be any feasible solution and $\hat{x} \in \mathcal{X}$ be an optimal solution. Then $R(x, \hat{x}) := f(x)/f(\hat{x}) \geq 1$ is called the *performance ratio* of x with respect to $\hat{x}$.

A polynomial time algorithm A for the problem is called an *$r(n)$-approximation algorithm*, if $R(A(I), \hat{x}) \leq r(|I|)$ for all instances I of the problem, where $A(I)$ is the solution found by A, $|I|$ denotes the size of the problem instance and $r : \mathbb{N} \to [1, \infty]$ is a function. $r(n) \equiv 1$ means that the problem is solvable in polynomial time by algorithm A. Note that $R(x, \hat{x}) = \varrho$ is equivalent to

$$\frac{f(x) - f(\hat{x})}{f(\hat{x})} = \varrho - 1.$$

We will investigate if it is possible to find *one* solution that is a good approximation for all efficient solutions of a multiobjective combinatorial optimization problem. In order to generalize the definition of approximation ratios, we will use norms. Let $\|\cdot\| : \mathbb{R}^p \to \mathbb{R}_{\geqq}$ be a norm. We say that $\|\cdot\|$ is monotone, if for $y^1, y^2 \in \mathbb{R}^p$ with $|y_k^1| \leq |y_k^2|$ for all $k = 1, \ldots, p$ we have $\|y^1\| \leq \|y^2\|$ (see Definition 4.19).

Now consider a multiobjective combinatorial optimization problem.

Definition 10.3. *Let $x \in \mathcal{X}$, and $\hat{x} \in \mathcal{X}_E$.*

1. *The* performance ratio R_1 *of x with respect to x^* is*

$$R_1(x, \hat{x}) := \frac{|\, \|f(x)\| - \|f(\hat{x})\| \,|}{\|f(\hat{x})\|}.$$

Algorithm A that finds a feasible solution of the MOCO problem is an $r_1(n)$-approximation algorithm if

$$R_1(A(I), \hat{x}) \leq r_1(|I|)$$

for all instances I of the problem and for all efficient solutions of that instance of the MOCO problem.

2. *The* performance ratio R_2 *of x with respect to $\hat{x}$ is*

$$R_2(x, \hat{x}) := \frac{|\, \|f(x) - f(x^*)\| \,|}{\|f(x^*)\|}.$$

Algorithm A that finds a feasible solution of the MOCO problem is an $r_2(n)$-approximation algorithm if

$$R_2(A(I), \hat{x}) \leq r_2(|I|)$$

for all instances I of the problem and for all efficient solutions of that instance of the MOCO problem.

We have some general results on approximation of efficient solutions.

Corollary 10.4. *An $r(n)$-approximation algorithm according to R_2 is an $r(n)$-approximation algorithm according to R_1.*

Proof. If $R_2(x, \hat{x}) \leq \varrho$ then also $R_1(x, \hat{x}) \leq \varrho$. □

Since we actually compare the norms of the objective vectors of a heuristic solution x and efficient solutions $\hat{x}$, a straightforward idea is to use a feasible solution whose objective vector has minimal norm as an approximate solution. This approach gives a performance ratio of at most 1.

Theorem 10.5. *Let $x^n \in \mathcal{X}$ be such that $\|f(x^n)\| = \min_{x\in\mathcal{X}} \|f(x)\|$ and let $\hat{x}$ be efficient. Then*

$$R_1(x^n, \hat{x}) \leq 1.$$

Proof.

$$R_1(x^n, \hat{x}) = \frac{|\ \|f(x^n)\| - \|f(\hat{x})\|\ |}{\|f(\hat{x})\|} = \frac{\|f(\hat{x})\| - \|f(x^n)\|}{\|f(\hat{x})\|} \leq 1.$$

□

Note that there always exists some x^n with minimal norm $\|f(x^n)\|$, which is also efficient optimal. This can be seen from Theorem 4.20, using $y^U = 0$ as reference point, which is possible because distances c_{ij}^p are nonnegative.

With Theorem 10.5 two questions arise: Is the bound tight and can x^n be computed efficiently? The answer to the first question is given by an example.

Example 10.6. Let $\mathcal{E} = \{e_1, e_2, e_3, 3_4\}$ and $\mathcal{X} = \{x \subset \mathcal{E} : |x| = 2\}$. The costs of all $e \in \mathcal{E}$ are $c(e_1) = (M, 0), c(e_2) = (0, M), c(e_3) = c(e_4) = (1, 1)$, where M is a large number.

The efficient solutions are $\{e_1, e_3\}, \{e_1, e_4\}, \{e_2, e_3\}, \{e_2, e_4\}$, and $\{e_3, e_4\}$. The solution with minimal norm is $x^n = \{e_3, e_4\}$. Computing performance ratios we obtain

$$R_1(\{e_3, e_4\}, \{e_1, e_3\}) = \frac{|\ \|(2,2)\| - \|(M+1,1)\|\ |}{\|(M+1,1)\|} \to 1$$

as $M \to \infty$ and

$$R_2(\{e_3, e_4\}, \{e_1, e_3\}) = \frac{|\ \|(2,2) - (M+1,1)\|\ |}{\|(M+1,1)\|} \to 1$$

as $M \to \infty$. This example shows that the bound of 1 for the approximation ratio cannot be improved in general. □

For the second question, we can state two sufficient conditions which guarantee the existence of polynomial time algorithms to compute x^n.

Proposition 10.7. *The problem $\min_{x\in\mathcal{X}} \|f(x)\|$ can be solved in polynomial time if one of the two conditions below is satisfied.*

1. $(\mathcal{X}, 1\text{-}\sum, \mathbb{Z})/\operatorname{id}/(\mathbb{Z}, <)$ *can be solved in polynomial time and* $\|\cdot\| = \|\ \|_1$.
2. $(\mathcal{X}, 1\text{-}\max, \mathbb{Z})/\operatorname{id}/(\mathbb{Z}, <)$ *can be solved in polynomial time and* $\|\cdot\| = \|\cdot\|_\infty$.

Proof. 1. In the first case,

$$\begin{aligned}
\min_{x\in\mathcal{X}} \|f(x)\| &= \min_{x\in\mathcal{X}} \sum_{k=1}^{p} f_k(x) \\
&= \min_{x\in\mathcal{X}} \sum_{k=1}^{p} \sum_{e\in x} c^k(e) \\
&= \min_{x\in\mathcal{X}} \sum_{e\in x} \sum_{k=1}^{p} c^k(e) \\
&= \min_{x\in\mathcal{X}} \sum_{e\in x} \hat{c}(e)
\end{aligned}$$

where $\hat{c}(e) = \sum_{k=1}^{p} c^k(e)$.

2. In the second case,

$$\begin{aligned}
\min_{x\in\mathcal{X}} \|f(x)\| &= \min_{x\in\mathcal{X}} \max_{k=1,\ldots,p} f_k(x) \\
&= \min_{x\in\mathcal{X}} \max_{k=1,\ldots,p} \max_{e\in x} c^k(e) \\
&= \min_{x\in\mathcal{X}} \max_{e\in x} \max_{k=1,\ldots,p} c^k(e) \\
&= \min_{x\in\mathcal{X}} \max_{e\in x} \hat{c}(e),
\end{aligned}$$

where $\hat{c}(e) = \max_{k=1}^{p} c^k(e)$.

Under the assumptions of the proposition, these problems are solvable in polynomial time. □

After these general results on approximability, we turn attention to the multicriteria TSP again. We consider two well known heuristic methods for the single objective problem, and analyze their performance for the multiobjective TSP. To apply these methods, we have to assume that the distances satisfy the triangle inequality and are symmetric, i.e. $c_{ij}^k \leq c_{il}^k + c_{lj}^k$ and $c_{ij}^k = c_{ji}^k$ for all i, j, k, l.

The first heuristic generalizes the tree heuristic, which generates a tour from a spanning tree via an Eulerian tour. A Eulerian tour of a graph $\mathcal{G}$ is an alternating sequence of nodes and edges with identical first and last node, which contains each edge of $\mathcal{G}$ exactly once. It is well known that a graph $\mathcal{G}$ is Eulerian (i.e. has a Eulerian tour) if and only if each node has even degree, see e.g. (Papadimitriou and Steiglitz, 1982, p. 412) for a proof.

Algorithm 10.2 (Tree heuristic for the TSP.)

Input: Distance matrices $C^k, k = 1, \ldots, p$.

Find $ST \in \operatorname{argmin}\{\|f(T)\| : T$ *is a spanning tree of* $\mathcal{K}_n\}$.

Define $\mathcal{G} := (\mathcal{V}(\mathcal{K}_n), \mathcal{E})$, *where* $\mathcal{E}$ *consists of two copies of every* $e \in \mathcal{E}(ST)$

Find a Eulerian tour embedded in $\mathcal{G}$, *and the corresponding TSP tour HC by eliminating duplicate nodes in the Eulerian tour.*

Output: A TSP tour HC.

Note that the graph $\mathcal{G}$, which has two copies of each edge of the spanning tree ST is Eulerian because each node is incident to an even number of edges. From the Eulerian tour a TSP tour HC can be constructed through "shortcuts". This is where the triangle inequality for the cost functions is important.

Example 10.8. We apply the algorithm to a TSP with three objectives. The distance matrices are

$$C^1 = \begin{pmatrix} - & 1 & 6 & 5 & 5 & 5 \\ 1 & - & 5 & 4 & 6 & 4 \\ 6 & 5 & - & 5 & 6 & 1 \\ 5 & 4 & 5 & - & 6 & 8 \\ 5 & 6 & 6 & 6 & - & 2 \\ 5 & 4 & 1 & 8 & 2 & - \end{pmatrix},$$
$$C^2 = \begin{pmatrix} - & 57 & 55 & 24 & 19 & 46 \\ 57 & - & 151 & 126 & 121 & 137 \\ 55 & 151 & - & 121 & 90 & 117 \\ 24 & 126 & 121 & - & 34 & 61 \\ 19 & 121 & 90 & 34 & - & 27 \\ 46 & 137 & 117 & 61 & 27 & - \end{pmatrix},$$
$$C^3 = \begin{pmatrix} - & 39 & 173 & 6 & 249 & 45 \\ 39 & - & 354 & 348 & 430 & 25 \\ 173 & 354 & - & 511 & 76 & 404 \\ 6 & 348 & 511 & - & 251 & 39 \\ 249 & 430 & 76 & 251 & - & 328 \\ 45 & 25 & 404 & 39 & 328 & - \end{pmatrix}.$$

Using either the l_2- or the l_1-norm, we get the tree of Figure 10.2 with objective vector $f(ST) = (23, 272, 339)$. Its l_1-norm is 634, the l_2-norm is 435.24. So $\mathcal{G}$ is the graph shown in Figure 10.3.

One possible TSP tour HC, $(1, 2, 4, 3, 5, 6, 1)$ is indicated by the broken lines. The objective function value is $f(HC) = (18, 467, 1879)$ and HC is indeed efficient, which can be verified in this small example by complete enumeration. □

The multicriteria tree heuristic has the theoretically best performance ratio of 1.

Proposition 10.9. *Algorithm 10.2 is a 1-approximation algorithm according to performance ratio R_1.*

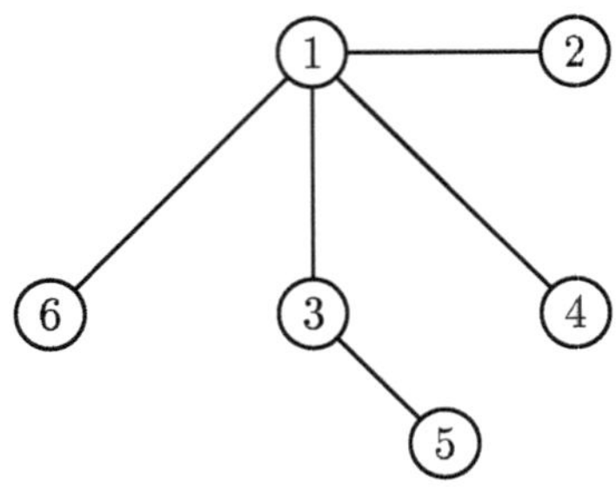

Fig. 10.2. Minimum norm spanning tree ST of $\mathcal{K}_6$.

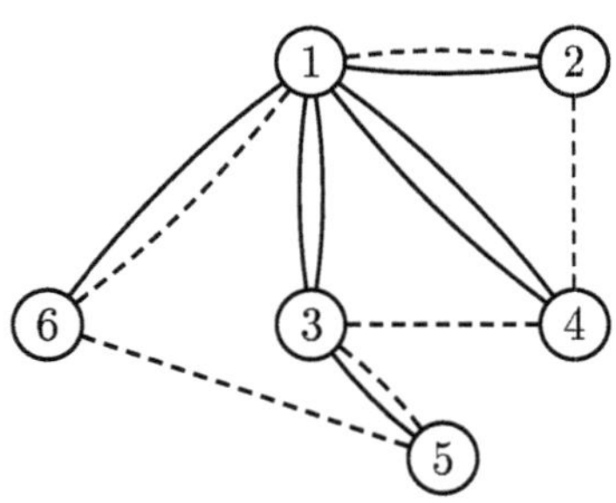

Fig. 10.3. $\mathcal{G}$ is obtained by duplicating all edges of ST.

Proof. Let HC be the tour found by Algorithm 10.2 and let $\hat{HC}$ be an efficient tour. We show that

$$-\|f(\hat{HC})\| \leq \|f(HC)\| - \|f(\hat{HC})\| \leq \|f(\hat{HC})\|. \tag{10.8}$$

The left inequality is trivial. To prove the right hand one, we first apply the the triangle inequality, which gives

$$f(HC) \leq 2f(ST) = f(G).$$

From monotonicity of the norm $\|\cdot\|$ we now conclude

$$\|f(C)\| \leq 2\|f(ST)\|. \tag{10.9}$$

By the choice of ST and since deleting an edge from $\hat{HC}$ gives a spanning tree we also have

$$\|f(ST)\| \leq \|f(\hat{HC})\|. \tag{10.10}$$

Combining (10.9) and (10.10) we conclude

$$\|f(HC)\| \leq 2\|f(\hat{HC})\|.$$

□

We can improve Algorithm 10.2 by adding copies of fewer edges in ST when creating $\mathcal{G}$. All we need is to add as many edges as necessary to be sure that all edges in $\mathcal{G}$ have even degree. Thus, we get a multiobjective version of Christofides' algorithm (Papadimitriou and Steiglitz, 1982).

Algorithm 10.3 (Christofides' heuristic for the TSP.)
Input: *Distance matrices* $C^k, k = 1, \ldots, p$.
Find $ST \in \text{argmin}\{\|f(T)\| : T$ *is a spanning tree of* $\mathcal{K}_n\}$.
Let $\mathcal{G}^* = (\mathcal{V}^*, \mathcal{E}^*)$, *where* $\mathcal{V}^* = \{v : v$ *has odd degree in* $ST\}$, *and* $\mathcal{E}^* = \{[u, v] : u, v \in \mathcal{V}^*\}$.
Find $PM \in \text{argmin}\{\|f(M)\| : M$ *is a perfect matching in* $\mathcal{G}^*\}$.
$\mathcal{G} = (\mathcal{V}, \mathcal{E}(ST) \cup \mathcal{E}(PM))$.
Find a Eulerian tour of $\mathcal{G}$, *and the embedded TSP tour* HC.
Output: *A TSP tour* HC.

Instead of duplicating all edges of ST only additional edges between those nodes that have odd degree in ST are added. There is always an even number of nodes with odd degree in a spanning tree (because the sum of the degrees of all nodes is even). Therefore $\mathcal{G}^*$ is a complete graph on an even number of nodes. Thus $\mathcal{G}^*$ has a perfect matching with $|\mathcal{V}^*|/2$ edges.

Example 10.10. We apply Algorithm 10.3 to the instance presented in Example 10.8. With the same spanning tree of minimal norm as before (Figure 10.2) nodes $2, 4, 5$, and 6 have odd degree. The (unique) perfect matching with

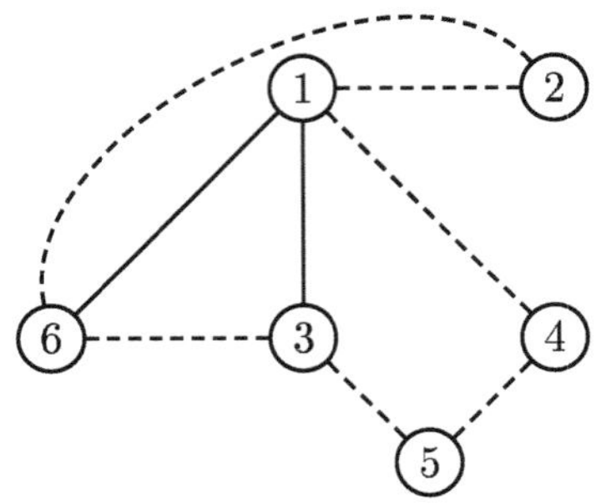

Fig. 10.4. Eulerian graph $\mathcal{G}$ obtained by Algorithm 10.3.

minimal norm for both the l_1 and l_2 norm is $PM = \{[2,6],[4,5]\}$. We get $\mathcal{G}$ as shown in Figure 10.4.

A TSP tour that can be extracted from $\mathcal{G}$ is $(1,2,6,3,5,4,1)$ with objective function values $(23,459,801)$, which is also an efficient TSP tour. □

The Christofides' heuristic has a performance ratio of 1/2 in the single objective case. This can no longer hold for multiple criteria. But it should not come as a surprise, that it also has the performance ratio 1.

Proposition 10.11. *Algorithm 10.3 is a 1-approximation algorithm for the multicriteria TSP according to performance ratio R_1.*

Proof. Let HC be the TSP tour found by Algorithm 10.3 and let $\hat{HC}$ be an efficient TSP tour. We show (10.8) again.

Let $\{i_1,\ldots,i_{2m}\}$ be the odd degree nodes in ST in the order they appear in $\hat{HC}$, i.e.

$$\hat{HC} = (\alpha_0, i_1, \alpha_1, i_2, \ldots, \alpha_{2m-1}, i_{2m}, \alpha_{2m}),$$

where α_i are sequences of other nodes.

$M_1 = \{[i_1,i_2],[i_3,i_4],\ldots,[i_{2m-1},i_{2m}]\}$ and $M_2 = \{[i_2,i_3],\ldots,[i_{2m},i_1]\}$ are two perfect matchings on the nodes $\{i_1,\ldots,i_{2m}\}$. By the triangle inequality we get $f(\hat{HC}) \geq f(M_1) + f(M_2)$ and by definition of PM $\|f(M_i)\| \geq \|f(PM)\|$ for $i = 1,2$. Therefore

$$\begin{aligned}\|f(\hat{HC})\| &\geq \|f(M_1) + f(M_2)\| \\ &\geq \max\{\|f(M_1)\|, \|f(M_2)\|\} \geq \|f(PM)\|. \qquad (10.11)\end{aligned}$$

On the other hand

$$\begin{aligned}\|f(C)\| \leq \|f(G)\| &= \|f(ST) + f(PM)\| \\ &\leq \|f(ST)\| + \|f(PM)\|. \end{aligned} \tag{10.12}$$

Putting (10.11) and (10.12) together we obtain

$$\|f(HC)\| \leq \|f(ST)\| + \|f(PM)\| \leq \|f(ST)\| + \|f(\hat{HC})\| \leq 2\|f(\hat{HC}\|,$$

because, as in the proof of Proposition 10.11,

$$\|f(\hat{HC})\| \geq \|f(ST)\|.$$

□

Note that if $p = 1$, $\|f(x)\| = f(x)$, and (10.11) can be strengthened to $f(\hat{HC}) \geq 2f(PM)$ which gives $f(HC) \leq 3f(\hat{HC})$ and $R_1(HC, \hat{HC}) = 1/2$. The reason why no better result is obtained in the multicriteria case is the maximum of $\|f(M_1)\|$ and $\|f(M_2\|$. We cannot replace this by the sum of the two terms in general.

To prove the approximation result for approximation ratio R_2, we restrict ourselves to l_p-norms

$$\|y\|_\mathsf{p} = \left(\sum_{k=1}^{\mathsf{p}} |y_k|^p\right)^{\frac{1}{\mathsf{p}}}.$$

Theorem 10.12 (Ehrgott (2000)). *Algorithms 10.2 and 10.3 are* $(2^\mathsf{p}+1)^{\frac{1}{\mathsf{p}}}$*-approximation algorithms according to performance ratio* R_2.

Proof. Let HC be the TSP tour found by either Algorithm 10.2 or Algorithm 10.3 and let $\hat{HC}$ be an efficient TSP tour.

$$\begin{aligned}
\frac{\|f(HC) - f(\hat{HC})\|}{\|f(\hat{HC})\|} &= \frac{\left(\sum_{k=1}^{p} |f_k(HC) - f_k(\hat{HC})|^{\mathrm{p}}\right)^{\frac{1}{\mathrm{p}}}}{\left(\sum_{k=1}^{p} (f_k(\hat{HC}))^{\mathrm{p}}\right)^{\frac{1}{\mathrm{p}}}} \\
&\leq \frac{\left(\sum_{k=1}^{p} ((f_k(HC))^{\mathrm{p}} + (f_k(\hat{HC}))^{\mathrm{p}})\right)^{\frac{1}{\mathrm{p}}}}{\left(\sum_{k=1}^{p} (f_k(\hat{HC}))^{\mathrm{p}}\right)^{\frac{1}{\mathrm{p}}}} && (10.13) \\
&= \left(\frac{\|f(HC)\|^{\mathrm{p}} + \|f(\hat{HC})\|^{\mathrm{p}}}{\|f(\hat{HC})\|^{\mathrm{p}}}\right)^{\frac{1}{\mathrm{p}}} \\
&\leq \left(\frac{2^{\mathrm{p}}\|f(\hat{HC})\|^{\mathrm{p}} + \|f(\hat{HC})\|^{\mathrm{p}}}{\|f(\hat{HC})\|^{\mathrm{p}}}\right)^{\frac{1}{\mathrm{p}}} && (10.14) \\
&= (2^{\mathrm{p}} + 1)^{\frac{1}{\mathrm{p}}} && (10.15)
\end{aligned}$$

For inequality (10.13) we used the crude estimate

$$\left(\sum_{k=1}^{p} |y_k^1 - y_k^2|^{\mathrm{p}}\right)^{\frac{1}{\mathrm{p}}} \leq \left(\sum_{k=1}^{p} ((y_k^1)^{\mathrm{p}} + (y_k^2)^{\mathrm{p}})\right)^{\frac{1}{\mathrm{p}}}.$$

Inequality (10.14) follows from the fact $\|f(HC)\| \leq 2\|f(\hat{HC})\|$ from the proofs of Proposition (10.9) and 10.11, which are true for any monotone norm. □

Even though the theoretical bounds for Algorithms 10.2 and 10.3 are the same, 10.3 will often yield better results in practice, see Exercise 10.4, which continues Examples 10.8 and 10.10. A more detailed analysis of Algorithms 10.2 and 10.3 can be found in Ehrgott (2000). For instance, the bound of Theorem 10.12 can be improved, when the l_1-norm is used. The reader is asked to obtain this better bound in Exercise 10.3

10.3 Notes

References on multiobjective versions of $\mathcal{NP}$-hard combinatorial optimization problems are fewer than for polynomially solvable ones.

The most popular is the knapsack problem. Apart from the branch and bound algorithm presented here algorithms based on dynamic programming are known (e.g. Eben-Chaime (1996); Klamroth and Wiecek (2000)). Heuristics and metaheuristics to approximate $\mathcal{X}_E$ are found in Gandibleux and

Fréville (2000); Hansen (1998); Safer and Orlin (1995); Salman *et al.* (1999). Metaheuristics have also been used to solve multi-constraint knapsack problems (Jaszkiewicz, 2001; Zitzler and Thiele, 1999).

Some further references on the TSP are Fischer and Richter (1982); Hansen (2000); Melamed and Sigal (1997). A few references must suffice to indicate that other problems have also been addressed, e.g. set partitioning problems in Ehrgott and Ryan (2002) and location problems in Fernández and Puerto (2003). For scheduling problems there is a vast amount of literature, see T'Kindt and Billaut (2002) and in Chapter 8 of Ehrgott and Gandibleux (2002b).

Exercises

10.1. Let x^1 and x^2 be two optimal solutions of the weighted sum knapsack problem

$$\min \lambda \sum_{i=1}^{n} c_i^1 x_i + (1-\lambda) \sum_{i=1}^{n} c_i^2 x_i$$
$$x_i \in \{0,1\}; \quad i = 1, \ldots, n$$

with $0 < \lambda < 1$. Let $x^{MT} \in [0,1]^n$ be a vector which attains the Martello-Toth bound (10.7) for this single objective knapsack problem. Show that $(\sum_{i=1}^n c_i^1 x_i^{MT}, \sum i = 1^n c_i^2 x_i^{MT})$ is an upper bound for all efficient solutions of the bicriterion knapsack problem with $\sum_{i=1}^n c_i^1 x_i^1 \leq \sum_{i=1} c_i^1 x_i \leq \sum_{i=1}^n c_i^1 x_i^2$ and $\sum_{i=1}^n c_i^2 x_i^2 \leq \sum_{i=1} c_i^2 x_i \leq \sum_{i=1}^n c_i^2 x_i^1$.

10.2. Solve the following bicriterion knapsack problem using Algorithm 10.1.

$$\begin{array}{lll}
\max & 10x_1 + 3x_2 + 6x_3 + 8x_4 + 2x_5 & \\
\max & 12x_1 + 9x_2 + 11x_3 + 5x_4 + 6x_5 & \\
\text{subject to} & 4x_1 + 5x_2 + 2x_3 + 5x_4 + 6x_5 \leq 17 & \\
& x_j \in \{0,1\}, \quad j = 1, \ldots, 5. &
\end{array}$$

10.3. Compute the approximation ratio $r_2(n)$ of Algorithms 10.2 and 10.3 explicitly when the l_1-norm or the l_∞-norm is used. For the l_1 norm you should obtain a better result than that of Theorem 10.12.

10.4. To see that the Christofides' heuristic (Algorithm 10.3) may yield much better results than the tree algorithm in practice, despite their having the same worst case approximation ratios, compute the actual deviations of all possible heuristic TSP tours from the efficient TSP tours. See Figures 10.3 and 10.4. There are seven efficient solutions.

Bibliography

Abhyankar, S., Morin, T., and Trafalis, T. (1990). Efficient faces of polytopes: Interior point algorithms, parametrization of algebraic varieties, and multiple objective optimization. *Contemporary Mathematics*, **114**, 319–341.

Aghezzaf, B. (1999). Second-order necessary conditions of the Kuhn-Tucker type in multiobjective programming problems. *Control and Cybernetics*, **28**(2), 213–224.

Aghezzaf, B. and Hachimi, M. (1999). Second order optimality conditions in multiobjective optimization problems. *Journal of Optimization Theory and Applications*, **102**(1), 37–50.

Arbel, A. (1997). An interior multiobjective primal-dual linear programming algorithm based on approximated gradients and efficient anchoring points. *Computers and Operations Research*, **24**(4), 353–365.

Armand, P. (1993). Finding all maximal efficient faces in multiobjective linear programming. *Mathematical Programming*, **61**, 357–375.

Armand, P. and Malivert, C. (1991). Determination of the efficient set in multiobjective linear programming. *Journal of Optimization Theory and Applications*, **70**(3), 467–489.

Arrow, K., Barankin, C., and Blackwell, D. (1953). Admissible points of convex sets. In H. Kuhn and A. Tucker, editors, *Contributions to the Theory of Games*, volume 2. Princeton University Press, Princeton New Jersey.

Azevedo, J., Santos Costa, M., Silvestre Madeira, J., and Martins, E. (1993). An algorithm for the ranking of shortest paths. *European Journal of Operational Research*, **69**, 97–106.

Balbás, A., Ballvé, M., and Jiménez Guerra, P. (1998). Density theorems for ideal points in vector optimization. Technical report, Universidad Carlos III Madrid. Presented at MOPGP 3, June 1998, Québec City.

Bazaraa, M., Jarvis, J., and Sherali, H. (1990). *Linear Programming and Network Flows*. John Wiley & Sons, New York, NY.

Bazaraa, M., Sherali, H., and Shetty, C. (1993). *Nonlinear Programming – Theory and Algorithms*. John Wiley & Sons, New York.

Behringer, F. (1977a). Lexicographic quasiconcave multiobjective programming. *Zeitschrift für Operations Research*, **21**, 103–116.

Behringer, F. (1977b). On optimal decisions under complete ignorance – A new criterion stronger than both Pareto and maxmin. *European Journal of Operational Research*, **1**, 295–306.

Bellman, R. (1958). On a routing problem. *Quarterly of Applied Mathematics*, **16**, 87–90.

Benson, H. (1978). Existence of efficient solutions for vector maximization problems. *Journal of Optimization Theory and Applications*, **26**(4), 569–580.

Benson, H. (1979). An improved definition of proper efficiency for vector maximization with respect to cones. *Journal of Optimization Theory and Applications*, **71**, 232–241.

Benson, H. (1981). Finding an initial efficient extreme point for a linear multiple objective program. *Journal of the Operational Research Society*, **32**, 495–498.

Benson, H. (1998a). Further analysis of an outcome set-based algorithm for multiple-objective linear programming. *Journal of Optimization Theory and Applications*, **97**(1), 1–10.

Benson, H. (1998b). Hybrid approach for solving multiple-objective linear programs in outcome space. *Journal of Optimization Theory and Applications*, **98**(1), 17–35.

Benson, H. (1998c). An outer approximation algorithm for generating all efficient extreme points in the outcome set of a multiple objective linear programming problem. *Journal of Global Optimization*, **13**(1), 1–24.

Benson, H. and Sun, E. (2000). Outcome space partition of the weight set in multiobjective linear programming. *Journal of Optimization Theory and Applications*, **105**(1), 17–36.

Bigi, G. and Castellani, M. (2000). Second order optimality conditions for differentiable multiobjective programming. *RAIRO Operations Research*, **34**, 411–426.

Bitran, G. and Magnanti, T. (1979). The structure of admissible points with respect to cone dominance. *Journal of Optimization Theory and Applications*, **29**(4), 573–614.

Bolintinéanu, S. and El Maghri, M. (1998). Second-order efficiency conditions and sensitivity of efficient points. *Journal of Optimization Theory and Applications*, **98**(3), 569–592.

Borwein, J. (1977). Proper efficient points for maximization with respect to cones. *SIAM Journal on Control and Optimization*, **15**(1), 57–63.

Borwein, J. (1983). On the existence of Pareto efficient points. *Mathematics of Operations Research*, **8**(1), 64–73.

Borwein, J. and Zhuang, D. (1991). Super efficiency in convex vector optimization. *Zeitschrift für Operations Research*, **35**(3), 175–184.

Borwein, J. and Zhuang, D. (1993). Super efficiency in vector optimization. *Transactions of the American Mathematical Society*, **338**(1), 105–122.

Brucker, P. and Hamacher, H. (1989). k-optimal solution sets for some polynomially solvable scheduling problems. *European Journal of Operational Research*, **41**, 194–202.

Brumbaugh-Smith, J. and Shier, D. (1989). An empirical investigation of some bicriterion shortest path algorithms. *European Journal of Operational Research*, **43**(2), 216–224.

Cambini, A., Martein, L., and Cambini, R. (1997). A new approach to second order optimality conditions in vector optimization. In R. Caballero, F. Ruiz, and R. Steuer, editors, *Advances in Multiple Objective and Goal Programming*, volume 455 of *Lecture Notes in Economics and Mathematical Systems*, pages 219–227. Springer Verlag, Berlin.

Cambini, R. (1998). Second order optimality conditions in multiobjective programming. *Optimization*, **44**(2), 139–160.

Camerini, P., Galbiati, G., and Maffioli, F. (1984). The complexity of multi-constrained spanning tree problems. In L. Lovasz, editor, *Theory of Algorithms*, pages 53 – 101. North-Holland, Amsterdam.

Carraresi, P. and Sodini, C. (1983). A binary enumeration tree to find k shortest paths. *Methods of Operations Research*, **45**, 177 – 188.

Chankong, V. and Haimes, Y. (1983). *Multiobjective Decision Making: Theory and Methodology*. Elsevier Science Publishing Co., New York, NY.

Charnes, A. and Cooper, W. (1961). *Management Models and Industrial Applications of Linear Programming*. John Wiley and Sons, New York.

Chegireddy, C. and H.W., H. (1987). Algorithms for finding k-best perfect matchings. *Discrete Applied Mathematics*, **18**, 155–165.

Choo, E. and Atkins, D. (1983). Proper efficiency in nonconvex multicriteria programming. *Mathematics of Operations Research*, **8**(3), 467–470.

Chung, S., Hamacher, H., Maffioli, F., and Murty, K. (1993). Note on combinatorial optimization with max-linear objective functions. *Discrete Applied Mathematics*, **42**, 139–145.

Climaco, J. and Martins, E. (1982). A bicriterion shortest path algorithm. *European Journal of Operational Research*, **11**, 399–404.

Corley, H. (1980). An existence result for maximization with respect to cones. *Journal of Optimization Theory and Applications*, **31**(2), 277–281.

Corley, H. (1985). Efficient spanning trees. *Journal of Optimization Theory and Applications*, **45**(3), 481–485.

Corley, H. and Moon, I. (1985). Shortest paths in networks with vector weights. *Journal of Optimization Theory and Applications*, **46**(1), 79–86.

Da Cuhna, N. and Polak, E. (1967). Constrained minimization under vector-valued criteria in finite dimensional spaces. *Journal of Mathematical Analysis and Applications*, **19**(1), 103–124.

Danilidis, A., Hadjisavvas, N., and Schaible, S. (1997). Connectedness of the efficient set for three-objective maximization problems. *Journal of Optimization Theory and Applications*, **93**(3), 517–524.

Dantzig, G. (1998). *Linear programming and extensions*. Princeton University Press, Princeton, NJ.

Dantzig, G. and Thapa, M. (1997). *Linear Programming – 1: Introduction*. Springer Verlag, New York.

Dantzig, G. and Thapa, M. (2003). *Linear Programming – 2: Theory and Extensions*. Springer Verlag, New York.

Dathe, H. (1978). Zur Lösung des Zuordnungsproblems bei zwei Zielgrößen. *Zeitschrift für Operations Research*, **22**, 105–118.

Dauer, J. (1993). On degeneracy and collapsing in the construction of the set of objective values in a multiple objective linear program. *Annals of Operations Research*, **46/47**(1-4), 279–292.

Dauer, J. and Gallagher, R. (1990). A combined constraint-space, objective-space approach for determining high-dimensional maximal efficient faces of multiple objective linear programs. *European Journal of Operational Research*, **88**, 368–381.

Dauer, J. and Liu, Y. (1990). Solving multiple objective linear programs in objective space. *European Journal of Operational Research*, **46**, 350–357.

Dauer, J. and Saleh, O. (1990). Constructing the set of efficient objective values in multiple objective linear programs. *European Journal of Operational Research*, **46**, 358–365.

Deb, K. (2001). *Multi-Objective Optimization using Evolutionary Algorithms*. Wiley-Interscience Series in Systems and Optimization. John Wiley & Sons, Chichester.

Dell'Amico, M. and Maffioli, F. (1996). On some multicriteria arborescence problems: Complexity and algorithms. *Discrete Applied Mathematics*, **65**, 191–206.

Dell'Amico, M. and Maffioli, F. (2000). Combining linear and non-linear objectives in spanning tree problems. *Journal of Combinatorial Optimization*, **4**(2), 253–269.

Eben-Chaime, M. (1996). Parametric solution for linear bicriteria knapsack models. *Management Science*, **42**(11), 1565–1575.

Ecker, H. and Hegner, N. (1978). On computing an initial efficient extreme point. *Journal of the Operational Research Society*, **29**(10), 1005–1007.

Ecker, J. and Kouada, I. (1975). Finding efficient points for linear multiple objective programs. *Mathematical Programming*, **8**, 375–377.

Ecker, J. and Kouada, I. (1978). Finding all efficient extreme points for multiple objective linear programs. *Mathematical Programming*, **14**(2), 249–261.

Ecker, J., Hegner, N., and Kouada, I. (1980). Generating all maximal efficient faces for multiple objective linear programs. *Journal of Optimization Theory and Applications*, **30**(3), 353–381.

Ehrgott, M. (1996). On matroids with multiple objectives. *Optimization*, **38**(1), 73–84.

Ehrgott, M. (1997). *Multiple Criteria Optimization – Classification and Methodology*. Shaker Verlag, Aachen.

Ehrgott, M. (1998). Discrete decision problems, multiple criteria optimization classes and lexicographic max-ordering. In T. Stewart and R. van den Honert, editors, *Trends in Multicriteria Decision Making*, volume 465 of *Lecture Notes in Economics and Mathematical Systems*, pages 31–44. Springer Verlag, Berlin.

Ehrgott, M. (2000). Approximation algorithms for combinatorial multicriteria optimization problems. *International Transcations in Operational Research*, **7**, 5–31.

Ehrgott, M. (2002). Location of rescue helicopters in South Tyrol. *International Journal of Industrial Engineering*, **9**, 16–22.

Ehrgott, M. (2005). A discussion of scalarization techniques for multiobjective integer programming. *Annals of Operations Research*, **Special Issue on Multiobjective Discrete and Integer Programming**.

Ehrgott, M. and Galperin, E. (2002). Min-max formulation of the balance number in multiobjective global optimization. *Computers & Mathematics with Applications*, **44**(7), 899–907.

Ehrgott, M. and Gandibleux, X. (2000). A survey and annotated bibliography of multiobjective combinatorial optimization. *OR Spektrum*, **22**, 425–460.

Ehrgott, M. and Gandibleux, X. (2002a). Multiobjective combinatorial optimization. In M. Ehrgott and X. Gandibleux, editors, *Multiple Criteria Optimization – State of the Art Annotated Bibliographic Surveys*, pages 369–444. Kluwer Academic Publishers, Boston, MA.

Ehrgott, M. and Gandibleux, X., editors (2002b). *Multiple Criteria Optimization – State of the Art Annotated Bibliographic Surveys*, volume 52 of *Kluwer's International Series in Operations Research and Management Science*. Kluwer Academic Publishers, Boston, MA.

Ehrgott, M. and Gandibleux, X. (2004). Approximative solution methods for multiobjective combinatorial optimization. *TOP*, **12**(1), 1–88.

Ehrgott, M. and Klamroth, K. (1997). Connectedness of efficient solutions in multiple criteria combinatorial optimization. *European Journal of Operational Research*, **97**, 159–166.

Ehrgott, M. and Nickel, S. (2002). On the number of criteria needed to decide Pareto optimality. *Mathematical Methods of Operations Research*, **55**(3), 329–345.

Ehrgott, M. and Ryan, D. (2002). Constructing robust crew schedules with bicriteria optimization. *Journal of Multi-Criteria Decision Analysis*, **11**(3), 139–150.

Ehrgott, M. and Skriver, A. (2003). Solving biobjective combinatorial max-ordering problems by ranking methods and a two-phases approach. *European Journal of Operational Research*, **147**, 657–664.

Ehrgott, M. and Tenfelde-Podehl, D. (2002). A level set method for multiobjective combinatorial optimization: Application to the quadratic assignement problem. Technical report, Universität Kaiserslautern.

Ehrgott, M. and Tenfelde-Podehl, D. (2003). Computation of ideal and nadir values and implications for their use in MCDM methods. *European Journal of Operational Research*, **151**(1), 119–131.

Ehrgott, M. and Wiecek, M. (2005). Multiobjective programming. In J. Figueira, S. Greco, and M. Ehrgott, editors, *Multicriteria Decision Analysis: State of the Art Surveys*, pages 667–722. Springer Science + Business Media, New York, NY.

Ehrgott, M., Hamacher, H., Klamroth, K., Nickel, S., Schöbel, A., and Wiecek, M. (1997). A note on the equivalence of balance points and Pareto solutions in multiple-objective programming. *Journal of Optimization Theory and Applications*, **92**(1), 209–212.

Ehrgott, M., Nickel, S., and Hamacher, H. (1999). Geometric methods to solve max-ordering location problems. *Discrete Applied Mathematics*, **93**, 3–20.

Emelichev, V. and Perepelitsa, V. (1992). On cardinality of the set of alternatives in discrete many-criterion problems. *Discrete Mathematics and Applications*, **2**(5), 461–471.

Eppstein, D. (1998). Finding the k shortest paths. *SIAM Journal on Computing*, **28**(2), 652–673.

Evans, J. and Steuer, R. (1973). A revised simplex method for linear multiple objective programs. *Mathematical Programming*, **5**, 375–377.

Fernández, E. and Puerto, J. (2003). Multiobjective solution of the uncapacitated plant location problem. *European Journal of Operational Research*, **145**(3), 509–529.

Figueira, J., Greco, S., and Ehrgott, M., editors (2005). *Multiple Criteria Decision Analysis: State of the Art Surveys*. Springer's International Series in Operations Research and Management Science. Springer Science + Business Media, New York, NY.

Fischer, R. and Richter, K. (1982). Solving a multiobjective traveling salesman problem by dynamic programming. *Mathematische Operationsforschung und Statistik, Series Optimization*, **13**(2), 247–252.

Fukuda, K. and Matsui, T. (1992). Finding all the perfect matchings in bipartite graphs. *Networks*, **22**, 461–468.

Gabow, H. (1977). Two algorithms for generating weighted spanning trees in order. *SIAM Journal of Computing*, **6**(1), 139–150.

Gal, T. (1977). A general method for determining the set of all efficient solutions to a linear vectormaximum problem. *European Journal of Operational Research*, **1**, 307–322.

Galperin, E. (1992). Nonscalarized multiobjective global optimization. *Journal of Optimization Theory and Applications*, **75**(1), 69–85.

Gandibleux, X. and Fréville, A. (2000). Tabu search based procedure for solving the 0/1 multiobjective knapsack problem: The two objective case. *Journal of Heuristics*, **6**(3), 361–383.

Garey, M. and Johnson, D. (1979). *Computers and Intractability – A Guide to the Theory of NP-Completeness.* Freeman, San Francisco, CA.

Geoffrion, A. (1968). Proper efficiency and the theory of vector maximization. *Journal of Mathematical Analysis and Applications*, **22**, 618–630.

Georgiadis, L., Georgatsos, P., Floros, K., and Sartzetakis, S. (2002). Lexicographically optimal balanced networks. *IEEE/ACM TRansactions on Networking*, **10**(6), 818–829.

Göpfert, A. and Nehse, R. (1990). *Vektoroptimierung (in German).* B.G. Teubner, Leipzig.

Guddat, J., Guerra Vasquez, F., Tammer, K., and Wendler, K. (1985). *Multiobjective and Stochastic Optimization Based on Parametric Optimization.* Akademie-Verlag, Berlin.

Haimes, Y., Lasdon, L., and Wismer, D. (1971). On a bicriterion formulation of the problems of integrated system identification and system optimization. *IEEE Transactions on Systems, Man, and Cybernetics*, **1**, 296–297.

Hamacher, H. (1995). K best network flows. *Annals of Operations Research*, **57**, 65–72. Special Volume "Industrial Systems".

Hamacher, H. and Queyranne, M. (1985). K best solutions to combinatorial optimization problems. *Annals of Operations Research*, **4**, 123–143.

Hamacher, H. and Ruhe, G. (1994). On spanning tree problems with multiple objectives. *Annals of Operations Research*, **52**, 209–230.

Hamacher, H., Labbé, M., and Nickel, S. (1999). Multicriteria network location problems with sum objectives. *Networks*, **33**, 79–92.

Hansen, M. (1998). *Metaheuristics for multiple objective combinatorial optimization.* Ph.D. thesis, Institute of Mathematical Modelling, Technical University of Denmark, Lyngby, Denmark. Report IMM-PHD-1998-45.

Hansen, M. (2000). Use of substitute scalarizing functions to guide a local search based heuristics: The case of moTSP. *Journal of Heuristics*, **6**(3), 419–431.

Hansen, P. (1979). Bicriterion path problems. In G. Fandel and T. Gal, editors, *Multiple Criteria Decision Making Theory and Application*, volume 177 of *Lecture Notes in Economics and Mathematical Systems*, pages 109–127. Springer Verlag, Berlin.

Hartley, R. (1978). On cone-efficiency, cone-convexity and cone-compactness. *SIAM Journal on Applied Mathematics*, **34**(2), 211–222.

Hazen, G. and Morin, T. (1983). Optimality conditions in nonconical multiple-objective programming. *Journal of Optimization Theory and Applications*, **40**, 25–60.

Helly, E. (1923). On sets of convex bodies with common points (in German). *Jahrbuch der Deutschen Mathematiker Vereinigung*, **32**, 175–176.

Henig, M. (1982). Proper efficiency with respect to cones. *Journal of Optimization Theory and Applications*, **36**(3), 387–407.

Henig, M. (1985). The shortest path problem with two objective functions. *European Journal of Operational Research*, **25**, 281–291.

Henig, M. (1986). The domination property in multicriteria optimization. *Journal of Mathematical Analysis and Applications*, **114**, 7–16.

Hiriart-Uruty, J. and Lemaréchal, C. (1993). *Convex analysis and minimization algorithms.* Springer-Verlag, Berlin.

Hirschberger, M. (2002). *Structure of Efficient Points in Quadratic Vector Optimization Problems (in German).* Ph.D. thesis, Katholische Universität Eichstätt-Ingolstadt.

Holder, A. (2004). Radiotherapy treatment design and linear programming. In M. Brandeau, F. Sainfort, and W. Pierskalla, editors, *Operations Research and Health Care*, pages 741–774. Kluwer Academic Publishers, Norwell MA.

Huarng, F., Pulat, P., and Shih, L. (1996). A computational comparison of some bicriterion shortest path algorithms. *Journal of the Chinese Institute of Industrial Engineers*, **13**(2), 121–125.

Hwang, C. and Masud, A. (1979). *Multiple Objective Decision Making – Methods and Applications*, volume 164 of *Lecture Notes in Economics and Mathematical Systems.* Springer Verlag, Berlin.

Isermann, H. (1974). Proper efficiency and the linear vector maximum problem. *Operations Research*, **22**, 189–191.

Isermann, H. (1977). The enumeration of the set of all efficient solutions for a linear multiple objective program. *Operational Research Quarterly*, **28**(3), 711–725.

Jahn, J. (1986). *Mathematical Vector Optimization in Partially Ordered Spaces*, volume 31 of *Methoden und Verfahren der mathematischen Physik*. Peter Lang, Frankfurt am Main.

Jahn, J. (2004). *Vector Optimization – Theory, Applications, and Extensions*. Springer Verlag, Berlin.

Jaszkiewicz, A. (2001). Comparison of local search-based metaheuristics on the multiple objective knapsack problem. *Foundations of Computing and Decision Sciences Journal*, **26**(1), 99–120.

Jimenez, B. and Novo, V. (2002). First and second order sufficient conditions for strict minimality in multiobjective programming. *Numerical Functional Analysis and Optimization*, **23**(3/4), 303–322.

Kaliszewski, I. (1987). A modified weighted Tchebycheff metric for multiple objective programming. *Computers and Operations Research*, **14**(4), 315–323.

Karmarkar, N. (1984). A new polynomial time algorithm for linear programming. *Combinatorica*, **4**(4), 373–395.

Karp, R. (1972). Reducibility among combinatorial problems. In R. Miller and J. Thatcher, editors, *Complexity of Computer Computations*, pages 85–103. Plenum Press, New York.

Katoh, N., Ibaraki, T., and Mine, H. (1981). An algorithm for finding k minimum spanning trees. *SIAM Journal of Computing*, **10**(2), 247–255.

Keeney, R. and Raiffa, H. (1993). *Decisions with Multiple Objectives: Preferences and Tradeoffs*. Cambridge University Press, Cambridge, GB.

Kellerer, H., Pferschy, U., and Pisinger, D. (2004). *Knapsack Problems*. Springer Verlag, Berlin.

Kirchhoff, G. (1847). On the resolution of the equations to which one is lead in the investigation of the linear distribution of galvanic currents (in German). *Annalen der Physik und Chemie*, **72**, 497–508.

Klamroth, K. and Wiecek, M. (2000). Dynamic programming approaches to the multiple criteria knapsack problem. *Naval Research Logistics*, **47**(1), 57–76.

Klamroth, K., Tind, J., and Wiecek, M. (2002). Unbiased approximation in multicriteria optimization. *Mathematical Methods of Operations Research*, **36**(3), 413–437.

Klinger, A. (1967). Improper solutions of the vector maximum problem. *Operations Research*, **15**(3), 570–572.

Korhonen, P. and Wallenius, J. (1988). A Pareto race. *Naval Research Logistics*, **35**, 615–623.

Korhonen, P., Salo, S., and Steuer, R. (1997). A heuristic for estimating Nadir criterion values in multiple objective linear programming. *Operations Research*, **45**(5), 751–757.

Kostreva, M. and Ogryczak, W. (1999). Linear optimization with multiple equitable criteria. *RAIRO Operations Research*, **33**, 275–297.

Kostreva, M. and Wiecek, M. (1993). Time dependency in multiple objective dynamic programming. *Journal of Mathematical Analysis and Applications*, **173**(1), 289–307.

Kruskal, J. (1956). On the shortest spanning subtree of a graph and the travelling salesman problem. *Proceedings of the American Mathematical Society*, **7**, 48–50.

Küfer, K. (1998). On the asymptotic number of efficient vertices in multiple objective linear programming. *Journal of Complexity*, **14**, 333–377.

Kuhn, H. and Tucker, A. (1951). Nonlinear programming. In J. Neyman, editor, *Proceedings of the Second Berkeley Symposium on Mathematical Statistics and Probability*, pages 481–492. University of California Press, Berkeley, California.

Kung, H., Luccio, F., and Preparata, F. (1975). On finding the maxima of a set of vectors. *Journal of the Association for Computing Machinery*, **22**(4), 469–476.

Lee, H. and Pulat, P. (1993). Bicriteria network flow problems: Integer case. *European Journal of Operational Research*, **66**, 148–157.

Lootsma, F., Athan, T., and Papalambros, P. (1995). Controlling the search for a compromise solution in multi-objective optimization. *Enginering Optimization*, **25**(1), 65–81.

Luc, D. (1989). *Theory of Vector Optimization*, volume 319 of *Lecture Notes in Economics and Mathematical Systems*. Springer Verlag, Berlin.

Malivert, C. and Boissard, N. (1994). Structure of efficient sets for strictly quasi convex objectives. *Journal of Convex Analysis*, **1**(2), 143–150.

Mangasarian, O. (1969). *Nonlinear Programming*. McGraw Hill, New York.

Marchi, E. and Oviedo, J. (1992). Lexicographic optimality in the multiple objective linear programming: The nucleolar solution. *European Journal of Operational Research*, **57**, 355–359.

Martello, S. and Toth, P. (1990). *Knapsack Problems-Algorithms and Computer Implementations*. John Wiley & Sons, Chichester.

Martins, E. (1984). On a multicriteria shortest path problem. *European Journal of Operational Research*, **16**, 236–245.

Martins, E., Pascoal, M., and Dos Santos, J. (1999a). Deviation algorithms for ranking shortest paths. *International Journal of Foundations of Computer Science*, **10**(3), 247–261.

Martins, E., Pascoal, M., and Santos, J. (1999b). Deviation algorithms for ranking shortest paths. *The International Journal of Foundations of Computer Science*, **69**, 97 - 106.

Melamed, I. and Sigal, I. (1997). The linear convolution of criteria in the bicriteria traveling salesman problem. *Computational Mathematics and Mathematical Physics*, **37**(8), 902–905.

Miettinen, K. (1999). *Nonlinear Multiobjective Optimization*, volume 12 of *International Series in Operations Research and Management Science*. Kluwer Academic Publishers, Dordrecht.

Mote, J., Murthy, I., and Olson, D. (1991). A parametric approach to solving bicriterion shortest path problems. *European Journal of Operational Research*, **53**, 81–92.

Murthy, I. and Her, S. (1992). Solving min-max shortest-path problems on a network. *Naval Research Logistics*, **39**, 669–683.

Naccache, P. (1978). Connectedness of the set of nondominated outcomes in multicriteria optimization. *Journal of Optimization Theory and Applications*, **25**, 459–467.

Nemhauser, G. and Wolsey, L. (1999). *Integer and Combinatorial Optimization*. John Wiley & Sons, New York.

Neumayer, P. (1994). Complexity of optimization on vectorweighted graphs. In A. Bachem, U. Derigs, M. Jünger, and R. Schrader, editors, *Operations Research 93*, pages 359–361. Physica Verlag, Heidelberg.

Noghin, V. (1997). Relative importance of criteria: A quantitative approach. *Journal of Multi-Criteria Decision Analysis*, **6**(6), 355–363.

Oxley, J. (1992). *Matroid Theory*. Oxford University Press.

Padberg, M. (1999). *Linear optimization and extensions*. Springer-Verlag, Berlin.

Papadimitriou, C. and Steiglitz, K. (1982). *Combinatorial Optimization*. Prentice Hall, Englewood Cliffs.

Pareto, V. (1896). *Manual d'économie politique (in French)*. F. Rouge, Lausanne.

Pascoletti, A. and Serafini, P. (1984). Scalarizing vector optimization problems. *Journal of Optimization Theory and Applications*, **42**(4), 499–524.

Philip, J. (1972). Algorithms for the vector maximization problem. *Mathematical Programming*, **2**, 207–229.

Philip, J. (1977). Vector maximization at a degenerate vertex. *Mathematical Programming*, **13**, 357–359.

Pisinger, D. (1997). A minimal algorithm for the 0-1 knapsack problem. *Operations Research*, **45**(5), 758–767.

Prim, J. (1957). Shortest connection networks and some generalizations. *Bell System Technics Journal*, **36**, 1389–1401.

Punnen, A. and Nair, K. (1996). An $O(m \log n)$ algorithm for the max + sum spanning tree problem. *European Journal of Operational Research*, **89**, 423–426.

Ramos, R., Alonso, S., Sicilia, J., and González, C. (1998). The problem of the optimal biobjective spanning tree. *European Journal of Operational Research*, **111**, 617–628.

Rockafellar, R. (1970). *Convex Analysis*. Princeton University Press, Pronceton, NJ.

Romero, C. (2001). Extended lexicographic goal programming: A unifying approach. *Omega: The International Journal of Management Science*, **29**, 63–71.

Roy, B. (1990). Decision-aid and decision-making. *European Journal of Operational Research*, **45**, 324–331.

Roy, B. (1996). *Multicriteria Methodology for Decision Aiding*, volume 12 of *Nonconvex Optimization and its Applications*. Kluwer Academic Publishers, Dordrescht.

Safer, H. and Orlin, J. (1995). Fast approximation schemes for multi-criteria flow, knapsack, and scheduling problems. Working paper 3757-95, Sloan School of Management, MIT, Cambridge, MA.

Salman, F., Kalagnanam, J., and Murthy, S. (1999). Cooperative strategies for solving the bicriteria sparse multiple knapsack problem. In *1999 Congress on Evolutionary Computation*, pages 53–60. IEEE Service Center, Piscataway, NJ.

Sawaragi, Y., Nakayama, H., and Tanino, T. (1985). *Theory of Multiobjective Optimization*. Academic Press, Orlando, FL.

Sayin, S. (1996). An algorithm based on facial decomposition for finding the efficient set in multiple objective linear programming. *Operations Research Letters*, **19**, 87–94.

Schechter, M. and Steuer, R. (2005). A correction to the connectedness of the Evans-Steuer algorithm of multiple objective linear programming. Technical report, Department of banking and Finance, University of Georgia.

Schönfeld, K. (1964). *Effizienz und Dualität in der Aktivitätsanlyse (German)*. Ph.D. thesis, Freie Universität Berlin, Germany.

Schweigert, D. (1990). Linear extensions and vector-valued spanning trees. *Methods of Operations Research*, **60**, 219–222.

Sedeño-Noda, A. and González-Martín, C. (2001). An algorithm for the biobjective integer minimum cost flow problem. *Computers and Operations Research*, **28**(2), 139–156.

Serafini, P. (1986). Some considerations about computational complexity for multi objective combinatorial problems. In J. Jahn and W. Krabs, editors, *Recent advances and historical development of vector optimization*, volume

294 of *Lecture Notes in Economics and Mathematical Systems*, pages 222–232. Springer Verlag, Berlin.

Skriver, A. and Andersen, K. (2000). A label correcting approach for solving bicriterion shortest path problems. *Computers and Operations Research*, **27**(6), 507–524.

Sniedovich, M. (1988). A multi-objective routing problem revisited. *Engineering Optimization*, **13**, 99–108.

Soland, R. (1979). Multicriteria optimization: A general characterization of efficient solutions. *Decision Sciences*, **10**, 26–38.

Sonntag, Y. and Zălinescu, C. (2000). Comparison of existence results for efficient points. *JOTA*, **105**(1), 161–188.

Steuer, R. (1985). *Multiple Criteria Optimization: Theory, Computation and Application*. John Wiley & Sons, New York, NY.

Steuer, R. (2000). ADBASE: A multiple objective linear programming solver. Technical report, Terry College of Business, University of Georgia, Athens, GA, 30602 USA.

Steuer, R. (2002). Example for degenerate MOLP. Private Communication.

Steuer, R. and Choo, E. (1983). An interactive weighted Tchebycheff procedure for multiple objective programming. *Mathematical Programming*, **26**, 326–344.

Stoer, J. and Witzgall, C. (1970). *Convexity and Optimization in Finite Dimensional Spaces I*. Springer Veralg, Berlin.

Strijbosch, L., Van Doorne, A., and Selen, W. (1991). A simplified MOLP algorithm: The MOLP-S procedure. *Computers and Operations Research*, **18**(8), 709–716.

Tamiz, M. and Jones, D. (1996). Goal programming and Pareto efficiency. *Journal of Information & Optimization Sciences*, **17**(2), 291–307.

Tamura, K. and Arai, S. (1982). On proper and improper efficient solutions of optimal problems with multicriteria. *Journal of Optimization Theory and Applications*, **38**, 191–205.

Thulasiraman, K. and Swamy, M. (1992). *Graphs: Theory and Algorithms*. John Wiley & Sons, New York, NY.

Tind, J. and Wiecek, M. (1999). Augmented Lagrangian and Tchebycheff approaches in multiple objective programming. *Journal of Global Optimization*, **14**, 251–266.

T'Kindt, V. and Billaut, J. (2002). *Multicriteria Scheduling – Theory, Models and Algorithms*. Springer Verlag, Berlin.

Tuyttens, D., Teghem, J., Fortemps, P., and Van Nieuwenhuyse, K. (2000). Performance of the MOSA method for the bicriteria assignment problem. *Journal of Heuristics*, **6**, 295–310.

Ulungu, E. (1993). *Optimisation combinatoire multicritère: Détermination de l'ensemble des solutions efficaces et méthodes interactives.* Ph.D. thesis, Faculté des Sciences, Université de Mons-Hainaut. Mons, Belgium.

Ulungu, E. and Teghem, J. (1994). The two-phases method: An efficient procedure to solve bi-objective combinatorial optimization problems. *Foundations of Computing and Decision Sciences*, **20**(2), 149–165.

Ulungu, E. and Teghem, J. (1997). Solving multi-objective knapsack problem by a branch-and-bound procedure. In J. Climaco, editor, *Multicriteria Analysis*, pages 269–278. Springer Verlag, Berlin.

Valiant, L. (1979a). The complexity of computing the permanent. *Theoretical Computer Science*, **8**, 189–201.

Valiant, L. (1979b). The complexity of enumeration and reliability problems. *SIAM Journal on Computing*, **8**(3), 410 –421.

Vincke, P. (1992). *Multicriteria Decision Aid.* J. Wiley & Sons, New York, NY.

Visée, M., Teghem, J., Pirlot, M., and Ulungu, E. (1998). Two-phases method and branch and bound procedures to solve the bi-obective knapsack problem. *Journal of Global Optimization*, **12**, 139–155.

Wang, S. (1991). Second-order necessary and sufficient conditions in multiobjective programming. *Numerical Functional Analysis and Optimization*, **12**(1-2), 237–252.

Warburton, A. (1983). Quasiconcave vector maximization: Connectedness of the sets of Pareto-optimal and weak Pareto-optimal alternatives. *Journal of Optimization Theory and Applications*, **40**, 537–557.

Warburton, A. (1987). Approximation of Pareto optima in multiple-objective shortest-path problems. *Operations Research*, **35**(1), 70 –79.

Weber, E., Rizzoli, A., Soncini-Sessa, R., and Castelletti, A. (2002). Lexicographic optimisation for water resources planning: The case of Lake Verbano, Italy. In A. Rizzoli and A. Jakeman, editors, *Integrated Assessment and Decision Support – Proceedings of the First Biennial Meeting of the the International Environmental Modelling and Software Society, Lugano, June 2002*, pages 235–240.

Welsh, D. (1993). *Complexity: Knots, Colourings and Counting.* Cambridge University Press, Cambridge.

White, D. (1988). Weighting factor extensions for finite multiple objective vector minimization problems. *European Journal of Operational Research*, **36**, 256–265.

Wiecek, M. (1995). Multicriteria optimization. Lecture Notes, University of Kaiserslautern.

Wierzbicki, A. (1980). A methodological guide to multiobjective optimization. In K. Iracki, K. Malanowski, and S. Walukiewicz, editors, *Optimization*

Techniques, Part 1, volume 22 of *Lecture Notes in Control and Information Sciences*, pages 99–123. Springer, Berlin.

Wierzbicki, A. (1986a). A methodological approach to comparing parametric characterizations of efficient solutions. In G. Fandel, M. Grauer, A. Kurzhanski, and A. Wierzbicki, editors, *Large-scale modelling and interactive decision analysis*, volume 273 of *Lecture Notes in Economics and Mathematical Systems*, pages 27–45. Springer Verlag, Berlin.

Wierzbicki, A. (1986b). On the completeness and constructiveness of parametric characterizations to vector optimization problems. *OR Spectrum*, **8**(2), 73–87.

Yamamoto, Y. (2002). Optimization over the efficient set – Overview. *Journal of Global Optimization*, **22**, 285–317.

Yu, P. (1974). Cone convexity, cone extreme points and nondominated solutions in decision problems with multiobjectives. *Journal of Optimization Theory and Applications*, **14**, 319–377.

Yu, P. (1985). *Multiple Criteria Decision Making: Concepts, Techniques and Extensions*. Plenum Press, New York, NY.

Yu, P. and Zeleny, M. (1975). The set of all nondominated solutions in linear cases and a multicriteria simplex method. *Journal of Mathematical Analysis and Applications*, **49**, 430–468.

Yu, P. and Zeleny, M. (1976). Linear multiparametric programming by multicriteria simplex method. *Management Science*, **23**(2), 159–170.

Zeleny, M. (1974). *Linear Multiobjective Programming*, volume 95 of *Lecture Notes in Economics and Mathematical Systems*. Springer Verlag, Berlin.

Zhou, G. and Gen, M. (1999). Genetic algorithm approach on multi-criteria minimum spanning tree problem. *European Journal of Operational Research*, **114**, 141–152.

Zionts, S. and Wallenius, J. (1980). Identifying efficient vectors: Some theory and computational results. *Operations Research*, **28**(3), 785–793.

Zitzler, E. and Thiele, L. (1999). Multiobjective evolutionary algorithms: A comparative case study and the strength Pareto approach. *IEEE Transactions on Evolutionary Computation*, **3**(4), 257–271.

List of Figures

1.1 Objective functions of Example 1.3. 3
1.2 Criterion space in Example 1.1. 4
1.3 Criterion space in Example 1.3. 6
1.4 Nondominated points in Example 1.3. 6
1.5 Min-max solution of Example 1.3. 8
1.6 Illustration of two cones. 12

2.1 Illustration of definitions of efficient solutions. 25
2.2 Feasible set of the original problem in Example 2.2. 26
2.3 Feasible set of the modified problem in Example 2.2. 26
2.4 Nondominated points of $\mathcal{Y}$ and $\mathcal{Y} + \mathbb{R}^p_{\geqq}$ are the same. 27
2.5 $\mathcal{Y}^0$ is a compact section of $\mathcal{Y}$. 29
2.6 Constructing a cover from a chain. 31
2.7 Efficient set, ideal, and nadir point. 35
2.8 Feasible set in Example 2.23. 36
2.9 $\mathcal{Y}_N$ is empty, $\mathcal{Y}_{wN}$ is not. 40
2.10 Nondominated and weakly nondominated points. 41
2.11 Level set and level curve in Example 2.29. 42
2.12 Level sets and efficiency. 42
2.13 Location problem of Example 2.31. 45
2.14 Properly nondominated point $\hat{y}$. 50
2.15 Conical hull and tangent cone. 53
2.16 Benson's proper efficiency. 54
2.17 Relationships among definitions of proper efficiency for the case $\mathcal{C} = \mathbb{R}^p_{\geqq}$. 59

3.1 A set $\mathcal{S}(\lambda, \mathcal{Y})$. 66
3.2 Properly separated sets $\mathcal{Y}_1$ and $\mathcal{Y}_2$. 68

3.3 Properly nondominated $\hat{y} \in \mathcal{Y}_N$. 73
3.4 Sets $\mathcal{Y}_N$ and $\mathcal{Y}_{pN}$. 75
3.5 Illustration of the first case in the proof of Theorem 3.17. 78
3.6 Illustration of the second case in the proof of Theorem 3.17. 79
3.7 The set $\mathcal{Y}$ in Example 3.19. 80
3.8 Objective functions in Example 3.30. 85
3.9 Connectedness of $\mathcal{Y}_N$. 87
3.10 The claim in the proof of Theorem 3.35. 88
3.11 Finding the sets $\mathcal{Y}(\alpha)$ in the proof of Theorem 3.35. 89

4.1 The weighted sum method fails for nonconvex problems. 98
4.2 Optimal solutions of ε-constraint problems (4.3). 99
4.3 Feasible set and objective function of Problem (4.8). 103
4.4 Illustration of Benson's problem (4.10). 107
4.5 Objective functions in Example 4.18. 110
4.6 Level sets $\{y : \|y - y^I\| \leq c\}$ for different norms. 112
4.7 ε-neighbourhoods of nondominated points in the l_∞-norm. 118
4.8 $\operatorname{cl} \mathcal{A}(\mathcal{Y})$ may contain dominated points. 120

5.1 Relationships between optimal solutions according to different MCO classes. 138
5.2 The sets $\mathcal{A}_i$ for three convex functions. 141

6.1 Feasible set and objective contours in Example 6.3. 155
6.2 Feasible set in objective space in Example 6.3. 155
6.3 Level sets of the objective of a weighted sum LP. 157
6.4 Feasible set and efficient set in Example 6.18. 168
6.5 Objective space and nondominated set in Example 6.18. 169

7.1 Adjacency of efficient bases and corresponding BFSs. 183
7.2 Feasible and efficient set in Example 7.13. 184
7.3 A polyhedron with extreme points and extreme rays. 186
7.4 $\mathcal{X}_E$ in Example 7.24. 188
7.5 Weights to obtain efficient faces in Example 7.13. 190

8.1 A graph $(\mathcal{V}, \mathcal{E})$ with edge costs $c(e) \in \mathbb{Z}^4$. 201
8.2 All spanning trees of the graph of Figure 8.1. 202
8.3 Illustration of Example 8.6. 204

9.1 Graph in the proof of Theorem 9.2. 222
9.2 A graph with exponentially many efficient paths. 223
9.3 Combining efficient paths. 225

9.4 Example graph for Algorithm 9.1. . . . 227
9.5 The eleven iterations of 9.1 in Example 9.9. . . . 228
9.6 Example graph for Algorithm 9.2. . . . 231
9.7 Initialization. . . . 232
9.8 First iteration. . . . 232
9.9 Second iteration. . . . 233
9.10 Minimal edges in cuts 237
9.11 ... and cycles. . . . 237
9.12 The converse of Theorem 9.14 does not hold. . . . 238
9.13 Example graph for Algorithm 9.4. . . . 239
9.14 Possible extensions of trees in $\mathcal{T}_1$. . . . 239
9.15 Possible extensions of Trees in $\mathcal{T}_2$. . . . 240
9.16 Example for the greedy algorithm. . . . 241
9.17 Finding $\mathcal{D}_T$ for the graph of Figure 9.16 and T as shown . . . 242
9.18 Connected Pareto graph of the graph in Figure 9.13. . . . 243
9.19 Graph $\mathcal{G}$ has nonadjacent efficient spanning trees. . . . 244
9.20 Showing $\mathcal{NP}$-completeness of the biobjective spanning tree problem. . . . 246
9.21 Pareto graph in Example 9.34. . . . 251
9.22 Example 9.38: Feasible set in objective space. . . . 255
9.23 New supported efficient solutions defining new extreme points. . . . 258
9.24 A new supported efficient solution on conv$\{f(x^r), f(x^s)\}$. . . . 258
9.25 Test 1 for the weighted sum of objectives. . . . 261
9.26 Test 2 for objective function 1. . . . 261
9.27 Test 3 for objective function 2. . . . 262

10.1 Branch and bound tree in Example 10.1. . . . 278
10.2 Minimum norm spanning tree ST of $\mathcal{K}_6$. . . . 284
10.3 $\mathcal{G}$ is obtained by duplicating all edges of ST. . . . 284
10.4 Eulerian graph $\mathcal{G}$ obtained by Algorithm 10.3. . . . 286

List of Tables

1.1 Criteria and alternatives in Example 1.1. 2
1.2 Some orders on $\mathbb{R}^p$. 12

2.1 Pay-off table and ideal point. 35
2.2 Single objectives and minimizers in Example 2.23. 37
2.3 Pay-off table in Example 2.23. 37
2.4 Terminology for efficiency and nondominance. 60

5.1 Feasible solutions and objective values in Example 5.17. 137

7.1 Number of efficient extreme points. 180
7.2 Criteria and alternatives in Example 7.30. 193

8.1 Parameters in (8.15) and resulting scalarizations. 217

9.1 Arcs and their costs in Theorem 9.3. 223
9.2 Efficient spanning trees of graph $\mathcal{G}$ in Figure 9.19. 244

Author Index

Authors appearing only in "Notes" sections are not listed.

Arai, S., 93
Arrow, K., 79
Atkins, D., 114

Balbás, A., 63
Ballvé, M., 63
Barankin, C., 79
Bazaraa, M., 80
Behringer, F., 140–142
Bellman, R., 229
Benson, H., 53, 55, 63, 106, 108, 179
Blackwell, D., 79
Boissard, N., 47, 48
Borwein, J., 29, 53

Camerini, P., 245
Castelletti, A., 127
Chankong, V., 98, 101
Choo, E., 114
Chung, S., 216
Climaco, J., 231
Corley, H., 30, 124, 229, 236

Dantzig, G., 158, 160, 179

Ehrgott, M., 16, 43, 49, 102, 127, 144, 198, 199, 210, 235, 243, 245, 251, 279, 287, 288
Emelichev, V., 279
Eppstein, D., 234
Evans, J., 173

Floros, K., 128
Fortemps, P., 262
Fukuda, K., 257

Galbiati, G., 245
Gandibleux, X., 199
Garey, M., 205, 215, 234, 248, 254
Geoffrion, A., 51, 57, 72, 76, 85
Georgatsos, P., 128
Georgiadis, L., 128
Göpfert, A., 23, 25
Guddat, J., 102, 216
Guerra Vasquez, F., 102, 216

Haimes, Y., 98, 101
Hamacher, H., 43, 212, 216, 236, 245, 252
Hansen, P., 204, 223
Hartley, R., 32, 77
Helly, E., 46
Hirschberger, M., 63, 91
Holder, A., 152

Isermann, H., 159, 191, 192, 195

Jiménez Guerra, P., 63
Johnson, D., 205, 215, 248, 254

Johnson, S., 234

Karmarkar, N., 179
Karp, R., 207
Keeney, R., 6
Kellerer, H., 272
Kirchhoff, G., 247
Klamroth, K., 43, 243, 245
Korhonen, P., 35
Kruskal, J., 205, 270
Küfer, K., 179
Kuhn, H., 57
Kung, H., 209, 230

Labbé, M., 212
Lasdon, L., 98
Luccio, F., 209, 230

Maffioli, F., 216, 245
Makowski, M., 215
Malivert, C., 47, 48
Mangasarian, O., 76, 82, 83
Martello, S., 272, 273
Martins, E., 225, 227, 231
Matsui, T., 257
Murty, K., 216

Naccache, P., 88
Nakayama, H., 23, 58, 116
Nehse, R., 23, 25
Nemhauser, G., 185, 248
Neumayer, P., 254
Nickel, S., 43, 49, 212

Oxley, J., 248

Padberg, M., 160
Papadimitriou, C., 254, 257, 282, 285
Pareto, V., 3
Perepelitsa, V., 279
Pisinger, D., 272
Preparata, F., 209, 230
Prim, J., 238

Raiffa, H., 6
Rizzoli, A., 127
Rockafellar, R., 68
Roy, B., 6
Ruhe, G., 236, 245, 252
Ryan, D., 102, 198

Salo, S., 35
Sartzetakis, S., 128
Sawaragi, Y., 23, 58, 116
Schöbel, A., 43
Serafini, P., 222, 241, 254
Sherali, H., 80
Shetty, C., 80
Soland, R., 125
Soncini-Sessa, R., 127
Steiglitz, K., 254, 257, 282, 285
Stephan T., 235
Steuer, R., 35, 154, 166, 173, 175
Stoer, J., 78
Swamy, M., 207

Tammer, K., 102, 216
Tamura, K., 93
Tanino, T., 23, 58, 116
Teghem, J., 255, 256, 262, 272, 276
Tenfelde-Podehl, D., 235
Thapa, M., 160, 179
Thulasiraman, K., 207
Toth, P., 272, 273
Tucker, A., 57
Tuyttens, D., 262

Ulungu, E., 255, 256, 272, 276

Valiant, L., 207, 254
Van Nieuwenhuyse, K., 262
Vincke, P., 6

Warburton, A., 90
Weber, E., 127
Welsh, D., 207
Wendler, K., 102, 216
Wessels, J., 215
Wiecek, M., 43, 94, 109, 180
Wierzbicki, A., 121, 215
Wismer, D., 98
Witzgall, C., 78
Wolsey, L., 185, 248

Subject Index

Bold face page numbers indicate definitions.

$<$, 12
$\leq$, 12
$\leq_{MO}$, 12
$\leq_{\text{lex}}$, 12, 17
$\leqq$, 12
$\mathbb{R}^p_{\geqq}$-bounded, 60, 62
$\mathbb{R}^p_{\geqq}$-closed, 60, 62, 116
$\mathbb{R}^p_{\geqq}$-compact, **31**, 62, 88
$\mathbb{R}^p_{\geqq}$-convex, **67**, 68–70, 74
$\mathbb{R}^p_{\geqq}$-semicompact, **30**, 31, 32
$\mathbb{R}^p_{\geqq}$-semicontinuous, **32**
$\prec$, 9
$\preceq$, 9
$\propto$, 206
$\propto_p$, 206
$\sim$, 9
z_{MO}, 140
$f^{\mathcal{P}}$, 46, 145
$\mathcal{A}(\lambda, \mathsf{p}, \mathcal{Y})$, 116, 125
$\mathcal{A}(\mathcal{Y})$, 116
$\mathcal{C}^{s\circ}$, 93
$\mathcal{K}^{\circ}$, 74
$\mathcal{K}^{\circ\circ}$, 74
$\mathcal{R}_{\mathcal{C}}$, 15
$\mathcal{S}(\mathcal{Y})$, 67–91, 93, 115
$\mathcal{S}_0(\mathcal{Y})$, 67–93
$T_{\mathcal{Y}}(y)$, 52
$\mathcal{X}^{\mathcal{Q}}$, 145
$\mathcal{X}_E$, **24**
$\mathcal{X}_{\text{lex-MO}}$, 136
$\mathcal{X}_{MO}$, 132
$\mathcal{X}_{\Pi}$, 131
$\mathcal{Y}(\alpha)$, 88
$\mathcal{Y}_N$, **24**
$\mathcal{Y}_{pN}$, 71
$\mathcal{Y}_{wN}$, 40
$\mathcal{Y}_{\text{lex-MO}}$, 136

achievement function, 121, **121**, 215
  increasing, 121
  strictly increasing, 121
  strongly increasing, 121
achievement function approach, 121
active index, 57
acyclic digraph, 222, 241
adjacent, **172**
  bases of LP, 172
  efficient basis, 173
  spanning trees, 243
algorithm
  approximation algorithm, 279, 280
  branch and bound, 274
  Christofides' heuristic, 285, 287, 290
  complexity, 205
  depth first strategy, 274
  exponential, 205, 230, 240, 253
  greedy strategy, 240

Hungarian method, 254
knapsack, 274
label correcting, 229, 269
label setting, 225–227, 269
level set, 235
polynomial time, 205, 209–211, 214, 281
tree heuristic, 282, 287
two phase method, 256
worst case running time, 205
alternatives, 1–8
antisymmetric, **9**
approximation, 279–288
approximation ratio, 290
assignment problem, 253–267
$\#\mathcal{P}$-complete, 253
$\mathcal{NP}$-complete, 253
intractable, 253
asymmetric, **9**

backtracking
branch and bound, 274
paths, 228, 231
bases
connected efficient bases, 174
basic feasible solution, 161, 164
efficient, 171
optimal, 168
basic solution, 161
basis, 160, 247
initial efficient, 178
supported efficient, 250
weakly efficient, 195
basis matrix, 160
Benson's method, 106, 217
Benson's problem, 124, 215
BFS, 161
binary relation, 8–11, 20
antisymmetric, **9**
asymmetric, **9**, 10, 11
compatible with scalar multiplication, 13
connected, **9**
irreflexive, **9**, 10, 11
negatively transitive, **9**
reflexive, **9**
strongly connected, **9**
symmetric, **9**
transitive, **9**, 10, 11
bipartite graph, 253
bottleneck problem, 199
boundary, 13, 28
branch and bound, 277
backtracking, 274
depth first, 274
fathom, 274

Christofides' heuristic, 286
class $\#\mathcal{P}$, 206
class $\mathcal{NP}$, 206
class $\mathcal{P}$, 205
classification, 65, 128, 136
closed convex set, 77, 78
combinatorial optimization problem, 197–288
instance, 200
size of an instance, 200
common index, 143
criteria for, 142
existence, 142
compact interval, 62
compact section, 29
compact set, 30, 39
compact, convex set, 90
complement, 30
complete bipartite graph, 253
complete set
maximal, 204
minimal, 204
compromise programming, 111–120, 132, 216
method, 124, 126
weighted problem, 113
computational complexity, 205–208
cone, 8, 15, 20
cone($\mathcal{Y}$), 52
acute, 21
closed, convex, 61
closed, convex, pointed, 63
convex, **13**, 14, 15

convex, pointed, 71
nontrivial, **13**
nontrivial, convex, pointed, closed, 61
pointed, **13**, 14, 15
polar, 93
proper, **13**
recession cone, 62
tangent cone, 52
conical hull, 52
cone($\mathcal{Y}$), 52
connected, **9**, **86**, **174**
extreme efficient bases, 251
Pareto graph, 243
connectedness
efficient bases, 175
linear program, 188
of $\mathcal{Y}_N$, 92
of $\mathcal{Y}_N$ and $\mathcal{X}_E$, 86–91
constraint
from objectives, 98
inactive, 57
KT qualification, 81
problem, 56
qualification, 57
constraint qualification, **57**
convex function, 46–49, 54, 62, 71, 76, 84, 90–91, 101, 108, 140
convex hull, 13
convex set, 88
Corley's method, 124, 126
counting problem, 206, 207
criteria, 1–8
conflicting, 2, 7
vectors, 7
criteria matrix, 153
criterion space, 4, 5
critical item, 272
critical value, 166
cycle, **221**
directed, 221

decision, 1
optimal, 5, 7
decision maker, 6
decision problem, 205, 207
decision space, 4
deviation variable, 106
differentiable curve, 57
dimension, 164
distance matrix, 279
distance measure, 111
divide and conquer, 209
dominating solution, 106, 109
domination, **24**

edge, 164
efficiency
test for, 260–261
efficient, **24**, 43, 68–71, 125
ε-constraint method, 100
$\mathcal{C}$-efficient, 94
basic feasible solution, 171
basis, 250
Benson's method, 107
compromise programming, 111, 113, 114
edge, 188
extreme point, 193
graphic illustration, 24
perfect matching, 253
properly efficient, 51, 53
spanning tree, 251
unbounded edge, 178
efficient basic feasible solution, 183
efficient basis, **172**, 174, 183, 191, 192, 195
adjacent, 191
initial, 177
efficient edge
unbounded, 191
efficient extreme point, 174, 195
average number, 179
expected number, 179
efficient face, **185**, 186, 189
maximal, 185
unbounded, 190
efficient nonbasic variable, **172**, 173, 174, 193
index set of, 192
maximal set of, 191

efficient path, 224, 233
  composition, 224
  simple, 224
  with at most k arcs, 229
efficient pivot, **172**, 173
efficient point
  existence, 28
efficient set, 20, **24**, 23–58
  connectedness, 86
  empty, 25
efficient solution, 8, 18, 20, 23–58
  assignment, 259
  characterization
    geometric, 43
  characterization by lex-MO solution, 139
  elastic constraint, 104
  existence, 32–61
  finite problem, 209
  knapsack, 271–277
  lexicographic solution, 129
  matroid, 248–252
  of a linear program, 158–193
  spanning tree, 236–247
elastic constraint method, 102
entering variable, 166, 173
ε-constraint
  method, 98, 115, 215
  problem, 99, 124
EQUI-PARTITION, 254
equivalence relation, **9**
Eulerian graph, 282
Eulerian tour, 282
externally stable, **33**
extreme efficient
  spanning tree, 270
extreme point, 164, 185
  efficient, 193
  initial, 177
extreme ray, 164, 185, 191, 192

face, 164
facet, 164
fathom, 275
feasible direction, 57
feasible pivot, 172, 173, 191
feasible set, 4
  finite, 203
  in objective space, 153
  of a linear program, 153
feasible solution
  finite set, 198
finite problem, 214
  efficient solution, 209
  lex-MO solution, 210
  lexicographic solution, 212
  max-ordering solution, 210
finite subcover, 30–32, 39
free variables, 272

global lexicographic solution, **131**
greedy algorithm, 240, 241, 243, 248

halfspace, 21
Helly's theorem, 46–48
heuristic method, 282
Hungarian method, 254, 257
hybrid method, 101
hyperplane, 67, 163
  supporting, 163

ideal point, 33, **34**, 63, 111, 133, 233
  proper, 63
identity map, 17, 128
identity matrix, 158
independence system, **247**
independent set, **247**
  maximal, 247, 248
  maximum, 247
inductively ordered set, **28**, 30, 31
inefficient nonbasic variable, 174
interior, 13
  relative interior, 13
intractable, **208**
  assignment problem, 253
  shortest path problem, 223
  spanning tree problem, 245
  TSP, 279
irreflexive, **9**
isolated node, 245

0-1 KNAPSACK, 207, 208, 222
KNAPSACK, 207, 246
knapsack problem, 207, 217, 271–277
  $\mathcal{NP}$-hard, 271
  bicriterion, 271
  optimal solution, 272
  upper bound, 290
  value to weight ratio, 271
k-PARTITION, 248

label correcting algorithm, 229
label setting algorithm, 225
Lagrangian relaxation, 217
leaving variable, 166
level curve, 41, **41**, 44, 234
level set, 41, **41**, 44, 133, 234
  convex, 46
  strict, 41
lex-MO problem, 201
lex-MO solution, 135
  finite problem, 210
lexicographic
  comparison, 7
  optimization, 128–131
lexicographic max-ordering, 135
lexicographic max-ordering solution, **136**
lexicographic problem, 20, 201
lexicographic solution, **129**
  global, 131
lexicographically optimal
  bases, 251
lexicographically optimal solution, 18, 128–131, 252
  finite problem, 212
  path, 231, 233
linear program
  biobjective, 169
  dual, 159, 177
  feasible, 162
  parametric, 165, 170
linear programming
  algebra, 163
  binary, 199, 253
  geometry, 163
  integer, 199
  optimality criterion, 165
  parametric, 164–169
local search, 86
location problem, 44
lower bound
  max-ordering, 135
lower bound, 28
  knapsack, 273
  max-ordering, 134
  on efficient set, 34
lower semicontinuous, 32, 146
LP(λ), 171–193

matching, **253**
matroid, **247**, 247–252
  efficient solution, 248–252
  graphic, 247
  matric, 247
  partition, 247, 251
  uniform, 247
matroid problem
  $\mathcal{NP}$-complete, 248
max-ordering optimal, 132–135
max-ordering optimization, 132–135
max-ordering problem, 20, 140, 201, 216
  matroid, 249, 250
max-ordering problem$\mathcal{NP}$-complete, 248
max-ordering solution, **132**, 252
  finite problem, 210
  geometric characterization, 133
  spanning tree, 270
maximal efficient face, 189, 191, 192
MCO class, **19**, 201
  lex-MO class, 145
  normalization property, 145
  reduction property, 145
  regularity property, 145
  relationships, 137
MCOP, 17
  data, 17
  feasible set, 17
  model map, 17
  objective function vector, 17

objective space, 17
ordered set, 17
metric, 111
minimal edge, 237, 242
minimal element, 28
minimal set, 31
minimax theorem, 79
model map, 17
monotone norm, **111**
multicriteria decision aid, 6
multicriteria optimization class, **19**
multicriteria optimization problem
classification, 16
multicriteria Simplex algorithm, 176, 178, 195
multiobjective problem
differentiable, 57

nadir point, 33, **34**, 36, 38, 61, 125, 233
estimate, 35
negative cycle, 224, 228, 229, 269
negative weight, 229
negatively transitive, **9**
neighbourhood search, 86, 243, 245
neighbours, 243
nonbasic variable
weakly efficient, 195
nonconvex problem, 73
nonconvex set, 116
nonconvexity
degree of, 119
nondominated, **24**
$\mathcal{C}$-nondominated, 62
nondominated frontier, **256**
nondominated point, 5, 23–58
existence, 61
nondominated set, 20, **24**, 23–58
connectedness, 86
properties, 27–28
norm, 21
l_∞-norm, 119
l_p-norm, 113, 122
monotone, 113
properties, 116
strictly monotone, 116
monotone, 111, 280
strictly monotone, 111
normalization property, **145**
not connected, 86
$\mathcal{NP}$-complete, 206
assignment problem, 253
CMOP, 207
matroid problems, 248
shortest path problem, 222
spanning tree problem, 245
TSP, 279
$\mathcal{NP}$-hard, 206
knapsack, 271
$\#\mathcal{P}$-complete, 206
assignment problem, 253
CMOP, 207
shortest path problem, 222

objective space, 4
open cover, 30, 32, 39
ophelimity, 3
optimal solution, **18**
existence for combinatorial problems, 203
global, 81
linear knapsack, 272
local, 81
optimal value, **18**
optimality condition, 189
Karush-Kuhn-Tucker, 84
optimality criterion, 165
optimization over the efficient set, 61
optimization problem
continuous, 2
discrete, 2
order, 8, 20
antisymmetric, 14, 15
canonical, 8, 17
componentwise order, 12, 240
lexicographic, 135
lexicographic order, 12, 136
max-order, 12
partial order, **11**, 21
preorder, **9**
quasi-order, **9**

reflexive, 14, 15
strict componentwise order, 12
strict partial order, **11**
strict weak order, **10**
topological, 240
total, 136
total order, **10**
total preorder, **10**
transitive, 14, 15, 39
weak componentwise order, 12

Pareto class, 63, 65
Pareto graph, 243, **243**, 250, 270
Pareto optimal, **24**
Pareto optimal solution, 4
parsimonious transformation, 206, 222
partial solution, 272, 273
completion, 272
PARTITION, 207, 248
path, **221**
directed, 221
simple, 221
simple directed, 221
pay-off table, 34–36
perfect matching, 253, 285, 286
#$\mathcal{P}$-complete, 254
counting, 207, 254
performance ratio, 279, **280**, 281, 283, 287
permutation
cyclic, 279
of objectives, 137
pivot column, 166
pivot element, 166
pivot row, 166
pivot step
exponential number, 179
polar cone, 93, 94
polyhedron, 163
polynomial time algorithm, 179
polynomial time transformation, 206
polytope, 163
power set, 199
predecessor, 275
preordered set, 9, 28
Prim's algorithm, 238
properly efficient, 36, 50–59, 61, 65, 71–85, 93
Benson, 53, 55, 56, 63, 74, 125
Borwein, **53**, 63
combinatorial problem, 203
Geoffrion, **51**, 55–57, 84, 85, 108
Kuhn-Tucker, 57, **57**, 58, 83–85
properly efficient set
dense in $\mathcal{X}_E$, 71
properly efficient solution, 61, 104
properly nondominated, 50–59, 61
Geoffrion, 51

quasi-convex, 91, 95

rank, **247**
of a set, 247
ranking
of items, 272
of matroid bases, 250
of objectives, 7, 17, 129
of paths, 234, 269
of solutions, 235
ray, 164
recession cone, 62
reduced cost, 161, 173
reduced cost matrix, 171
nonbasic part, 171
reduced problem, 145
reduction property, **145**
reduction result, 47, 61
reference point, 33, 111, 114, 216
reflexive, **9**
regularity property, **145**
relative interior, 68
of a polyhedron, 186
root node, 275

scalarization
ε-constraint, 215
ε-constraint method, 98
achievement function approach, 121, 215
Benson's method, 106, 215

compromise programming, 111
elastic constraint method, 102
hybrid method, 101, 216
of MOIP, 214–218
weighted sum, 215
second-order necessary condition, 92
separation, 88
proper, 68
theorem, 68
set of optimal solutions, 18, 144
set of optimal values, 18
shortest path problem, 221–234
$\#\mathcal{P}$-complete, 222
$\mathcal{NP}$-complete, 222
k-best, 234
intractable, 223
simple path, 224
Simplex algorithm
multicriteria, 176, 178
parametric, 165
Simplex method, 194
Simplex tableau, 166
Sion-Kakutani minimax theorem, 77, 79
slack variable, 180
sort(y), 135
spanning tree, **200**, 282, 285
dominated, 238
efficient, 236–247
spanning tree problem, 200, 202, 236–248, 251
$\mathcal{NP}$-complete, 245
counting problem, 206, 247
decision problem, 205
efficient solution, 203
instance, 201
intractable, 245
lex-MO solution, 203
lexicographically optimal solution, 203
max-ordering solution, 203
strictly efficient, **38**, 38–50, 63, 109
lex-MO solution, 138
lexicographic solution, 130
strictly efficient solution
geometric characterization, 43
strictly increasing, 125
strictly monotone norm, 111
strictly quasiconvex, 146
strongly connected, **9**
substitute problem, 106
sum problem, 199
supported efficient
basis, 250
supported efficient solution, **204**, 218
extreme, 205
supported nondominated point, 204
extreme, 205
symmetric, **9**

tangent cone, 52, **52**, 54
theorem of the alternative
Motzkin, 81
Tucker, 83
topological order, **240**, 242
total unimodularity, 254, 256
totally orderd set, 28
trade-off, 50, 72
transitive, **9**
travelling salesperson problem, 206, 279–288
1-tree, 270
tree heuristic, 283
triangle inequality, 111, 282
TSP, 279–288
$\mathcal{NP}$-complete, 279
intractable, 279

UMOCO, 208, 216
unbounded edge, 185
unique optimal solution, 8, 39, 100, 111, 113, 114, 130
upper bound
knapsack problem, 273
max-ordering, 134
on efficient set, 34
utopia point, **114**, 139

weak efficiency
of max-ordering solution, 132

weakly efficient, 36, **38**, 38–49, 71, 99, 125
  compromise programming, 111, 114
weakly efficient set
  nonempty, 40
  structure, 47
weakly efficient solution
  characterization
    by compromise programming, 115
    by max-ordering, 132
    geometric, 43
weakly nondominated, **38**
weight function, 199
  vector valued, 200
weighted distance, 44
weighted sum
  problem, 101, 215, 249
  scalarization, 51, 65–91, 101, 113, 126
weighted sum problem, 256
weighting vector, 189

Zorn's lemma, 28, 30, 39

GPSR Compliance
The European Union's (EU) General Product Safety Regulation (GPSR) is a set of rules that requires consumer products to be safe and our obligations to ensure this.

If you have any concerns about our products, you can contact us on

ProductSafety@springernature.com

In case Publisher is established outside the EU, the EU authorized representative is:

Springer Nature Customer Service Center GmbH
Europaplatz 3
69115 Heidelberg, Germany

www.ingramcontent.com/pod-product-compliance
Ingram Content Group UK Ltd.
Pitfield, Milton Keynes, MK11 3LW, UK
UKHW020238290926
15470UKWH00002B/599
9783540213987